RELIABILITY OF RADIOACTIVE
TRANSFER MODELS

Proceedings of the Workshop on "Methods for Assessing the Reliability of Environmental Transfer Model Predictions" organized by the Commission of the European Communities in collaboration with the Office of Health and Environmental Research of the Department of Energy (U.S.A.), and the National Research Centre for Physical Sciences "Demokritos" Greece, and in co-operation with the International Atomic Energy Agency (Vienna).

PROGRAMME COMMITTEE

N. CADELLI	CEC, Belgium
M. HILL	NRPB, UK
O. HOFFMAN	Oak Ridge, USA
G. JOHANSSON	SIRP, Sweden
J. KOLLAS	NRCPS Demokritos, Greece
G.S. LINSLEY	IAEA, Austria
C. MYTTENAERE	CEC, Belgium
H.G. PARETZKE	GSF, Fed. Rep. Germany
D. ROBEAU	CEA, France
R.G. SCHRECKHISE	Pac. Northw. Labs., USA
J. SINNAEVE	CEC, Belgium
P. VAN DORP	NAGRA, Switzerland

SCIENTIFIC SECRETARIAT

G. DESMET	CEC, Belgium
F. LUYKX	CEC, Luxembourg
E.G. SIDERIS	NRCPS Demokritos, Greece

LOCAL ORGANIZING COMMITTEE

J. BARTZIS	NRCPS Demokritos, Greece
N. CATSAROS	NRCPS Demokritos, Greece
E. OLYMPIOS	NRCPS Demokritos, Greece
G. PANTELIAS	NRCPS Demokritos, Greece

CONFERENCE SERVICE

E. OLYMPIOS
National Research Centre for Physical Sciences Demokritos
4-15310 Aghia Paraskevi, Attiki
ATHENS, Greece

RELIABILITY OF RADIOACTIVE TRANSFER MODELS

Edited by

G. DESMET

Commission of the European Communities, Brussels, Belgium

ELSEVIER APPLIED SCIENCE

LONDON and NEW YORK

ELSEVIER APPLIED SCIENCE PUBLISHERS LTD
Crown House, Linton Road, Barking, Essex IG11 8JU, England

Sole Distributor in the USA and Canada
ELSEVIER SCIENCE PUBLISHING CO., INC.
52 Vanderbilt Avenue, New York, NY 10017, USA

WITH 122 ILLUSTRATIONS

© 1988 ECSC, EEC, EAEC, BRUSSELS AND LUXEMBOURG
© CROWN COPYRIGHT—pp. 241–249

Softcover reprint of the hardcover 1st edition 1988

British Library Cataloguing in Publication Data

Reliability of radioactive transfer models.
1. Environment. Radioactivity. Measurement.
Mathematical models
I. Desmet, G.
363.7'38

ISBN-13: 978-94-010-7110-9   **e-ISBN-13: 978-94-009-1369-1**
DOI: 10.1007/978-94-009-1369-1

Library of Congress CIP data applied for

Publication arrangements by Commission of the European Communities, Directorate-General Telecommunications, Information Industries and Innovation, Scientific and Technical Communications Service, Luxembourg

EUR 11367

Assessment of the radiological impact of planned or existing practices involving the (actual or potential) release of radionuclides to the environment are largely based on the use of modelling techniques which allow prediction of the relationship between environmental levels and releases and the associated radiation dose to man. Models are imperfect means of representing environmental transfer processes, and it is essential to know the reliability which can be associated with the predictions of these models for each and every assessment situation. Such information is necessary in order to establish confidence in model predictions and, in particular, to allow adequate safety margins to be set in the design of nuclear facilities. This knowledge is also a prerequisite to determine release limits or to decide whether further research is justified in order to improve predictive accuracy.

Therefore a number of distinguished papers have been presented during this workshop which focused both on practical aspects of variability of observations of facts occuring in nature, but also on learned aspects of the science of statistics. It is not very clear, however, whether much insight in mechanisms is gained by such an approach. This insight is probably rather reached by a straightforward judgment of the quality of the primary data and by the willingness to think over carefully the experiments and measurements before doing them.

The book is composed such as to give the reader the chance to quietly study the presented papers in good order.

It is also hoped that modellers and scientists doing research in the laboratory and the field will find points of agreement which could then lead to a better cooperation and exchange of opinion between both usually very remote groups of scientists.

G. DESMET

CONTENTS

SESSION 3: TRANSFER IN TERRESTRIAL ENVIRONMENT
(Chairman: H. PARETZKE)

SESSION 4a: TRANSFER IN THE AQUATIC ENVIRONMENT
FRESH WATER
(*Chairman:* D. ROBEAU)

SESSION 4b: TRANSFER IN THE AQUATIC ENVIRONMENT
MARINE ECOSYSTEM
(*Chairman:* A. AARKROG)

SESSION 5: TRANSFER IN THE BIOSPHERE FROM
WASTE REPOSITORIES
(*Chairman:* M. HILL)

SESSION 6: UNCERTAINTY ANALYSIS
(*Chairman:* F. O. HOFFMAN)

OPENING ADDRESS

E. Economou, Professor of Theoretical Physics, Demokritos, National Research Center for Physical Sciences.

Dear participants,

It is with great pleasure that I am adressing you, the participants of the workshop on "Methods for Assessing the Reliability of Environmental Transfer models Predictions".

The subject of your workshop has proved to be among the most interesting areas of Accident Analysis in Radiation Protection the aftermath of the Chernobyl accident. Our Ministry considers the subject to be of great importance. In order to improve predictions of the consequences of accidental releases of radioactivity the specialists should improve the accuracy of their models used to assess the atmospheric transport of radionuclides, their deposition on terrestrial surfaces such as soil and vegetation and their transfer through the terrestrial environment, food chains, surface waters, urban environment etc... To improve de reliability of such models is necessary in order to set up safety margins in the design of nuclear installations.

It is also necessary to the regulatory bodies, and this is of particular importance for a country like Greece, in order to determine release limits and intervention levels and to implement emergency plans and procedures. Finally this information is necessary to the research policy makers in order to determine the areas where further research efforts are necessary.

I am sure that you will have a very fruitful exchange of experiences and informations. I am also convinced that the results and conclusions of your seminar will be of great help for the scientific community as a whole, they will certainly contribute to make nuclear energy more safe.

Inclosing my remarks I wish you every success in your workshop. Thank you.

OPENING ADDRESS

S. Finzi, Director XII-D, Directorate General for Science, Research adn Development, Joint Research Centre, Commission of the European Communities
G. Desmet, Radiation Protection Programme, Commission of the European Communities

Mr. Secretary General, Mr. Director, Ladies and Gentlemen,

It is a real pleasure for us to address, on behalf of the European Commission and particularly on behalf of the Directorate General XII, this distinguished audience which has come together in the National Research Centre of Demokritos for this radioecology meeting on environmental transfer modelling.

It may be useful to show where the Radiation Protection Programme is now integrated.

The Nuclear Safety Research Directorate includes 3 divisions :

- Safety of Nuclear Installations
- Fuel Cycle
- Radioprotection

Task of the Directorate is to manage the programmes on S.C.A., and to coordinate the Nuclear Safety programmes of research of the Commission.

The present programme of S.C.A. are :

Radiation Protection

Radioactive Waste Management

Decommissioning of nuclear installations at the end of this life

In addition, an intense work on harmonization of safety criteria, and codes and standards for nuclear reactor is pursued from several years.

In the 1987-1991 framework programme of research just the Radiation Protection programme is under the item : quality of life, and the other programmes are under the item : nuclear energy, of which a part belongs to J.R.C. activities.

In this new context, the Radiation Protection activities, will be well coordinated with the nuclear accident management and the fuel cycle safety.

However, this should not at all lead to neglect the fact that the goal of the Radioprotection programme is the protection of man against radiation and the safeguard of his wellbeing.

The building of radioecological models requires an adequate and useful synthesis of previous and ongoing researches. Allow me to formulate some comments on this practice.

In the environment, radionuclides have many sources, and they follow various pathways to reach eventually their target, which is man.
There is the different sources such as the natural radioactivity in the environment, the controlled releases from nuclear power plants and reprocessing plants, the uncontrolled accidental releases, and the releases from deposited waste. Despite the manifold efforts and studies, uncertainty is still inherent to these source terms. Modelling is going to suffer from this uncertainty still in the next and further future.
A next step which is a subject of consideration is the dispersion after release from whatever source.
Three main dispersion pathways may be described : atmospheric dispersion, transfer through the geosphere after any deposition, and finally the dispersion through the biosphere including the terrestrial and aquatic environments. In general this workshop is going to focus to the uncertainty occurring in the biosphere, the two others topics being the subject of either the Methods for Assessing the Radiological Impact of Accidents (called MARIA) programme or the Migration of Radionuclides through the Geosphere (called MIRAGE) programme and finally the ongoing European Performance Assessment of Geological Isolation Systems, called PAGIS. The main objectives of the latter study, dealing with the disposal of nuclear waste, are :

-        a best estimate of the performance of various geological formations for the normal evolution scenario ;
-        a best estimate of the performances for all the altered scenarios ;
-        a sensitivity analysis and an uncertainty analysis are performed on the models.

Anyway, in all these different systems a reshuffle takes places in the primary source terms mainly by alterations of concentrations and chemical forms (speciation), when touching the boundaries of the adjacent individual compartments. So to say, source terms of the second order are created.
For many years, this reshuffle has almost been exclusively investigated experimentally. Much work was done in both the aquatic environment and terrestrial environment and long lasting exercises have been undertaken such as the study of absorption phenomena in sediments, plants and animals in river and sea ecosystems, the determination of transfer factors from soils to plants and to animals, the role of physical and chemical properties and the presence of microbial life on the speciation of radionuclides. These exercises have culminated in a library of information, unfortunately sometimes hidden on dusty shelves.

It is worthwhile to give value to these data, and to restructure and integrate them in comprehensive models, which lead to a reliable estimation of the dose to man.

Two sources of uncertainty have to be removed ; from one side the basic knowledge must be improved ; on the other hand, an effort of interpretation of the presently available information is needed. In general terms these two uncertainties are summarised by the expression "knowledge uncertainty".

Of great help may be to follow, from nearby, progress made in neighbouring areas such as soil sciences, plant sciences, animal sciences and agricultural and ecological studies in general.

For part of this audience however, uncertainty of models predictions refers to the uncertainty due to stochastic variability of quantities within the model, and is at present the subject of many different statistical exercises which should lead to an improvement of the output data. Most of the techniques in use here are of general application in many scientific disciplines.

From a general radiation protection point of view it may be good to bear in mind the ultimate goals of all these scientific efforts. A few of them may be recalled here.

1.   Reliable description of the sequential transfer of radionuclides to the biosphere.

2.   Adequate estimation of the amount of contamination eventually taken up by man after inhalation or ingestion of radionuclides via different pathways.

3.   Reliable estimation of dose to man and human population brought about by different pathways.

These first three goals comprise forward calculations and have been practised for a while, resulting in environmental models.

Two other goals correspond to the reliable calculation of derived levels in the backward direction of the usual modelling practice.

4.   Establishment of derived reference levels in cases of more or less uncontrolled releases.

5.   Setting up of schemes of scientifically founded countermeasures which may help to bring down contamination levels in the environment.

They are of extreme importance considering the health and economic consequences of establishing such levels. The modellers should be asked therefore to approach this exercise with much care, closely inspecting the real important input parameters in order to come to surveyable calculations of the derived levels. One should also not neglect the less known exposure pathways such as those occurring in poor upland areas and heathlands used for grazing animals like goats and sheep. They have been shown to be of importance for local populations. In general, establishing derived levels should be done with much criticism and common sense.

In the case of setting up countermeasure schemes, the situations is twofold. Several actions have been taken after major technical failures in power plants, not only to diminish the direct consequences to human health by simple administration of analogs or by preventing the public from coming in contact with the contamination. Also indirectly has it been tried to lower the concentrations of radionuclides is various compartments of the environment. Both from an experimental point of view and a modelling point of view a big intellectual effort may be requested here, and a search of more basic knowledge may turn out to be necessary.

The scientists have a big responsibility to reach a sound consensus on scientific evidence of this issue and to offer this evidence to the decision makers.

Hopefully this meeting will also give another opportunity for scientists of the Greek peninsula and other C.E.C. members states to establish, to renew or to strengthen personal contacts with scientists from many countries all over the world and to lie the germs of new initiatives, a prerequisite for a fruitful collaboration.

Finally, we wish to emphasise the appreciation we already have for the co-organization of this workshop with the Greek Atomic Energy Commission, and also with the U.S. Department of Energy and International Atomic Energy Agency. To the attendees, remains us to wish a fruitful week of discussions and exchange of opinions, but also a good time in this cradle of civilisation, Athens.

An Overview of the IAEA Safety Series on Procedures for Evaluating the Reliability of Predictions Made by Environmental Transfer Models[*]

F. Owen Hoffman

Environmental Sciences Division
Oak Ridge National Laboratory
P.O. Box X
Oak Ridge, TN 37831

and

Eduard Hofer

Gesellschaft für Reakorsicherheit, mbH
D-8046 Garching
Federal Republic of Germany

ABSTRACT

The International Atomic Energy Agency (IAEA) is preparing a Safety Series publication on practical approaches for evaluating the reliability of predictions made by environmental radiological assessment models. This IAEA document discusses factors that affect the reliability of model predictions and describes methods for quantifying uncertainty. Emphasis is placed on distinguishing between: (1) uncertainty due to stochastic variability and (2) uncertainty due to lack of knowledge. The document states that the best method for evaluating the accuracy in model predictions is the process of testing against data sets that are independent from those used to develop the model (model validation). For situations in which validation results are not available, analytical and numerical methods are presented for propagating the uncertainty in model parameters into a quantitative statement of uncertainty about the model prediction (parameter uncertainty analysis). The strengths and weaknesses of model intercomparison exercises are also discussed. It is recognized that quantitative statements about the reliability of model predictions must rely upon the use of expert judgment when models are applied to situations that are different from those under which they have been tested.

---

[*] Research sponsored by the International Atomic Energy Agency and the Office of Health and Environmental Research, U.S. Department of Energy under contract DE-AC05-840R21400 with Martin Marietta Energy Systems, Inc., Publication No. 3095, Environmental Sciences Division, ORNL.

INTRODUCTION

The International Atomic Energy Agency is preparing a new Safety Series on practical approaches for evaluating the reliability of predictions made by environmental radiological assessment models (IAEA, in preparation). The need for such a document is obvious. Mathematical models are used extensively to evaluate the acceptability of planned or accidental releases of radioactivity. Such models typically produce only a single prediction of concentration, dose or risk to an individual or population group, and until recently, statements about the reliability of these predictions have been merely qualitative.

The new Safety Series publication introduces methods for obtaining quantitative reliability statements. When compared to limits of concern (i.e., dose limits or derived concentration limits) quantitative reliability statements provide a measure of confidence with which model predictions can guide decisions and indicate the situations in which this confidence must be improved.

Numerous individuals have contributed to the IAEA document. Special recognition is given to: H. G. Paretzke and R. H. Gardner, who along with the authors served the IAEA as special consultants, and the IAEA Scientific Secretaries, G. S. Linsley and I. Savolainen.

GENERAL CONCEPTS

Factors Affecting Reliability

Five groups of factors are recognized that affect the reliability of model predictions (Fig. 1). Among these, the most important is the formulation of the assessment question (specification of the problem). The assessment question affects the scale of time and space as well as the processes and

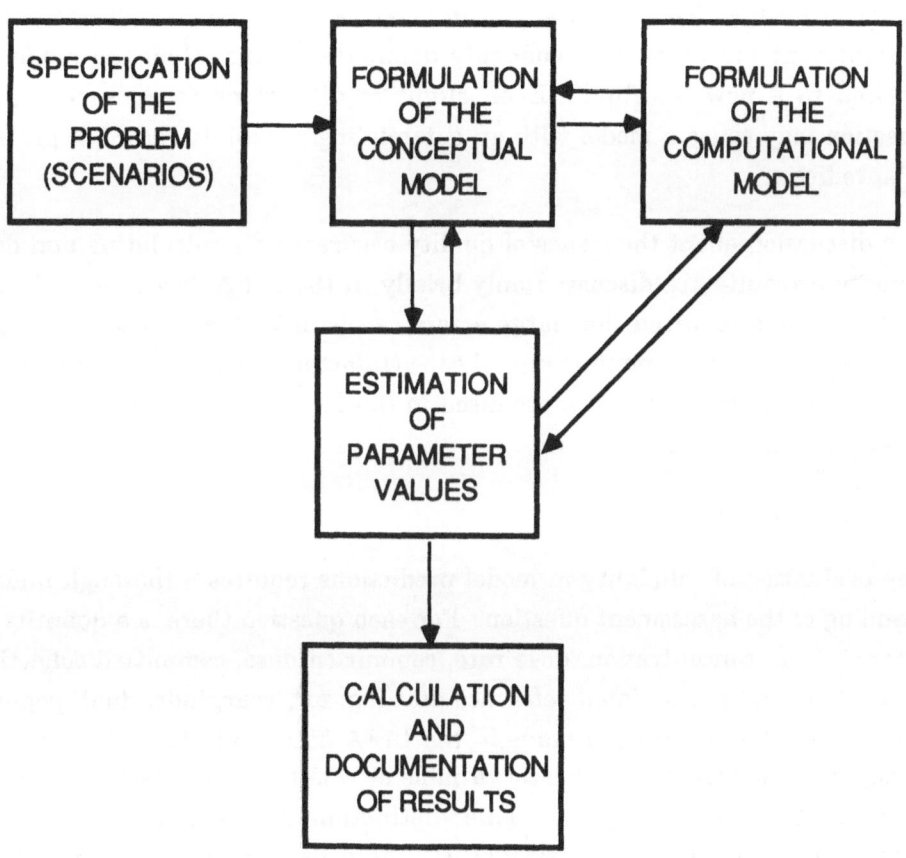

Figure 1. The five classes affecting the reliability of model predictions

mechanisms that should be taken into account by the model. Changes in the assessment question may enhance the significance of previously assumed minor processes. Thus, errors and uncertainties associated with previously unimportant processes may suddenly dominate results when the model is applied to a new question. Even slight modifications to an assessment question may cause a model with an outstanding reliability record to grossly mispredict.

The discussion about the issues of quality assurance for calculating and documenting results are discussed only briefly in the IAEA document. Nevertheless, it is recognized that large errors can be introduced by simple typographical and programming mistakes. Unfortunately, quality assurance procedures are frequently compromised in the face of pressing deadlines and emergencies.

Important Distinctions

The evaluation of reliability in model predictions requires a thorough understanding of the assessment question. For each question there is a quantity of interest (e.g., concentration, dose rate, committed dose, committed collective dose, etc.) and an associated reference unit (kg, m^3, year, individual, population, etc.). Distinctions are made in the IAEA document about whether the quantity of interest is a stochastic variable or whether it is invariant with respect to the reference unit. This distinction determines whether the assessment question has a probabilistic or a deterministic answer. The document therefore discriminates between two fundamentally different types of uncertainty: Type A and Type B.[1]

Type A uncertainty is due to stochastic variability with respect to the reference unit of the question asked of the model.

Type B uncertainty is due to a lack of knowledge about items that are invariant with respect to the reference unit of the question asked of the model.

---

[1] In previous drafts of the IAEA document and in papers summarizing its contents (Hofer and Hoffman 1987), the terms Type 1 and Type 2 uncertainties have been used. These terms have been changed to Type A and B to avoid confusion with Type I and Type II errors in statistical hypotheses testing.

1. Both Type A and Type B uncertainties are negligible

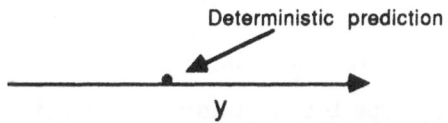

2. Type A is negligible but Type B is not

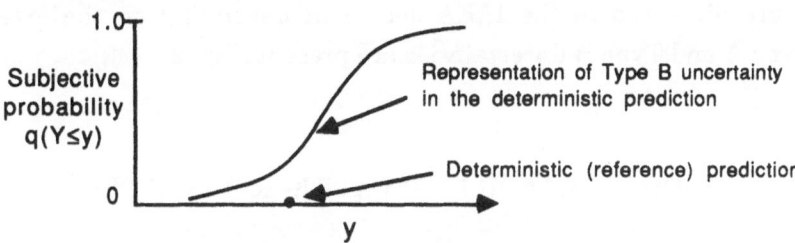

3. Type B is negligible but Type A is not

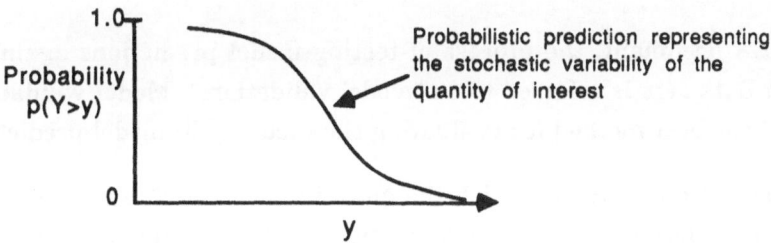

4. Neither Type A nor Type B are negligible

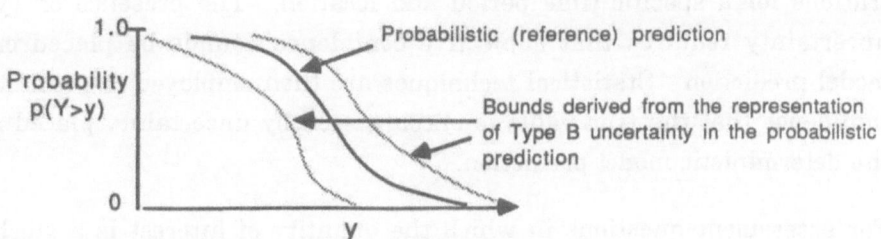

Figure 2. Prediction formats accounting for the presence of Type A and Type B uncertainties.

The presence of Type A uncertainty in the quantity of interest requires a probabilistic answer to the question. A probabilistic answer represents stochastic variability in the quantity of interest in the form of a distribution.

In practice, deterministic as well as probabilistic answers can be determined only imprecisely because of Type B uncertainties. Type B uncertainties suggest a range, not of variability but of alternative, possibly true deterministic or probabilistic answers to the assessment question. Four possible prediction formats are discussed in the IAEA document depending on the extent to which Type A and Type B uncertainties are present (Fig. 2).

APPROACHES FOR EVALUATING RELIABILITY

Model Validation

In the IAEA document, the process of testing model predictions against independent data sets is referred to as "model validation." Model validation is considered the best method for evaluating the accuracy in model predictions.

For assessment questions in which the quantity of interest is invariant with respect to its reference unit, a single predicted value is compared with a single value which is often an estimate obtained from a number of measurements. Examples are average concentrations or time-integrated concentrations for a specific time period and location. The presence of Type B uncertainty requires that subjective confidence bounds be placed on the model prediction. Statistical techniques are then employed to evaluate the confidence that the true value is encompassed by uncertainty placed about the deterministic model prediction.

For assessment questions in which the quantity of interest is a stochastic variable, comparisons are made of the empirical distribution of observed values with the predicted distribution (assuming the use of a probabilistic model to address these questions). The presence of Type B uncertainty requires placement of subjective confidence bounds on the predicted distribution. Statistical techniques are then employed to evaluate the confidence that certain characteristic values (i.e., mean, median, 95th percentile, etc.) of the true distribution are encompassed by the uncertainty placed on those of the predicted distribution.

For both types of questions, the validation process is repeated as many times as necessary to encompass the range of conditions over which the model may be applied. Analyses are made of the potential for the model to over- or underpredict the true values and the extent to which these results confirm quantitative standards of performance that have been assigned to the model (e.g., "the model should not underpredict by more than a factor of three nor overpredict by more than one order of magnitude"). For quantities of interest that vary as a function of time and space, the validation process analyzes the temporal and spatial trends in the over- or underpredictions. These trends indicate the capacity of the model to simulate the dynamics and spatial resolution of a system as well as to produce accurate results for a specific time and location.

Parameter Uncertainty Analysis

Validation data are not likely to ever be available for the full range of conditions in which models may be applied. In the absence of results from validation tests, an analysis of the reliability of model predictions must rely on methods that translate uncertainty in individual model parameters into an estimate of uncertainty in the model prediction. The IAEA document discusses several methods, among which are variance propagation, moment matching, and numerical methods involving the use of a computer. For the types of models used in environmental radiological assessment, numerical methods are most widely applicable. The numerical methods given the most detailed discussion are Simple Random Sampling and Latin Hypercube Sampling. Both methods are "Monte Carlo" procedures .

In the IAEA document, emphasis is given to the treatment of Type B uncertainties as Type A uncertainties are assumed to be properly addressed using a probabilistic model. Among Type B uncertainties, most attention in the past has been given to uncertainty in the estimation of parameter values.

As a consequence, present methods are particularly suited for the analysis of the effect of parameter uncertainties on the model prediction.

The steps recommended for conducting parameter uncertainty analyses are as follows:

1) List all potentially important uncertain parameters.

2) Within the context of the assessment question, specify the maximum conceivably applicable range for each parameter.

3) Over this range, specify a subjective probability distribution.

   (This step usually requires a considerable amount of judgment, given that data are seldom collected from experiments explicitly designed to address specific assessment questions. The probability distribution is to represent the degrees of belief for the appropriate parameter value to lie within any specific part of the the range specified in step 2).

4) Account for depedencies between parameters by introducing suitable restrictions, quoting conditional degrees of belief, or by estimating correlation coefficients, respectively.

5) Use analytical or numerical procedures to generate a "subjective probability distribution" of predicted values from the joint "subjective probability density function" over the combined range of parameter values (Fig. 3).

6) Derive quantitative statements ("subjective confidence intervals") representing the effect of parameter uncertainty on the model prediction.

7) Rank the parameters with respect to their contribution to the uncertainty in the model prediction. (This step provides incentive for additional research and gives priority to those parameters that have the greatest influence on the uncertainty in the model prediction).

8) Present and interpret the results of the analysis.

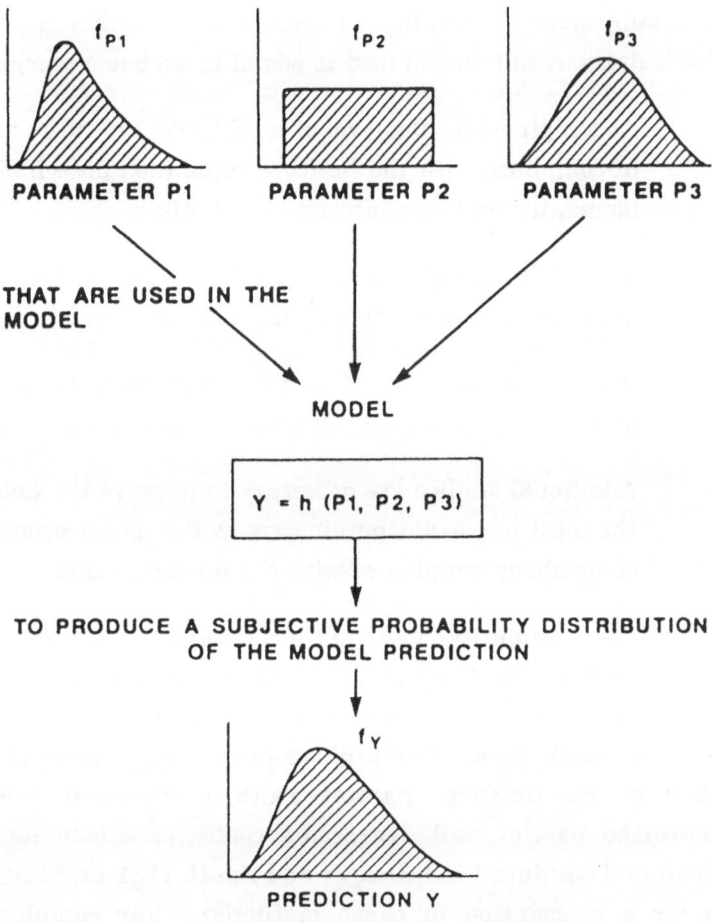

ORNL-DWG 87-15510

THE STATE OF KNOWLEDGE ABOUT UNCERTAIN
PARAMETERS OF A MODEL IS DESCRIBED BY
SUBJECTIVE PROBABILITY DISTRIBUTIONS F_p OR
THEIR DENSITY FUNCTIONS $f_p = dF_p/dp$

f_{P1} f_{P2} f_{P3}

PARAMETER P1   PARAMETER P2   PARAMETER P3

THAT ARE USED IN THE
MODEL

MODEL

$$Y = h \, (P1, \, P2, \, P3)$$

TO PRODUCE A SUBJECTIVE PROBABILITY DISTRIBUTION
OF THE MODEL PREDICTION

f_Y

PREDICTION Y

Fig. 3 An illustration of a subjective probability distribution of a deterministic model prediction. The distribution is produced from subjective probability distributions specified for each uncertain model parameter.

The results of a parameter uncertainty analysis determine a subjective level of confidence (degree of belief) that the value to be predicted is within a specified range or, that it is in compliance with specified limiting values, provided that all uncertainties not quantified are judged to be neglible.

Once a subjective probability distribution of predicted values has been established, the result should lead to one of three basic conclusions:

1) At a high subjective level of confidence, the value to be predicted is in compliace with the limiting value (i.e., dose limit, source upper bound, derived concentration limit, etc.);

2) At a high subjective level of confidence, the value to be predicted is not in compliance with the limiting value, or

3) The subjective levels of confidence for violation of and for compliance with the limiting value are of the same order of magnitude.

Additional studies are necessary to improve the knowledge base for the most important parameters in the model prior to making decisions about compliance with the limiting value.

Analytical and numerical methods for ranking important parameters are discussed in detail in the IAEA document. Among the measures used for ranking are: ratios of variances, simple correlation coefficients, partial correlation coefficients, standardized partial regression coefficients, rank correlation coefficients, partial rank correlation coefficients, and standardized partial rank regression coefficients (see Iman et al. 1981, Hoffman and Gardner 1983, Draper and Smith 1981 and Freund and Minton 1979 for a description of these methods). For complex models, it is recommended that several of these methods be used as no one method provides the best results for all model types and applications.

It is important to note that the results of a parameter uncertainty analysis account for only those uncertainties that have been quantified in the analysis. Therefore, additional terms must be added to the model to represent uncertainty beyond those associated with estimation of parameter values (i.e., suspected structural deficiencies in the conceptual and computational model).

Different quantitative uncertainty statements and parameter rankings may be expected depending on the question asked of the model. Therefore, different quantitative uncertainty statements and parameter rankings should be associated with the assessment of dose to members of critical population subgroups than for assessments involving large populations. Differences are also expected for uncertainty statements associated with generic assessments than for assessments that are site-specific. For this reason, the uncritical acceptance of published distributions for model parameters should be avoided. Caution must be exercised to ensure that the quantifications of parameter uncertainty conform to the conditions defined by the assessment question.

Because subjective judgment must be used to specify ranges and probability distributions for uncertain parameters, the results of a parameter uncertainty analysis will change with the state of knowledge of the experts involved. For this reason, the IAEA document recommends that individuals be consulted who have intimate familiarity with the model, the data available for the uncertain parameters, and the type of question being asked of the model. In the absence of such expertise, the individuals performing the analysis should specify distributions of sufficient width to reflect their lack of knowledge about these parameters and to ensure that their subjective confidence intervals will encompass the true value. The analysis can then be refined through acquisition of additional information when these subjective confidence intervals approach a level of concern (i.e., dose or concentration limit).

Model Intercomparison

Another procedure for evaluating the reliability of model predictions is the intercomparison of different models. This method is often criticized because it offers no measure of accuracy without independent test data. When discrepancies exist between model predictions, the question arises as to which set of results is closest to being correct. This question is extremely difficult to answer without independent test data.

Intercomparison exercises, however, are very useful for identifying obvious errors and weaknesses and for sharing information. They also provide a forum for learning about alternative modelling approaches and on how different model structures, parameterizations and numerical characteristics influence predictions. They are especially useful when each of the models has been subjected to a parameter uncertainty analysis and when at least one of the models has been tested against relevant data. Care, however, must be taken to ensure that similarity in the predictions of different models is not simply due to similarity in the data bases for estimating parameter values. It is not at all uncommon to find a variety of different models using similar references for their parameters; thus, despite differences in their structure, their results may be nearly identical (Hoffman et al. 1984). These references, however, may not be directly relevant to a given assessment situation.

CONCLUSIONS

The IAEA document recognizes that models, at the very best, are only approximations of real systems; therefore, truly "valid" models rarely exist. Models can only be considered "valid" when sufficient testing has been performed to ensure an "acceptable" level of accuracy. The level of accuracy considered "acceptable" is determined subjectively and will vary from case to case. Reliability is defined as a measure of confidence in model predictions. This measure of confidence can be expressed quantitatively or qualitatively. Although many of the procedures recommended in the IAEA document are still in a state of development and refinement, they should facilitate the transition from a qualitative to a quantitative assessment of reliability.

In the absence of relevant test data, or data obtained from experiments designed to answer a specific assessment question, the uncertainty associated with model predictions must be determined. Well established procedures from probability calculus and statistics can be used to propagate subjective estimates of parameter uncertainty into quantitative uncertainty statements about the model prediction. However, these uncertainty statements must still be recognized as being subjective. Quantitative estimates of uncertainty will change depending on the assessment question and the "state of knowledge" of the experts involved.

Once quantitative uncertainty statements are given, the following questions arise:

1) Is this uncertainty acceptable?

2) What are the most important sources that contribute to this uncertainty?

3) Can this uncertainty be effectively reduced?

Quantitative uncertainty statements are necessary in the process of defending decisions about the acceptability of potential and actual releases of radioactivity and provide indications of priorities for future research. Through such research, the predictive accuracy of environmental transfer models will increase as well as our overall understanding of natural systems.

REFERENCES

Draper, N. R. and H. Smith. 1981. *Applied Regression Analysis.* John Wiley and Sons, Inc., N. Y.

Freund, R. J. and P. D. Minton. 1979. *Regresion Methods - A Tool for Data Analysis.* Marcel Dekker, Inc., New York and Basel.

Hofer, E. and F. O. Hoffman. 1987. Selected examples of practical approaches for the assessment of model reliability - Parameter uncertainty analysis. In: Proceedings of a Workshop on Uncertainty Analysis for Systems Performance Assessments, OECD/NEA, Seattle, U.S.A., 24th-26th February 1987.

Hoffman, F. O. and R. H. Gardner. 1983. Evaluation of uncertainties n environmental radiological assessment models. In: *Radiological Assessment: A Textbook on Environmental Dose Assessment.* J.E. Till and H. R. Meyer, eds. U. S. Nuclear Regulatory Commission, Washington, D. C. NUREG/CR-3332, ORNL-5968, Ch. 11, pp. 11-1 to 11-55 (1983).

Hoffman, F. O., U. Bergström, C. Gyllander and A.-B. Wilkens. 1984. Comparison of predictions from internationally recognized assessment models for the transfer of selected radionuclides through terrestrial food chains. *Nuclear Safety 25*: 533-546.

Iman, R. L., J. C. Helton, and J. E. Campbell. 1981. An appraoch to sensitivity analysis of computer models: Part II - Ranking of inpur variables, response surface validation, distribution effect and technique synopsis. *Journal of Quality Technology 13*: 232-239

International Atomic Energy Agency (draft report 1987-09-29). Procedures for evaluating the reliability of predictions made by environmental transfer models. IAEA, Division of Nuclear Fuel Cycle. (to be published as an IAEA Safety Series).

VERIFICATION AND VALIDATION OF NRPB MODELS FOR RADIONUCLIDE TRANSFER THROUGH THE ENVIRONMENT

Marion D Hill

National Radiological Protection Board
Chilton, Oxon, OX11 ORO, UK

ABSTRACT

Some of the models available at NRPB for predicting radionuclide transfer through the environment are briefly described and the problems encountered in verifying and validating them are identified. It is suggested that models for transfer through the various parts of the environment can be ranked in terms of whether they can be validated quantitatively, or whether all that can be achieved is quantitative validation of their conceptual basis. Priorities for further NRPB work on model verification and validation are given.

INTRODUCTION

Since the organisation was established in 1970 the National Radiological Protection Board (NRPB) has developed models for radionuclide transfer through all sectors of the geosphere and biosphere, and has also implemented models developed by other organisations, and by national and international expert groups. During the course of this work, verification and validation have gradually become more important and more formal procedures. They have also become more difficult, because as knowledge of radionuclide transfer through the environment has increased, so has the complexity of the models used to predict transfer patterns and rates. The main purpose of this paper is to identify the problems which have been encountered by NRPB in verifying and validating models for radionuclide transfer through various parts of the environment, to compare these models in terms of ease of verification and validation, and to suggest the areas to which most effort might be directed in the future for assessing model reliability.

MODELS AVAILABLE AT NRPB

Brief descriptions of all the models available at NRPB for predicting radionuclide transfer through the environment are given in reference [1], together with summaries of the procedures which have been used to verify and validate them. Table 1 shows the models which are most relevant to the scope of this workshop; two of them, FARMLAND and BIOS, are discussed in other papers in these proceedings [2,3]. Apart from GESAMP 6, NUM 6, and GESAMP 7, all the models shown in Table 1 are of the compartment type. GESAMP 6 and GESAMP 7 are analytical models; NUM 6 is the numerical version of GESAMP 6 and is a finite difference model. FARMLAND, TRIMFOOD, RIVER-GEN, LAKE-GEN, SEVERN, NOCEAN 2, MINIBOX, NUM 6 and BIOS were developed by NRPB. MARIN-1 has been developed jointly by NRPB and the Riso National Laboratory as part of the CEC MARINA project. COMMA was developed by the Fisheries Laboratory of the UK Ministry of Agriculture, Fisheries and Food, and NRPB. GESAMP 6 and GESAMP 7 are models recommended by the United Nations Group of Experts on the Scientific Aspects of Marine Pollution [4].

VERIFICATION

Verification is the process of showing that a mathematical model is a proper representation of the conceptual model on which it is based, and of checking that the mathematical equations involved have been solved correctly. For models which are implemented on computers, as all those at NRPB are, verification includes quality assurance of the computer code.

Verification procedures vary from one organisation to another, and also depend on the purpose for which the model is to be used. If the model is primarily a research tool, it may be acceptable to be somewhat less vigorous about verification than if the results of the modelling are to form an input to decisions on authorisations for effluent discharges or to licensing of a nuclear facility. Most of the models listed in Table 1 are used for the latter type of work, and have therefore been subject to fairly vigorous verification. This is achieved through internal NRPB procedures which include:

TABLE 1

Models for Radionuclide Transfer in the Terrestrial
and Aquatic Environments

Model Name	Radionuclide Transfer Process Modelled
FARMLAND (Food Activity from Radionuclide Movement on LAND)	Transfer through terrestrial foodchains (milk, meat, green vegetables, grain, root crops).
TRIMFOOD	Tritium transfer through terrestrial foodchains
RIVER-GEN	Dispersion and transfer in rivers
LAKE-GEN	Turnover and transfer in lakes
SEVERN	Dispersion and transfer in Severn Estuary
NOCEAN 2	Dispersion in N European coastal seas and interactions with sediments
MARIN-1	Dispersion in N European coastal seas and interactions with sediments
COMMA (COMpartment Model of the Atlantic) MINIBOX GESAMP 6 and NUM 6 GESAMP 7	Dispersion in the deep ocean and interactions with sediments
BIOS	Transport in the biosphere following release from waste repositories

i)   review of model structure and basic equations by staff other than
     those involved in developing the model;

ii) checking computer codes to ensure that programming is correct;

iii) comparing computed results with problem solutions obtained from other
     models.

These internal procedures are supplemented by taking part in UK and
international model-model comparison exercises (see, for example
references [5] - [7]), and by submitting model descriptions and example
results to other organisations and experts for external peer review.

The NRPB experience with model verification has been that it is very straightforward when the model in question is a relatively simple one and when other similar models exist and results can be directly compared. When the model is complex, or no comparable model exists, verification is best carried out as an integral part of development and computer coding, rather than retrospectively. In such cases more planning of test procedures is required and more documentation is needed in order to demonstrate that the model has been verified.

VALIDATION

Validation consists of showing that a conceptual model and the computer code derived from it provide an adequate representation of radionuclide transfer processes in the real environment. The definition of what constitutes an adequate representation necessarily involves some subjective judgement and may depend on whether the model is intended for general or site specific applications, and on the purpose for which model results will be used (for example, whether they are to be used in demonstrating compliance with a limit or are to be an input to optimisation exercises). Although it is rarely possible to define precisely what will be judged to be an adequate representation in advance of carrying out a validation exercise for a particular model, it is preferable to have an outline definition in mind.

Ideally, validation is carried out by comparing model calculations with sets of field observations and experimental measurements other than those which were used in developing the model. In practice it is not possible to fully validate any environmental transfer model in this way because independent data sets do not exist for all the radionuclides, time scales and environmental conditions to which the model is likely to be applied. However, some models, or parts of them, can be partially validated by comparisons of calculations with measurements. This procedure has been applied to parts of the NRPB FARMLAND models for radionuclide transfer through terrestrial foodchains [2], to the NOCEAN 2 and MARIN 1 models for radionuclide dispersion in coastal seas [1], and to the radionuclide-sediment interaction components of the deep ocean models listed in Table 1 [8,9]. In addition, data on salinity and temperature in the ocean have been used to check that the flow field which is implicit in COMMA is realistic [8].

In cases where quantitative validation of the type described above is not feasible, recourse must be made to more qualitative techniques which mainly aim to check that a conceptual model is adequate, rather than that both this and the computer code derived from it are valid. These techniques include the use of data on the environmental behaviour of chemically analogous natural or artificial radionuclides or stable elements, use of data from laboratory experiments which simulate environmental conditions, and external peer review of the assumptions used in developing a model. In NRPB work, the tendency has been to use the last of these techniques for all models, to rely primarily on environmental data for models for radionuclide transfer in the aquatic environment, and to use a combination of environmental and laboratory data in validation of terrestrial foodchain models [1]. The validation of BIOS presents particular problems, which are discussed in reference [3].

From experience gained at NRPB, it is apparent that models for radionuclide transfer through the various parts of environment can be ranked in terms of whether they can be validated in a largely quantitative way or whether validation must be mainly qualitative. Table 2 shows a suggested ranking for the types of model listed in Table 1. The entire ranking may not be generally applicable because parts of it are based on availability of data for radionuclide movement in the environment in and around the UK (in particular, the abundance of data from Sellafield discharges, and the lack of data for freshwater bodies). However, the broad features of the ranking are believed to apply to environmental transfer models in general.

TABLE 2

Suggested Ranking of Models in Terms of Quantitative/Qualitative Validation

Ranking	Type of Model
Quantitative ↓ Qualitative	Terrestrial foodchain Dispersion in coastal seas Dispersion in the deep ocean Transfer in freshwater bodies Transport in the biosphere following release from waste repositories

DIRECTIONS FOR FUTURE WORK

Mathematical modelling of radionuclide transfer through the environment will always be an important aspect of radiological protection. It is an essential component in predicting doses and risks to individuals and populations, and hence in providing an input to decisions on the siting, design and operational procedures for all types of facility which will or could release radionuclides into the environment. It is also the only way in which present doses from some radionuclides which have already been released into the environment can be assessed. For all these purposes it is necessary to have models which are both verified and validated. However, since resources are limited, there is a need to decide which areas of model verification and validation should receive the highest priority. In the case of validation, such decisions should take into account the practical problems involved.

As far as NRPB's own work is concerned, the main priorities for future effort in model verification and validation are as follows:-

i) increased emphasis on verification during model development;

ii) quantitative validation of terrestrial food chain models and models for radionuclide dispersion in coastal seas;

iii) qualitative validation of the BIOS model for predicting radionuclide transfer following release from waste repositories.

In the quantitative validation work the intention is to use uncertainty analysis techniques to generate probability distributions of modelling results for comparisons with field measurements.

Other organisations will doubtless have differing priorities, depending on their own interests and capabilities. Nevertheless, during the course of this workshop it might be possible to reach agreement on the general directions for future work, particularly that to be carried out at as part of international projects.

REFERENCES

1. Hill, M D (ed). Verification and validation of NRPB models for calculating rates of radionuclide transfer through the environment. National Radiological Protection Board, Chilton (to be published).

2. Brown, J, Haywood, S M and Wilkins, B T. Validation of the FARMLAND models for radionuclide transfer through terrestrial foodchains. These proceedings.

3. Smith, G M. Modelling the radiological impact of release of radionuclides into the biosphere from solid waste disposal facilities. These proceedings.

4. IAEA. An oceanographic model for the dispersion of wastes disposed of in the deep sea. IAEA Technical Report 263, IAEA, Vienna 1986.

5. Meekings, G F and Walters, B. Dynamic models for radionuclide transport in agricultural ecosystems: summary of results from a UK code comparison exercise. J. Soc. Radiol. Prot., 1986, 6 (2) 83-

6. Brown, J, Haywood, S M, Simmonds, J R, Pröhl, A, Friedland, W, and Paretzke, H G. A comparison of two dynamic foodchain models following accidental releases, CEC, Luxembourg (to be published).

7. Mobbs, S F, Hill, M D, Koplik, C, and Demuth, C. A preliminary comparison of models for the dispersion of radionuclides released into the deep ocean, NRPB-R194, London, HMSO, 1986.

8. NEA. Review of the continued suitability of the dumping site for radioactive waste in the North East Atlantic, OECD/NEA, Paris, 1985.

9. Delow, C E, and Mobbs, S F. Modelling the interactions between radionuclides and particles in the ocean for assessment of sea disposal of radioactive waste, NRPB report for the Dept of Environment (to be published).

BIOMOVS: AN INTERNATIONAL MODEL VALIDATION STUDY

C. Haegg and G. Johansson

National Institute of Radiation Protection

Stockholm - SWEDEN

ABSTRACT

BIOMOVS (BIOspheric MOdel Validation Study) is an international study where models used for describing the distribution of radioactive and nonradioactive trace substances in terrestrial and aquatic environments are compared and tested. The main objectives of the study are to compare and test the accuracy of predictions between such models, explain differences in these predictions, recommend priorities for future research concerning the improvement of the accuracy of model predictions and act as a forum for the exchange of ideas, experience and information.

INTRODUCTION

Up to now, model testing has focused on aspects of the prediction of the physical dispersion of contaminants in the atmosphere, surface water and ground water. Only limited attention has been given to the testing of models used to predict the accumulation of radionuclides in the biosphere, e.g. soils or terrestrial and aquatic food chains.

BIOMOVS (BIOspheric MOdel Validation Study) is an international cooperation study initiated in 1985 by the Swedish National Institute of Radiation Protection (NIRP). The study concentrates on the testing of models which predict the accumulation and remobilisation of radionuclides and other toxic trace substances in the biosphere. Emphasis is placed on terrestrial and aquatic pathways of importance in the assessments of exposure to human populations.

The study has two approaches. Approach A aims to test model predictions against independent sets of data, i.e. validation of the models. Approach B, on the other hand, aims to compare model predictions and associated uncertainties for some test scenarios that cannot be validated or are impracticable to validate, e.g. scenarios for waste disposals.

Results obtained from this study may be used in conjunction with results obtained from other model testing studies in order to enhance the reliability of model predictions. This is particularly beneficial when the predicted accumulation of stable and non-stable nuclides constitute the major uncertainty in the evaluation of human exposure and health risk.

The BIOMOVS study is directed by a coordinating group. Each organisation that formally participates in the study appoints one member to the group. The managing organisation is the NIRP and the NIRP together with Kemakta Consultants Co are acting as the project secretariat. At present, 22 organisations from 14 different countries (cf. Table 1) have joined the study. Of these, 12 organisations are represented in the BIOMOVS coordinating group (indicated with an asterix in Table 1). The International Atomic Energy Agency and the Nuclear Energy Agency of the Organisation for Economic Cooperation and Development participate as observers. Organisations that provide economic support, beside the participating organisations and the NIRP, are the International Union of Radioecologists and the Nordic Liaison Committee for Atomic Energy.

TABLE 1
Organisations represented in the BIOMOVS study

Organisation	Country
National Research Institute for Radiobiology and Radiohygiene (*)	Hungary
Comitato Nazionale per l'Energia Atomica-Disp (*)	Italy
National Cooperative for the Storage of Radioactive Waste (*)	Switzerland
National Radiological Protection Board (*)	United Kingdom
National Institute of Radiation Protection (*)	Sweden
Risoe National Laboratory (*)	Denmark
Technical Research Centre of Finland (*)	Finland
Tokai Research Establishment (*)	Japan
Department of Energy (*)	USA
Atomic Energy of Canada Limited (*)	Canada
Studiecentrum voor Kernenenergie, SCK/CEN (*)	Belgium
Department of Environment/ANS	United Kingdom
Ministry of Agriculture, Fisheries and Food	United Kingdom
Institute of Radiation Hygiene	West Germany
Central Electricity Generating Board	United Kingdom
Swiss Federal Institute for Reactor Research	Switzerland
Gesellschaft Pur Strahlen- und Umwelt Forschung	West Germany
Japan Atomic Energy Research Institute	Japan
Oak Ridge National Laboratory	USA
Laboratory of Radiation Research	The Netherlands
Empresa Nacional de Residuos Radioactivos(*)	Spain
Studsvik Energiteknik AB	Sweden

(*) member of the BIOMOVS coordinating group.

In this paper we will give a short background of why the BIOMOVS study was initiated and how it will be carried out. We also briefly discuss the present status of the study.

BACKGROUND AND SOME GENERAL CONSIDERATIONS

Mathematical models are extensively used for evaluating the environmental impact of releases of radionuclides from nuclear power plants, nuclear waste disposal, uranium ore mining etc. However, what can be studied by means of mathematical methods is not the reality, but more or less simplified and abstract ideas of the real situation. Thus, we study the behaviour of mathematical models which only approximate real world conditions. For reliable decisions to be made, especially when optimization is being carried out, the extent of uncertainty associated with model predictions must be quantified by some means.

The primary objectives of the BIOMOVS study are to estimate the extent of the uncertainty associated with model predictions of the transfer, accumulation and remobilisation of environmentally significant radionuclides or trace substances in the biosphere and to recommend procedures for improving predictive accuracy. These estimates and recommendations will be made for a variety of scenarios relevant to the release of some radionuclides and trace substances in the biosphere following, e.g. the operation of nuclear power plants and nuclear waste disposal.

We need a deep and genuine understanding of the processes in nature which in fact constitute the reality behind our models. The better the understanding, the less uncertainty there is in model predictions. A lack of understanding of natural processes may also lead to very uncertain predictions when extrapolating in space and time, a problem that arises in for example safety assessment studies. However, no fixed procedures exist for ensuring that all processes, or even all relevant processes, have been considered.

Thus, the representation of reality in a simplified form for modelling purposes is by no means easy and we can rarely specify a model which explicitly considers all processes. However, many processes can be taken into account implicitly. An example is the use of Kd-values when describing adsorption/desorption phenomena. Nevertheless, the best we can do is to try to specify a model that considers all relevant processes. In this case conceptual uncertainties in the model are reduced as much as is possible and the remaining significant uncertainties are linked to the mathematical formulation and those in the controlling parameters of the model. Unfortunately, it is difficult to achieve this situation and therefore the BIOMOVS study considers the two general types of uncertainties:

- uncertainties in the controlling parameter values and

- uncertainties in the model structure.

Uncertainties in the parameter values are the types of uncertainty that are generally considered, while uncertainties in the model structure are somewhat more difficult to consider. The best method for assessing these uncertainties is to test model predictions against independent sets of

data, i.e. validate the models. This also gives the possibility of
finding out if our scenario identification and conceptual understanding
of nature are relevant.

Another way to improve our understanding is model intercomparision by
subjecting scenarios to different experts in the field for analysis.
Another method for improving our understanding of the uncertainties
associated with model predictions are parameter uncertainty and
sensitivity analysis. Parameter uncertainty analysis and model
intercomparisions could be regarded as complementary methods. These
methods both play a significant role when validation is impracticable or
impossible.

It is being increasingly recognised that model predictions are of
limited value unless accompanied by some indication of the associated
uncertainty. An estimate of this uncertainty may be obtained by varying
the control parameter values; either by random or stratified sampling
techniques. Here, the parameter values are subjected to some distribution
that generally is unknown and has to be chosen by the model user.
Moreover, the degree of correlation between different types of parameters
may also be unknown. As a consequense of this, a modelling study not only
includes the choice of model structure and parameter values, but also
many elements of judgement. Validation should cover all these aspects.

In the procedure of validation we must realise that the applicability
of the model is limited by the validity of the scientific principles
underlying the model. Therefore, a question we may ask ourselves is:

    - to what degree shall the model predictions agree with
      experimental data before we can say that the model is validated?

It seems that no well defined distinction between a model that is said to
be validated or not be validated exists. This decision has finally to be
made by responsible decision makers, but it is of great importance from a
scientific point of view to give guidance on how this distinction should
be made. This is one question that will be discussed in the final report
of the BIOMOVS study.

Another question that may be raised is:

    - is it possible to apply a validated model to a system other than
      the one for which it is validated and state that the outcome of
      the model reflects the status of the new situation?

The situation is analogous to opening a door with a key. If the key is
properly constructed it may open several doors, but we cannot forsee
which door it will open. The BIOMOVS study aims to reduce this
uncertainty and hence extend the applicability and reliability of
biospheric assessment models. This special task is also to be discussed
in the final report.

IMPLEMENTATION OF THE STUDY

The best method for testing models is, as discussed above, to compare
their predictions for specific scenarios with field data. These field
data-sets must be independent of the data used for developing non-site

specific parameter values in the model. Examples of parameters which often are treated as non-site specific (or scenario independent) in biosphere models are, e.g.:

(i) element specific transfer factors such as solid/water distribution coefficients, milk and meat transfer factors, plant/soil concentration ratios and water/fish concentration factors and

(ii) element independent parameters such as deposition velocity, the effective mass per unit area of the root zone of soil and of the exchangable sediments of aquatic systems.

When model predictions are compared with independent sets of data, the model is attempting to reproduce values that are already known. In the BIOMOVS study, the known values will not be revealed until after the calculations have been carried out in order to avoid premature calibration of the model.

A comparision of the model predictions with an independent set of data provides a measure of the accuracy of the prediction with respect to the scenario and location from which the field data set was derived. This type of testing is generally referred to as "model validation" and within this study constitutes approach A.

On the other hand, when it is not possible to test models against independent sets of data, predictions and related uncertainty estimates can be obtained from different models for specific test scenarios. This procedure is suitable for clarifying discrepancies and similarities among model predictions. Conclusions may also be drawn about the extent of the expected uncertainty associated with these predictions. The disadvantage of this procedure is that no estimate of the accuracy of model predictions and associated uncertainties is possible.

When formulating specific assessment questions for the purpose of comparing model predictions, it is necessary to provide a standard set of data for input parameters that are normally determined on a site-specific basis. These parameters include such factors as meteorology, length of growing season, turnover time and flowrate of water etc. If such detailed information is supplied the discrepancies in model predictions should be affected primarily by differences in the site independent data adopted for use with the model, the model structure and in the modellers' estimates of uncertainty about the prediction of the model.

For each scenario a site specific source term must be quantified, e.g. concentration of radionuclides in air surface water or ground water. It may also be of interest for some test scenarios to use a given concentration of a radionuclide in sediment or surface soil to simulate a situation that follows an accidental release or decommissioning of a nuclear power plant or an intrusion into a radioactive waste repository.

This alternative test procedure is referred within the study as approach B. This approach is capable of addressing a much larger number of scenarios than approach A.

PRESENT STATUS OF THE STUDY

In this section only a brief overview of the progress of the BIOMOVS study is given. A more comprehensive presentation of some of the results will be given by other participants at this symposium (ref. 1-4).

Figure 1 shows schematically the time schedule for the study. It can be seen that four work shops have been held so far. The results from these work shops are summarised in progress reports nos 1 - 4 (ref. 5). For approach A there is a short time-lag when compared with approach B owing to the time needed for compiling data sets suitable for validation.

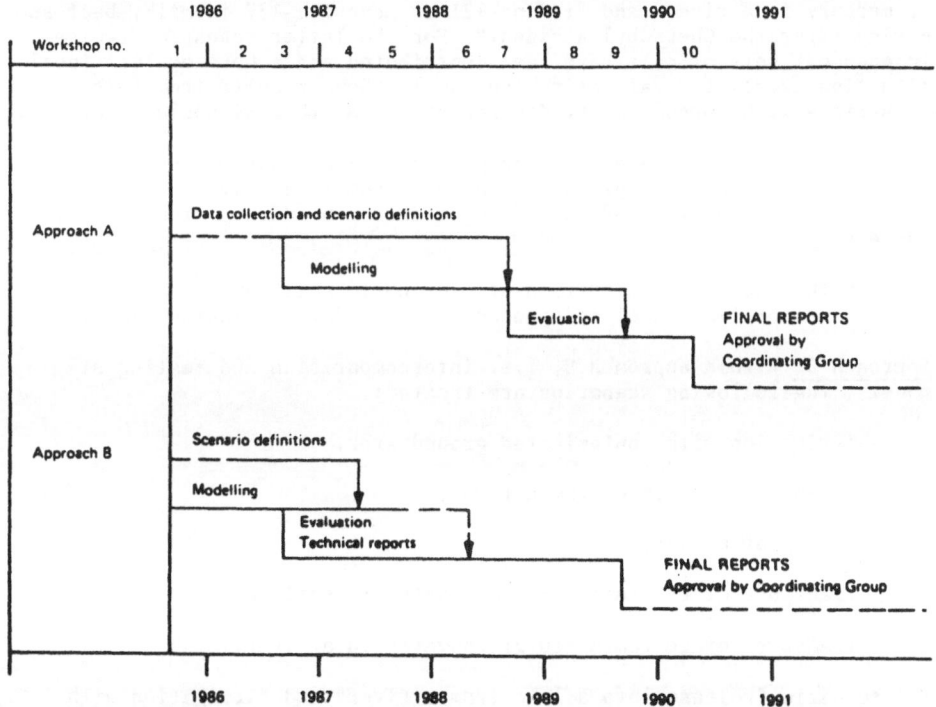

Figure 1. Approximate time schedule for the BIOMOVS study.

However, the great activity among scientists after the Chernobyl accident has led to a large body of suitable data. In order to take advantage of this situation, several new scenarios have been proposed.

The scenarios in the study have been chosen to assure that pathways of importance will be addressed. The choice of scenarios should also make it possible to critically test and compare different models. A thorough description of the scenarios is given in progress reports nos 1 - 4 (ref. 5).

Approach A. Within approach A, i.e. validation of models, five scenarios are treated. They are:

- release of mercury into a river,

- radioactivity in the environment around Sellafield (postponed),

- iodine-131 and caesium-137 in milk, beef and barley after the Chernobyl accident,

- migration of caesium in soil and

- "the Chernobyl lake".

Two scenarios are already in the "phase of validation", namely; "release of mercury in a river" and "iodine-131 and caesium-137 in milk, beef and barley after the Chernobyl accident". For the latter scenario, twelve independent sets of data have been contributed and a more or less ideal situation exists for validating the models. Some results from this scenario will be presented by Koehler et al. at this symposium (ref. 1).

The scenario "release of mercury into a river" has recently been extended to include independent data sets from three rivers. This has been done in order to find out if the models can address other similar systems.

For the scenarios "migration of caesium in soil" and "the Chernobyl lake", the compilation of independent sets of data is taking place.

Approach B. Within approach B, i.e. intercomparision and testing of models, the following scenarios are treated:

- irrigation with contaminated groundwater,

- release into a lake from a river,

- ageing of a lake,

- transport of contaminated groundwater to soil and

- transport of contaminated groundwater to a river.

The scenario "release into a lake from a river" and "irrigation with contaminated groundwater" are the first to be completed in the study and some results will be presented at this meeting by Bergstroem (ref. 2) and Grogan et al. (ref. 3), respectively. For the other B-scenarios, only preliminary results exist. The scenario "release into a lake from a river" and "ageing of a lake" treat waterborne radionuclides that reach water reservoirs and are related to assessment studies for radioactive waste and mining facilities. However, in these scenarios the interface between the biosphere and geosphere was not modelled. In order to address this problem, the two scenarios "transport of contaminated water to soil and to a river" were formulated. These two scenarios deal with groundwater borne radionuclides that reach the biosphere via surface soil or sediment. Some preliminary results will be presented at this meeting (ref. 4).

REFERENCES

1. Koehler, H. and Nielsen S., From model intercomparision to model validation - an example from the BIOMOVS study, "Methods for Assessing the Reliability of Environmental Transfer Models Predictions", Athens, October 5-9, 1987.

2. Bergstroem U., Intercomparision of model calculation of the turnover of Ra-226 within an aquatic ecosystem, ibid.

3. Grogan H. and van Dorp F., The reliability of environmental transfer models applied to waste disposal, ibid.

4. Bergstroem, U., Sundblad, B., Argarde, A.-C. and Ericsson, A.-M., Comparison of two model approaches for the geosphere/biosphere interface, ibid.

5. BIOMOVS progress reports. Can be ordered from the BIOMOVS project secretariat: Kemakta Consultants Co, Pipersgatan 27, S-112 28 Stockholm, Sweden.

A COMPARISON OF MODEL PREDICTIONS OF GASEOUS DISPERSION
WITH ENVIRONMENTAL MEASUREMENTS OF ^{14}C AROUND SELLAFIELD, UK

M J Fulker
Environmental Protection Group, B433
British Nuclear Fuels plc, Sellafield,
Seascale, Cumbria CA20 1PG

R L Otlet and A J Walker
Low Level Isotope Laboratory, Harwell
Didcot, Oxfordshire OX11 ORA

Abstract

The Sellafield nuclear fuel reprocessing plant, sited 0.5 km from the coast, emits ^{14}C to atmosphere primarily from two 120m high stacks. Measurements of ^{14}C in hawthorn berries (Crataegus) at a distance of 4 km over six years shows significant variations in the net ^{14}C per kg carbon with direction from the site. The predicted variation with direction, taking into account carbon uptake rate as well as the pattern of wind bearing and dispersion conditions through the growing season, shows reasonable agreement in most cases. The ^{14}C distributions are largely accounted for by angular wind frequency during the daylight hours growing season, which for this site is markedly affected by sea breezes. For the years when there is a marked disagreement between the measured and predicted profiles the most likely explanation is considered to be curvature of the plume due to the influence of the hilly terrain.

Introduction

The doses from ^{14}C to people living close to a nuclear plant are mainly

from ingestion of food and may be estimated by using an activity

concentration method assuming that the isotopic carbon composition in the

food is the same as in local air. Any differences due to different rates

of incorporation or transfer of different carbon isotopes are likely to be

small. Ingestion of locally grown food is then assumed to give rise to

^{14}C incorporation in the body in the same ratio of ^{14}C to total carbon as

that originally in the air at the point and time of production of the

food. This simple procedure is however, subject to a number of

complications. There is a need for further work on the possibility of

uptake of ^{14}C from soil for some crops. Also, if the activity

concentration method is to be used, the concentrations of ^{14}C in air must

be closely related to the period during which photosynthesis operates. In

this respect seasonal and diurnal variations in atmospheric dispersion can

result in large differences between annual average concentrations in air and that which is present during the growth period.

Killough and Rohwer [1] have considered the effects of dispersion during the daylight hours growing season for an inland site at Knoxville, Tennessee. In that case they found that during the period of ^{14}C uptake by plants the frequency of unstable dispersion categories was significantly increased, and for a high release point, the uptake doses to food consumers were increased by a factor of three compared with 24 hour average dispersion conditions.

For the more equitable climate of the west coast of the United Kingdom the effect of unstable dispersion conditions is likely to be less marked, but sea breezes tend to occur during the CO_2 uptake period. It is therefore necessary to examine the directional wind frequency for the CO_2 uptake period, since this may be very different from the annual average.

The uptake of ^{14}C by different crops, even grown at the same site, may be complicated by differences in growing season. The directional wind frequency may vary significantly for growing seasons which differ by only a few weeks. For this reason the choice of plant material, for a study of directional variations in ^{14}C, was constrained by the need to find something which was available at a wide range of sites and in which ^{14}C would be assimilated by photosynthesis within a predictable growing season. Hawthorn berries (Crataegus) were thus chosen on the basis of these criteria. Additional uptake of ^{14}C from soil, which may be a possibility for some plants (the subject of a separate study), was considered unlikely to be a significant process for hawthorn berries where incorporation of carbon in the fruit should be solely from CO_2 assimilated from the atmosphere by photosynthesis during the daylight hours growing season.

In an earlier study [2] ^{14}C was measured in hawthorn berries collected in the autumn at sites on a 4 km arc around the Sellafield reprocessing plant and the relationship between the ^{14}C activity concentration (Bq ^{14}C per kg C), elevation above sea level and wind direction was examined. For the year 1982 there was evidence of a general correlation between wind frequency (weighted during the growing period by carbon uptake rate) and the ^{14}C activity concentration. A study of the ^{14}C in hawthorn berries over a wider area also showed evidence of topographical effects, the contours of ^{14}C activity concentration being distorted into the openings of seaward facing valleys in the Lake District fells [2].

The present study examines in more detail the dispersion aspects of ^{14}C uptake in hawthorn berries over several years, taking into account both dispersion category and directional wind frequency.

Experimental Methods

Hawthorn berries were collected at sites on the 4 km arc in the years 1981 to 1986 inclusive. In the year 1986 samples were collected from one tree at intervals throughout the growing season. Analysis of carbon content showed that carbon uptake started at about the third week of June and followed an approximately exponential half build-up time of three weeks (Figure 1). This was similar to the uptake profile determined in 1983 in Oxfordshire, but beginning about one week later in the more northerly latitude.

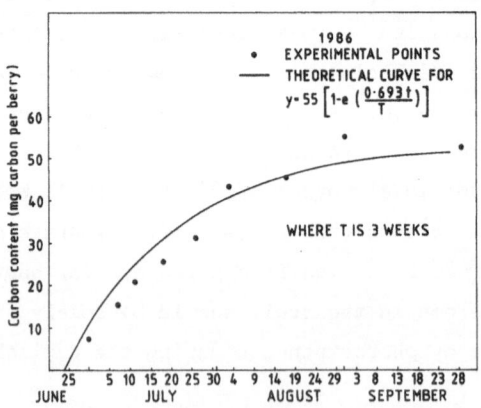

FIGURE 1 - Uptake of ^{14}C in hawthorn berries in Cumbria 1986

For the ^{14}C determinations berries were stripped of stalks, treated with 3M HCl to remove surface contamination (eg lime), washed in distilled water and dried in a hot cupboard. Samples were then converted to C_6H_6

for counting by the normal procedures used for standard precision ($\pm$1%) low-level radiocarbon measurements, ie, bomb combustion conversion to CO_2, CO_2 to C_2H_2 via lithium carbide and polymerisation to C_6H_6 over a V_2O_5 catalyst [4], [5].

The predicted ^{14}C distribution was determined from meteorological data and dispersion category estimates considered throughout the growing season. The dispersion category was assessed three times each day during daylight hours, assigned to the relevant 30° sectors and weighted according to the ^{14}C uptake rate for each week of the growing season. The ^{14}C uptake for the whole growing season, for each sector, was then determined using the dispersion factors for each dispersion category for an emission height of 120m, appropriate to the main emission sources, and a distance of 4 km.

Results and Discussion

Using the above procedure to estimate the variation of ^{14}C uptake as a function of plume direction it was found that the predicted peak ^{14}C concentration was mainly due to the high wind frequency in this direction. During the CO_2 uptake period the wind frequency is influenced by the occurrence of sea breezes. From the dispersion model, the unstable categories A and B resulted in somewhat higher concentrations at 4 km than the neutral conditions C and D. Although the frequency of unstable conditions was higher for the growing season than for the complete year, this had only a small effect on the profile of ^{14}C concentrations expected at 4 km. This is in contrast to the dispersion predictions of Killough and Rohwer [1] for an inland site in Tennessee where as stated earlier the high frequency of unstable conditions during the growing season had a marked effect on expected concentration.

Figures 2 to 7 show the measured net activity concentration of ^{14}C in hawthorn berries (Bq/kg C) as a function of direction from the source for the years 1981 to 1986, and compares the angular distribution with that predicted from analysis of the meteorological data. No attempt has been made here to predict the magnitude of ^{14}C in hawthorn berries but for clarity, the theoretical curves have been normalized to give the same peak values as the measured results.

34

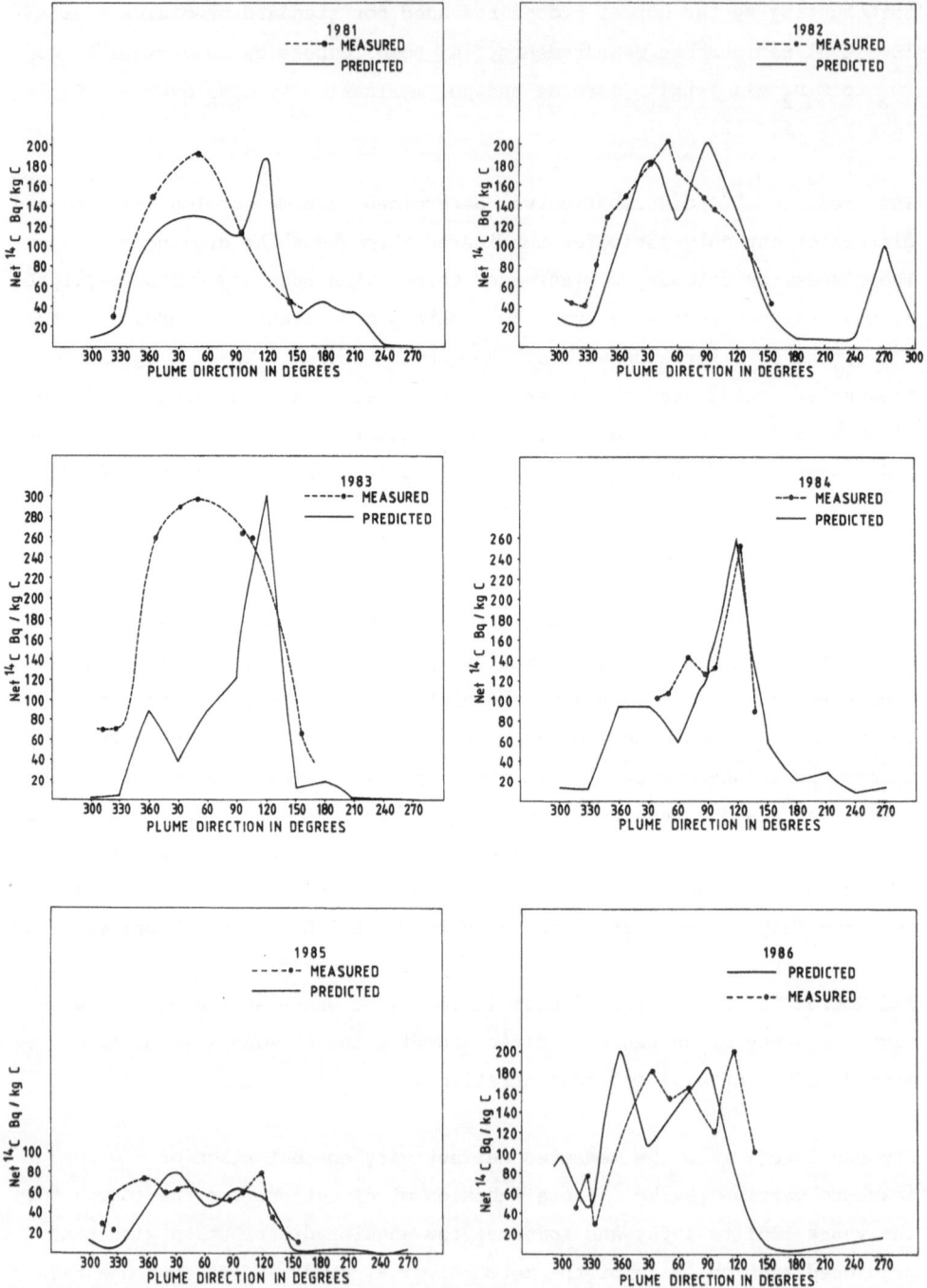

Figures 2-7 - Net ^{14}C in hawthorn berries at 4 km compared with predicted angular distribution, for the years 1981 to 1986.

For some years the predicted angular distribution is in general agreement
with that measured. In 1982, 1985 and 1986 the agreement is reasonable
although the position of the peak concentrations are not precisely
reproduced in the predicted curves (Figures 3, 6 and 7). For 1984 both
measured and predicted curves show a pronounced peak for a plume direction
of 120° (Figure 5), due primarily to the high frequency of wind blowing
towards the 120° sector throughout the daylight hours growing season in
that year. Figure 8 shows the directional wind frequency for day time,
June, July and August 1984. Compared with the average for the whole year,
the growing period winds are almost all on-shore with the highest
frequency in the 120° sector where the highest ^{14}C values in hawthorn
berries were found.

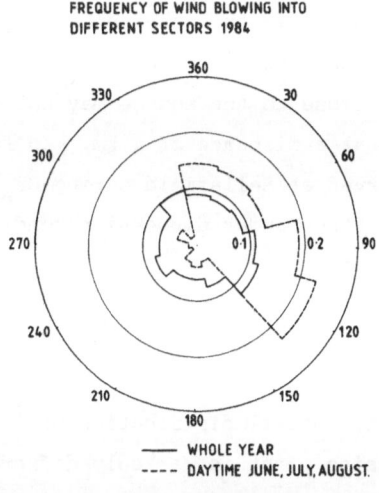

Figure 8 - Frequency of wind into different sectors 1984, daylight hours in
June, July and August, compared with the whole year.

The two years 1981 and 1983 show similar differences between measured and
predicted ^{14}C distribution, with measured peak concentrations for a plume
direction of about 50°. For these years the predicted peaks are at about
120°, as it is for the year 1984 (Figures 2 and 4). The shape of the
predicted curve is dependent on the assumed growth period for the hawthorn
berries, and, although this has been determined for the year 1986, there

may be some variation from year to year depending on differences in early summer weather. It was found, however, that for 1981 and 1983, the prominent predicted peak at 120° persisted even if the assumed start of the growth period was delayed by several weeks because of the high frequency of winds into this sector over much of the growing period. Since the measured peak concentrations in 1981 and 1983 are anti-clockwise of the predicted directions, it is unlikely that the difference can be explained by a wind shear effect. It is noted that the direction of the measured peak concentrations in these two years is more closely related to the land elevation topographical profile which rises to around 170m above sea level at about 3.5 km distance. However, since the gradients involved are relatively gentle it is also unlikely that plume compression effects, or decrease in wind speed, which are not considered in the simple dispersion model, could account for the observed differences. It is felt that a more likely hypothesis is that the topography over the first few kilometers creates plume curvature, so that the plume direction derived from the 10m anemometer close to the source may not be adequate for predicting its position at a distance of 4 km. In fact such plume curvature has been observed at Sellafield during SF_6 tracer release experiments from a high stack under dispersion conditions typical of daylight hours growing season.

Conclusion

This study shows that the spatial distribution of ^{14}C in hawthorn berries grown around a reprocessing plant is markedly different from the spatial distribution which might be obtained from annually averaged directional wind frequency data. The asymmetric distribution results primarily from the dispersion conditions relevant to the CO_2 uptake period when, for a coastal site, on-shore breezes dominate the directional wind frequency distribution. The agreement demonstrated by the detailed wind frequency/category analyses over the relevant growing period in some years gives confidence to the applicability of this technique more generally. In some years, however, differences in the angular position of the measured and predicted peak concentrations suggests the need to consider more complex factors such as plume curvature induced by topography. Theoretical prediction of ^{14}C activity concentrations in food crops is therefore uncertain and confirmation by crop sampling is necessary to provide good estimates of doses to consumers of locally grown food.

Acknowledgements

The financial support of British Nuclear Fuels plc, the Atomic Energy Reserach Establishment and the Department of the Environment is gratefully acknowledged. This research was conducted by agreement with these agencies, as part of the joint comprehensive study of radioactive materials in the Cumbrian environment until 1985, and through direct support from BNFL since then.

References

1   Killough, G G; Rohwer, P S. A new look at the dosimetry of ^{14}C released to the atmosphere as carbon dioxide. Health Physics 1978 34 141-159.

2   Walker, A J; Otlet, R I; Longley, H. Application of the use of hawthorn berries in monitoring ^{14}C emissions from a UK nuclear establishment over an extended period. Radiocarbon 1986 28-2A, p681-688.

3   Clarke, R H. A model for short and medium range dispersion of radionuclides released to the atmosphere, 1979 NRPB-R91. National Radiological Protection Board, Chilton, Didcot, Oxfordshire, UK.

4   Otlet, R L; Slade, B S. Harwell Radiocarbon Measurements I: Radiocarbon 1974 16, 2, 178-191.

5   Otlet, R L; Warchal, R M. Liquid Scintillation Counting of Low-Level Carbon 14, in Crook, M A and Johnson, P (eds) 1978. Liquid Scintillation Counting 5, Heyden and Sons Ltd, London, 210-218.

CALCULATION MODEL FOR THE IMPACT ON SOIL DUE TO TRITIUM EMISSION INTO THE ATMOSPHERE DURING SNOWFALL

D. Papadopoulos, L.A. König, K.-G. Langguth
Kernforschungszentrum Karlsruhe GmbH,
Hauptabteilung Sicherheit/Radioökologie,
Postfach 36 40, D-7500 Karlsruhe,
Federal Republic of Germany

ABSTRACT

In January 1985, after each of three periods of snowfall, snow samples were collected repeatedly at 33 sampling locations in the vicinity of the Karlsruhe Nuclear Research Center and the concentration of tritium in the snow-water was measured. The results of measurement have been published (Radiation Protection Dosimetry, Vol. 16, Nos. 1-2, pp. 95-100 (1986), Nuclear Technology Publishing). By parallel measurements of the intensity of snowfall it was possible to determine the tritium impact on the soil due to snowfall, expressed in Bq/m^2. The values obtained are considered here as the results of measurement of the tritium impact on the soil due to snowfall.A calculation model which takes account of the intensity of emissions of tritium into the atmosphere as well as of the meteorological situation was used to determine the tritium impacts on the soil at each of the 33 sampling locations and for each of the three periods of snowfall. the measured and calculated results are in good agreement.

INTRODUCTION

In January 1985 snow-water samples were collected repeatedly after snowfalls at 33 sampling points (Fig. 1) and their tritium concentrations were measured [1]. The background concentration during the period of sampling is known from measurements on snow-water collected at the Augustenberg reference sampling point located about 12 km south-east of the Karlsruhe Nuclear Research Center (KfK). The tritium emission by the main emitter, the heavy water moderated multipurpose research reactor MZFR (200 MW_{th}), decommissioned on May 3, 1984 (Fig. 1), is in the form of HTO; the daily emissions were likewise measured (Table 1).

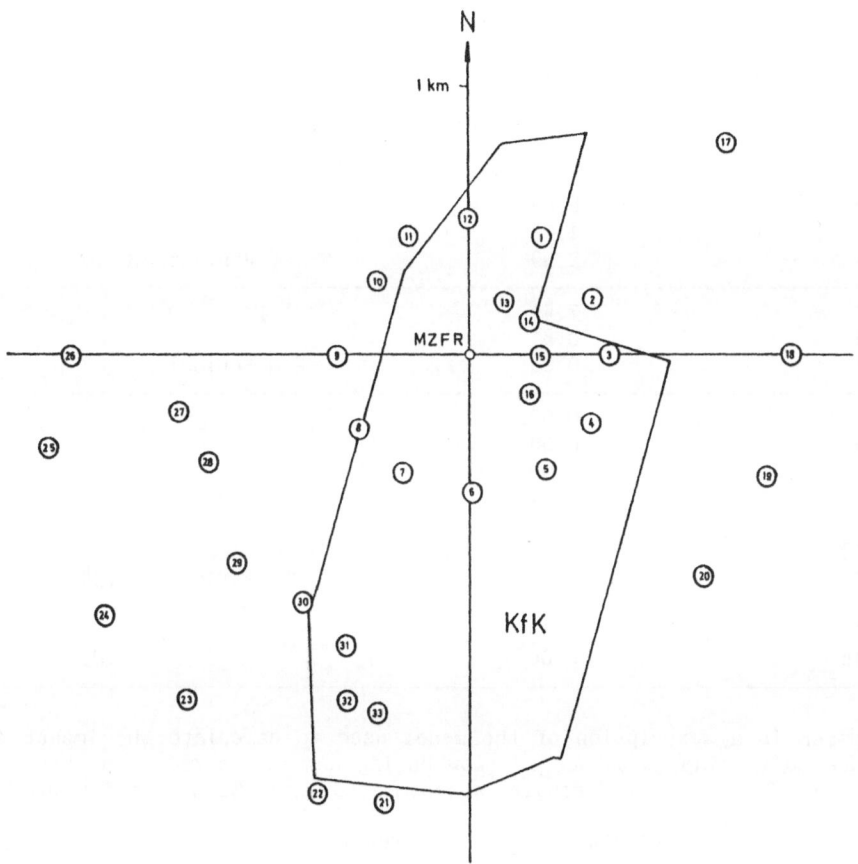

Figure 1. Map of sampling points around the emitter MZFR

TABLE 1
Tritium released with the effluent air from the MZFR stack from
Jan. 1 until 14, 1985

1985	Tritium emission (GBq/d)	1985	Tritium emission (GBq/d)
Jan. 1	60	Jan. 8	49
Jan. 2	53	Jan. 9	49
Jan. 3	60	Jan. 10	46
Jan. 4	51	Jan. 11	46
Jan. 5	49	Jan. 12	44
Jan. 6	47	Jan. 13	49
Jan. 7	44	Jan. 14	43

The meteorological data for the said period were measured at the meteorological tower and summarized as daily statistics with the corresponding intensities of precipitation indicated (Table 2).

TABLE 2
Intensity of precipitation due to snowfall

1985	Intensity (1/m^2-snow-water)	Intensity for the period (1/m^2-snow-water)
Jan. 1	0.00	
Jan. 2	6.88	
Jan. 3	1.03	
Jan. 4	2.99	1st Period: 10.90
Jan. 5	7.41	
Jan. 6	0.81	
Jan. 7	0.00	2nd Period: 8.22
Jan. 8	0.00	
Jan. 9	0.00	
Jan. 10	0.00	
Jan. 11	0.21	
Jan. 12	0.00	
Jan. 13	0.18	
Jan. 14	4.15	3rd Period: 4.54
Jan. 15	0.00	
Jan. 16	0.00	0.00

This paper is a description of the model used to calculate the impact on soil due to tritium contained in snow during continuous emission into the atmosphere. The calculated results are compared with the measured results.

MEASUREMENTS OF THE IMPACT ON SOIL

According to samples collected of fresh fallen snow three periods of snowfall can be considered:

- snowfalls from January 1-4, 1985: 1st period
- snowfalls from January 4-7, 1985: 2nd period
- snowfalls from January 7-14, 1985: 3rd period.

In Table 2 the sums have been entered of the intensities of precipitations in the periods mentioned above. As shown by Table 2, no precipitation occured from January 7 until 9, 1985. Therefore, the measurements made on the samples collected on Januar 9, 1985 (sampling points 21 to 33, Table 5 in [1]) have been included in the second period. This is justified by the fact that no significant change in tritium concentration in the snow-water was observed during periods without snowfall [1]. This gives Table 3 where also the HTO activity concentrations in the snow-water collected at the Augustenberg reference point have been entered. Multiplying the measured HTO concentration (Table 3) by the intensities of precipitation (Table 2) gives the measured impacts on soil, M, expressed in Bq/m^2. They have been entered in Table 4.

TABLE 3
Results of tritium measurement in snow samples

Sampling location	Coordinates from MZFR (m)		Measured tritium activity concentration (Bq/l) in fresh fallen snow		
	X (W-E)	Y (S-N)	1st Period Jan. 1-4, 1985	2nd Period Jan. 4-7, 1985	3rd Period Jan. 7-14, 1985
1	250	440	9.0 ± 6.2	14.6 ± 6.5	6.3 ± 6.1
2	440	250	< 5.7	25.5 ± 6.8	14.0 ± 6.4
3	500	0	7.6 ± 6.1	10.9 ± 6.3	7.1 ± 6.0
4	440	-265	< 5.7	22.4 ± 6.8	10.4 ± 6.4
5	265	-440	14.9 ± 6.6	21.7 ± 6.6	< 5.9
6	0	-500	17.8 ± 6.5	27.4 ± 6.8	< 5.8
7	-250	-440	< 5.7	8.0 ± 6.1	7.4 ± 6.1
8	-440	-260	< 5.8	7.9 ± 6.1	< 5.8
9	-500	0	< 5.9	9.5 ± 6.1	7.8 ± 6.1
10	-325	270	< 5.8	7.3 ± 6.1	12.7 ± 6.4
11	-190	490	< 5.8	< 5.8	< 5.8
12	0	500	< 5.8	7.3 ± 6.1	6.2 ± 6.2
13	120	220	6.6 ± 6.1	20.0 ± 6.6	< 5.8
14	210	130	6.4 ± 6.1	25.9 ± 6.8	11.0 ± 6.2
15	250	0	< 5.8	9.0 ± 6.3	8.5 ± 6.2
16	210	-145	< 5.9	11.1 ± 6.3	< 5.9
17	970	750	17.3 ± 6.5	17.2 ± 6.2	36.1 ± 7.2
18	1200	0	< 5.9	20.3 ± 6.6	8.7 ± 6.2
19	1100	-450	< 5.8	15.7 ± 6.4	7.5 ± 6.1
20	880	-840	< 5.8	27.5 ± 6.9	9.9 ± 6.3
21	-330	-1580	-	7.3 ± 6.1	< 5.9
22	-590	-1560	-	9.8 ± 6.2	6.1 ± 6.1
23	-1080	-1270	-	< 5.9	26.6 ± 6.9
24	-1360	-960	-	9.0 ± 6.3	22.1 ± 6.7
25	-1620	-350	-	10.6 ± 6.2	8.8 ± 6.2
26	-1530	0	-	10.4 ± 6.3	8.5 ± 6.1
27	-1160	-210	-	8.8 ± 6.2	< 5.9
28	-1000	-400	-	13.3 ± 6.3	26.9 ± 6.8
29	-890	-740	-	7.5 ± 6.1	22.6 ± 6.6
30	-650	-920	-	11.4 ± 6.5	18.6 ± 6.5
31	-490	-1010	-	7.5 ± 6.2	9.2 ± 6.2
32	-495	-1210	-	13.5 ± 6.4	8.5 ± 6.1
33	-410	-1280	-	11.7 ± 6.3	13.2 ± 6.3
Reference point	Augustenberg		8.5 ± 6.2	8.5 ± 6.2	8.5 ± 6.2

TABLE 4
HTO depositions on the ground due to snowfall during continuous emission
into the atmosphere, expressed in Bq/m^2
M: measured; T: calculated
1st period: snowfalls from Jan. 1-4, 1985
2nd period: snowfalls from Jan. 4-7, 1985

Sampling location	Coordinates from MZFR (m)		1st Period			2nd Period		
	X (W-E)	Y (S-N)	M	T	M/T	M	T	M/T
1	250	440	98	251	0.39	120	102	1.18
2	440	250	62	223	0.28	210	118	1.78
3	500	0	83	193	0.43	90	98	0.92
4	440	-265	62	179	0.35	184	72	2.56
5	265	-440	162	119	1.36	178	121	1.47
6	0	-500	194	127	1.53	225	543	0.41
7	-250	-440	62	102	0.61	66	123	0.54
8	-440	-260	63	93	0.68	65	70	0.93
9	-500	0	64	93	0.69	78	70	1.11
10	-325	270	63	93	0.68	60	70	0.86
11	-190	490	63	93	0.68	48	70	0.69
12	0	500	63	102	0.62	60	73	0.82
13	120	220	72	409	0.18	164	133	1.23
14	210	130	70	362	0.19	213	168	1.27
15	250	0	63	293	0.22	74	126	0.59
16	210	-145	64	256	0.25	91	75	1.21
17	970	750	189	150	1.26	141	89	1.58
18	1200	0	64	135	0.47	167	82	2.04
19	1100	-450	63	133	0.47	129	73	1.77
20	880	-840	63	143	0.44	226	73	3.10
21	-330	-1580	-	-	-	60	162	0.37
22	-590	-1560	-	-	-	81	113	0.72
23	-1080	-1270	-	-	-	49	71	0.69
24	-1360	-960	-	-	-	74	70	1.06
25	-1620	-350	-	-	-	87	70	1.24
26	-1530	0	-	-	-	86	70	1.23
27	-1160	-210	-	-	-	72	70	1.03
28	-1000	-400	-	-	-	109	70	1.56
29	-890	-740	-	-	-	62	70	0.89
30	-650	-920	-	-	-	94	77	1.22
31	-490	-1010	-	-	-	62	106	0.58
32	-495	-1210	-	-	-	111	118	0.94
33	-410	-1280	-	-	-	96	145	0.66

TABLE 4 (Continued)
HTO depositions on the ground due to snowfall during continuous emission
into the atmosphere, expressed in Bq/m^2
M: measured; T: calculated
3rd period: snowfalls from Jan. 7-14, 1985
1st + 2nd + 3rd period: snowfalls from Jan. 1-14, 1985

Sampling location	Coordinates from MZFR (m)		3rd Period			1st + 2nd + 3rd Period		
	X (W-E)	Y (S-N)	M	T	M/T	M	T	M/T
1	250	440	29	40	0.73	247	393	0.63
2	440	250	64	63	1.02	336	404	0.83
3	500	0	32	44	0.73	205	335	0.61
4	440	-265	47	42	1.12	293	293	1.00
5	265	-440	27	40	0.68	367	280	1.31
6	0	-500	26	47	0.55	445	717	0.62
7	-250	-440	34	193	0.18	162	418	0.39
8	-440	-260	26	52	0.50	154	215	0.72
9	-500	0	35	38	0.92	177	201	0.88
10	-325	270	58	38	1.53	181	201	0.90
11	-190	490	26	38	0.68	137	201	0.68
12	0	500	28	38	0.74	151	213	0.71
13	120	220	26	42	0.62	262	584	0.45
14	210	130	50	89	0.56	333	619	0.54
15	250	0	39	50	0.78	176	469	0.38
16	210	-145	27	45	0.60	182	376	0.48
17	970	750	164	47	3.49	494	286	1.73
18	1200	0	40	41	0.98	271	258	1.05
19	1100	-450	34	40	0.85	226	246	0.92
20	880	-840	45	41	1.10	334	257	1.30
21	-330	-1580	27	57	0.47	-	-	-
22	-590	-1560	28	73	0.38	-	-	-
23	-1080	-1270	121	73	1.66	-	-	-
24	-1360	-960	100	46	2.17	-	-	-
25	-1620	-350	40	39	1.03	-	-	-
26	-1530	0	39	38	1.03	-	-	-
27	-1160	-210	27	39	0.69	-	-	-
28	-1000	-400	122	40	3.05	-	-	-
29	-890	-740	103	60	1.72	-	-	-
30	-650	-920	84	103	0.82	-	-	-
31	-490	-1010	42	104	0.40	-	-	-
32	-495	-1210	39	89	0.44	-	-	-
33	-410	-1280	60	77	0.78	-	-	-

CALCULATION OF THE IMPACT ON SOIL

According to [1] the tritium activity concentrations measured in snow-water are roughly similar to those of rainwater, with the emission rate and diffusion conditions approximately identical, and no significant changes in tritium concentrations were observed during the periods without precipitation.

Therefore, the surface loading by tritium in snow-water has been calculated using the same Equation 1 as in [1] for rainwater

$$B_{F,i} = A \frac{N}{2 \pi x} \sum_{jmt} q_{ijmt} \frac{\Lambda_t}{U_{jm}}$$ (1)

(corresponding to Equation (4.12) in [2]).

In Equation 1, the symbols used have the following meanings:

$B_{F,i}$ = surface loading per unit area (Bq/m^2) on the day under consideration
A = activity emitted (Bq) on that day
N = number of sectors of wind directions
x = distance from emitter (m)
q_{ijmt} = daily frequencies of precipitations (i, sector of wind direction; j, diffusion category; m, wind velocity level; t, precipitation intensity level)
u_{jm} = average of wind velocity level (m/s)
Λ_t = wash-out constant (s)

If the wash-out constant, Λ, is directly proportional to the precipitation intensity, θ [3], the following relation holds

$$\Lambda = s\,\theta$$ (2)

where s is a proportionality constant (y mm^{-1} s^{-1}), and θ is the precipitation intensity (mm y^{-1}) on the day under consideration.

Except for s, all quantities needed to calculate $B_{F,i}$ are known from measurements conducted at the exhaust stack of the emitter and from those conducted at the meteorological tower.

To calculate the surface loading as a result of wash-out in the ISOLA-III program [4], s = 3×10^{-9} y mm^{-1} s^{-1} was used.

For each period of snowfall the sum, T', was formed of the daily values $B_{F,i}$. In order to allow a comparison to be made of the measured values with the calculated values the background level caused by nuclear weapons tests has to be taken into account in addition. This has been done for each period of snowfall by adding to the T' values the surface loads measured at the reference point. The T-values obtained in this way have also been entered in Table 4.

COMPARISON OF THE CALCULATED VALUES WITH THE MEASURED VALUES

For comparison of the measured values M with the calculated values T the ratios M/T were composed for each period of snowfall and for each sampling point and entered in Table 4. Besides, the sums of the calculated and of the measured values were added up for all periods of snowfall and they were considered as the values applicable to the entire period of snowfall lasting from January 1 until 14, 1985. Also for these values the ratios M/T have been composed and entered in Table 4.

For each period of snowfall the cumulative sums have been calculated for all sampling points available and entered in Table 5. The steps of frequencies selected in Table 5 are the steps of the geometric series (1.5^n; with n = -5, -4, ..., 4) for a semi-logarithmic representation.

TABLE 5
Frequency distribution of the M/T ratios
Symbols: M measurement, in Bq/m^2, T calculation, in Bq/m^2

M/T step of frequencies	1st period	2nd period	3rd period	1st + 2nd + 3rd period
1/7.59 ≤...< 1/5.06	2	0	1	0
1/5.06 ≤...< 1/3.38	3	0	0	0
1/3.38 ≤...< 1/2.25	4	2	3	2
1/2.25 ≤...< 1/1.50	4	4	6	6
1/1.50 ≤...< 1/1.00	4	9	12	8
1/1.00 ≤...< 1/1.50	2	11	5	3
1/1.50 ≤...< 1/2.25	1	5	4	1
1/2.25 ≤...< 1/3.38	0	2	1	0
1/3.38 ≤...< 1/5.06	0	0	1	0
total	20	33	33	20

It is evident from Table 5 that the calculated results agree well with the measured results. The model Equation 1 used in the ISOLA III computer code and Equation 2 are satisfactory for calculating the impact on soil by HTO in snowfalls during emission into the atmosphere.

REFERENCES

1. Papadopoulos, D., König, L.A., Langguth, K.-G. and Fark, S., Contamination of Precipitation due to Tritium Release into the Atmosphere. Rad. Prot. Dosim., 1986, 16, 95-100.

2. Der Bundesminister des Innern, Allgemeine Berechnungsgrundlage für die Strahlenexposition bei radioaktiven Ableitungen mit der Abluft oder in Oberflächenwasser. BMBl, 1979, 369-436.

3. Brenk, H.D. and Vogt, K.J., Konzeption für eine praxisnahe Berechnung der Ablagerung radioaktiver Stoffe aus der Abluft kerntechnischer Anlagen durch Niederschlag. Jül-1328, August 1976.

4. Hübschmann, W. and Nagel, D., ISOLA-III - Ein FORTRAN-IV-Programm zur Berechnung der langfristigen Dosisverteilung in der Umgebung kerntechnischer Anlagen. KfK-2698, Dezember 1978.

CONFIRMATION OF LABORATORY RESULTS BY HT RELEASES UNDER FIELD CONDITIONS: HT DEPOSITION VELOCITY AND REEMISSION RATE.

H. Förstel, H. Trierweiler and K. Lepa
Nuclear Research Centre Jülich (KFA)
Radioagronomy, POB 1913, 5170 Jülich
Fed. Rep. Germany

ABSTRACT

The two main forms of tritium, elementary gas HT and tritiated water HTO do not only differ in their radiotoxicity but also in their radioecological behaviour. To predict the radioecological pathway of tritium in a specific local situation a technique easy to apply was tested. Mechanically undisturbed soil cores were collected in the field and exposed to HT in the laboratory. The deposition velocity sufficiently describes the first step of the radioecological pathway of HT after its distribution in the air. Numerous results of deposition velocities in field, pasture and forest soil have been obtained in Jülich, but a tritium release experiment only enabled us to confirm the technique proposed. Soil cores have been exposed to the HT plume in the field and the HT deposition velocity of soil samples from the exposure site has been determined in the laboratory. The field data confirm the applicability of our method and one may be able to propose a way of collecting representative data under accidental and under chronic conditions.
Another important aspect of radioprotection is the reemission of the reaction product HTO from the soil. Laboratory measurements demonstrate a very rapid HTO reemission during the first hour directly after the HT/HTO turnover in the soil. Thereafter the HTO reemission decreases distinctly.

INTRODUCTION

Tritium is used, generated or released by fission and fusion techniques. Therefore it's necessary to know the radioecological pathway exactly. One has to distinguish between different chemical forms of tritium which differ not only in their radiotoxicity |1| but also in their radioecological behaviour. Tritium gas HT is taken up by man, animals and plants. On the other side tritiated water HTO is quickly absorbed in the lungs. Organically bound tritium OBT can be incorporated via the digestive tract and may act as a precursor of biomass.
It has been demonstrated that tritium is taken up by the soil very quickly |2-5|. HT is nearly exclusively converted to HTO (not to OBT) by the enzymatic activity of soil organisms. The mobile gaseous form is changed to a liquid one. But due to an exchange between the liquid form of soil water and the gaseous form of water vapour in the air (air humidity) a reemission of the more radiotoxic form HTO must be taken into account.
The ability of a soil to oxidaize HT to HTO is demonstrated by simple experiments, but obtaining data usable for model calculations requires another procedure. Soil contains a variety of microorganisms. A small-scale stratification indicates that the state of organisms, living in certain microsites, may be important. Therefore soil must not be mechani-

cally treated. We have tested a procedure to collect small cores of soil directly in the field into steel tubes, which can be closed for storage and can be attached to a gas circuit in the laboratory. In spite of the fact that the atmospheric distribution of HT after release can be predicted quite well, field studies were necessary to demonstrate the role of natural soil as a sink of elementary tritium under field conditions.

The HT uptake into the soil is described by the deposition velocity:
$$V_d = k * V/q$$
(k: HT decrease constant from the well-mixed headspace, V: volume of the total gas space above the soil core, q: surface area of the soil core). The deposition velocity obtained in the manner does not depend on the specific experimental conditions. The relation between the deposition velocity and the amount of HT taken up by the soil is given by the equation:
$$F = V_d * c_{HT}$$
(F: flux of HT into the soil, c_{HT}: concentration of HT in the gas space above the soil).

MATERIAL AND METHODS

Soil Samples
During the HT release experiment in France, which will be summarized by an official report |6|, 20 soil cores (50 mm diameter) collected in the vicinity of the Nuclear Research Centre Jülich (KFA Jülich) were exposed to the plume on October 15, 1986. They were kept in two containers, arranged in a randomly selected manner. After exposure the soil cores were kept frozen and allowed to thaw only overnight before each measurement in the laboratory. Originally 30 containers were brought to the release site, but because of the long waiting period two cores of each soil type were used to measure the actual state of the soil on the release day.
The soil cores represent both different soil types and various uses (Table 1).

TABLE 1
Soil cores exposed in France, October 15, 1986.

sign	use	soil type	vegetation
MR	field	parabraunerde	sugar-beets
MW	field	parabraunerde	wheat
W	pasture	parabraunerde	grass
S	forest	sand	pine/grass
L	forest	pseudogley	beech

The results from the laboratory and the field exposure cannot be compared directly, but only relatively, for the exact position of the containers during the release is not known. The site theoretically expected as the exposure centre was not hit direktly by the plume, but was situated at its edge. The sharp decrease of the HT concentration in the air at that point necessitated a very exact localisation of the containers exposed.

Parallely to this procedure soil cores were collected across the whole exposure area to detect the local variability of the HT deposition. At three sites, in the field, in a meadow, and in a forest, samples were routinely taken during the whole waiting period. They were to be used to characterize the state of the soil on the release day in comparison with the whole waiting period. The soil cores are characterized by their profiles of soil water content and the profile of the conversion product HTO.

The exposure area, situated north-east of Bruyères-le-Châtel consists of arable land grown with maize, corn and various vegetables. Its common agricultural use enables one to handle the data in other studies, too.

Tritium Measurements

The deposition velocity is calculated from the decrease of the HT concentration in the head space which is in contact with the surface of a soil core. The volume of the gas space and the surface area of the soil core must be known. About 1.8 MBq were injected into the circuit to start each test run. The gas is circulated by a pump in a closed loop. Before the gas is fed into the ionisation chamber, the air is freed from HTO and water vapour by the passage through cooling traps kept at the temperature of a dry ice/isopropanol mixture. The application of the ionisation chamber allows a continuous record of the HT activity in the air. Experimental details have been given in a previously published paper |5|.

Preliminary reemission data can be obtained from the amount of HTO which is trapped from the air stream before entering the ionisation chamber. All of the HTO released from the soil is trapped there. The air flowing back into the reaction vessel does not contain HTO. Therefore the situation is similar to that which we expect under natural conditions, especially after an accidental release, too. The HTO which comes out of the soil will be quickly carried away into the well-mixed atmosphere.

For laboratory tests we have used soil cores of 100 mm diameter taken from the beech forest in our centre. The upper layer of litter was removed carefully to ensure defined experimental conditions. First the soil samples were exposed to HT gas in order to label them. As a result the HTO is distributed in a manner which would result during an HT release under natural conditions, too. Afterwards the soil surface was brought into contact to an HTO free air stream of defined humidity at room temperature. No significant changes of the soil water content were observed. During the series of increasing air turnover rates (from 0 to about 11 l min^{-1}) the relative humidity of the air was kept close to saturation. Changing the relative humidity the flow rate was kept constant at 7.7 l min^{-1}.

RESULTS

Exposure of known soil samples

Samples of five soil types were exposed to the HT plume. Simultaneoulsly the deposition velocity was measured in the laboratory at home, too. **Figure 1** summarizes the results. The soil from a field near Jülich which has been continuously observed for years is set as reference. One may recognize that the kind of soil (both the soil texture and the treatment by agricultural methods) governs the HT deposition velocity. Neither the soil water content nor the penetration profile of HT are able to explain the differences observed. In most of the cases the maximum of the reaction product HTO was not observed at the surface directly but 2-3 cm below it. This result confirms the idea, that not only the biological state of the soil but mainly the physical condition of diffusion determine the deposition velocity.

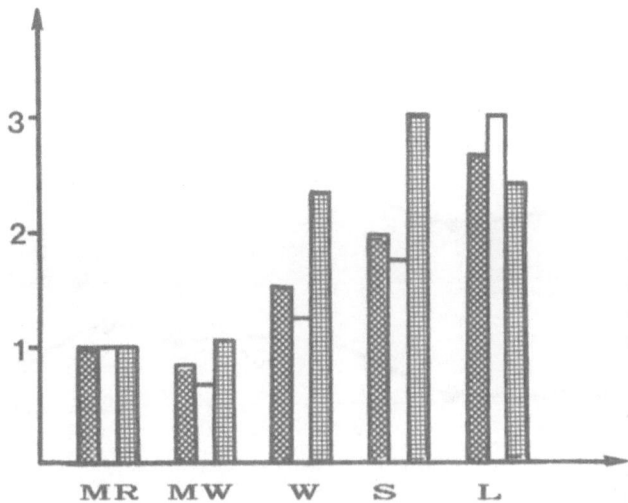

Figure 1   Comparison of deposition velocities measured in the laboratory
            and in the field. The first column is set 1.0 and is measured
            in the lab. The other two are the mean values of deposition
            velocities calculated from the HT concentration during release
            and the HTO in the soil cores. Abbreviations see Table 1.

Soil Cores from the Release Site

The mean deposition velocity of a first set of samples collected in spring
1986 at different points of the release site was: $6.4 \pm 0.7 \times 10^{-4}$ m s^{-1}.
The most surprising result was the observation that the depositon velocity
does not depend on the temperature, exept an increase to above 40°. Under
these conditions the enzymes are inhibited irreversibly.

The deposition velocities of the sampes collected from September to October 1986 **(Figure 2)** demonstrate that the typical sites (field, meadow and
forest) distinctly differ in their ability to take up HT. According to our
earlier results the forest soil is the most active one, its deposition
velocity does not depend on the soil water content. The water content of
the field does not change drastically, therefore the depositon velocity
also does not change. In contrast to these findings the water content of
the meadow increased and consequently the deposition velocity decreased.
About 50 % of the HTO was formed in the upper soil layer between 0.8 and
1.8 cm. The reemission rate calculated from the HTO activity of the traps
ranged between 2.4 and 7.3 % h^{-1}, including extreme situations.

The deposition velocities of soil samples taken on the day of release and
during the next one demonstrate the pattern reported above **(Figure 3)**. The
deposition velocities increase from field to meadow and forest. The 50 %
depth data of the formation product HTO range from 0.7 to 3.1 cm, reemission rates grown 1.4 and 7.8 % h^{-1}.

Reemission Rate Studies

The results show that the reemission rate depends on the air turnover in
the reaction vessels **(Figure 4)**. A turnover of 1 l min^{-1} corresponds to a
wind velocity of about 0.4 m s^{-1}. One can see that the reemission is most
intensive during the first hour and diminishes thereafter. This decrease
is not the result of the HTO loss during reemission, but might be caused

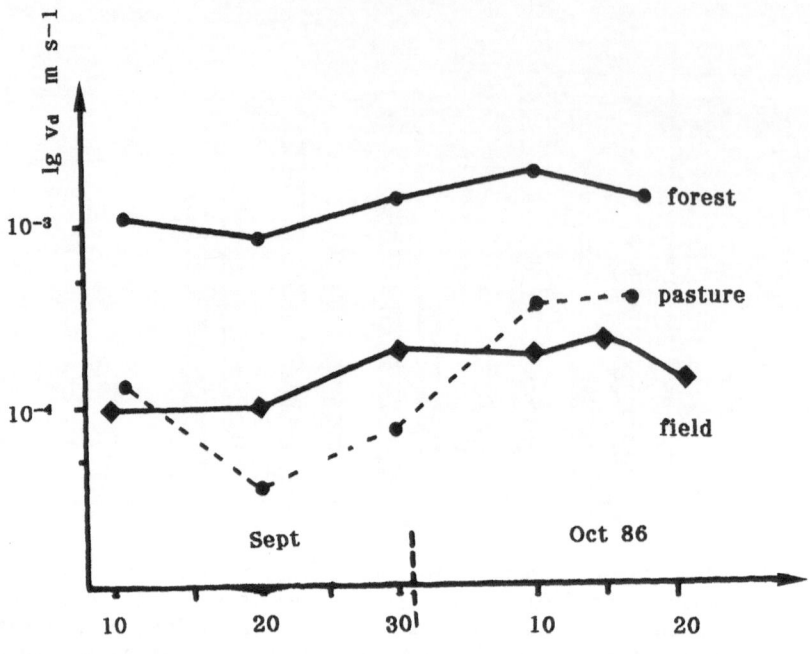

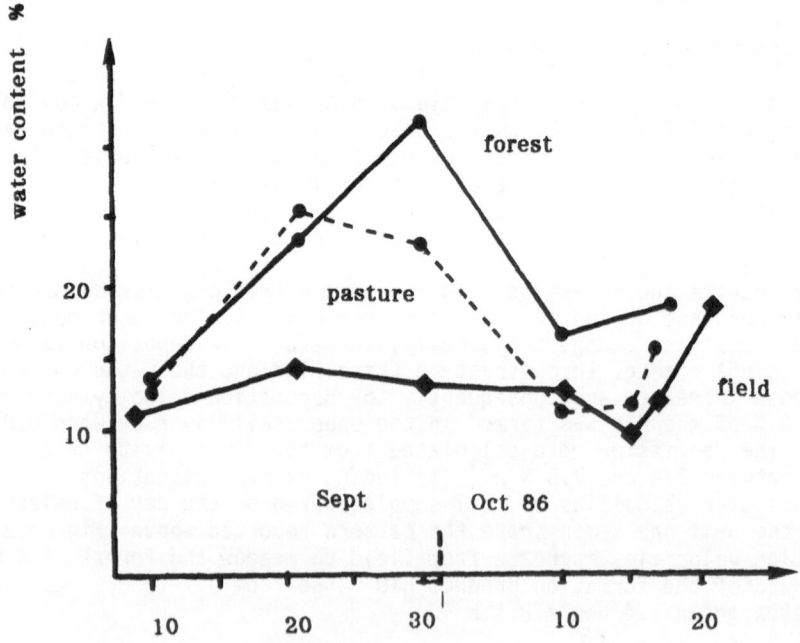

Figure 2   Time course of deposition velocities (upper graph) and soil
water content (lower graph) at three points of the release site
(field, meadow and forest).

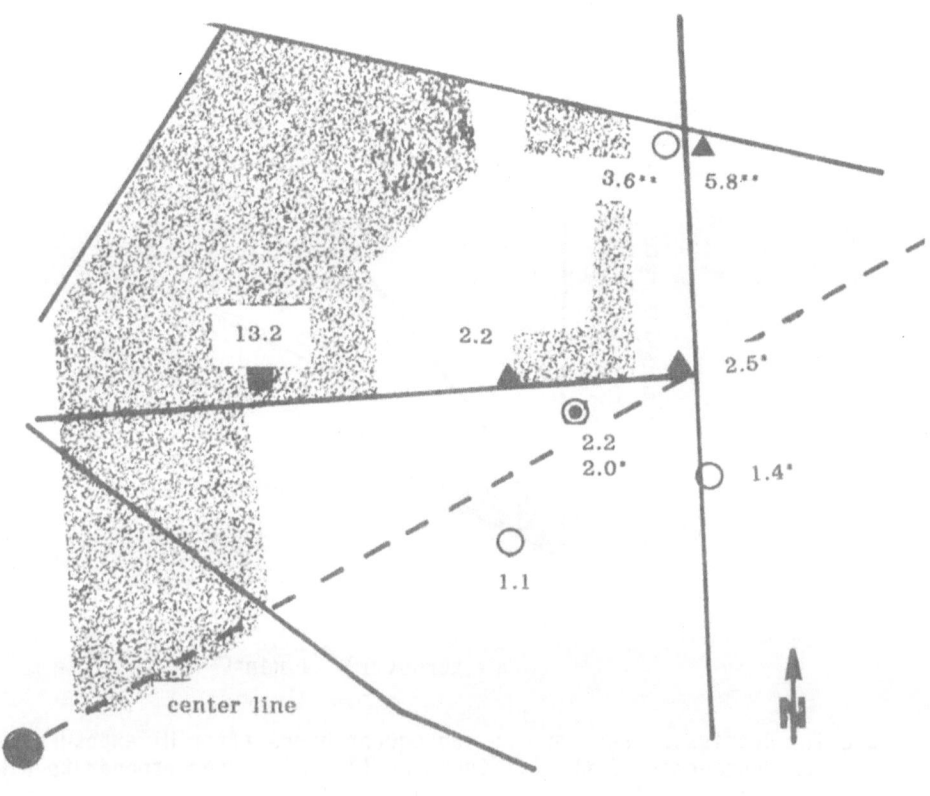

Figure 3   Deposition velocity on the day of release and the next ones. The lines represent streets, the grey areas forest. Circles represent field sites, triangles meadow and squares forest. The data are reported as 10^{-4} m s^{-1}. The asteric symbolizes the time of soil collection in days after release.

by internal transport processes in the soil. The decrease of the reemission rate is more at higher air turnover rates. The change of the relative air humidity does not influence the result (two test runs at 0.28 and 96 % each).

DISCUSSION

The deposition velocities measured are in good agreement. Compared to our results of laboratory measurements (2.2 * 10^{-4} m s^{-1}) at point A the CEA Fontenay found 5.5 * 10^{-4} m s^{-1} (also chamber result). From the HT concentration in the air and in the soil water profile the NIR Hannover calculated deposition velocities between 1-2 * 10^{-4} m s^{-1}. At line B our results of field soil range between 2.0 and 2.5 * 10^{-4} m s^{-1}, that of CEA between 1.7 and 3.6 * 10^{-4} m s^{-1}.

52

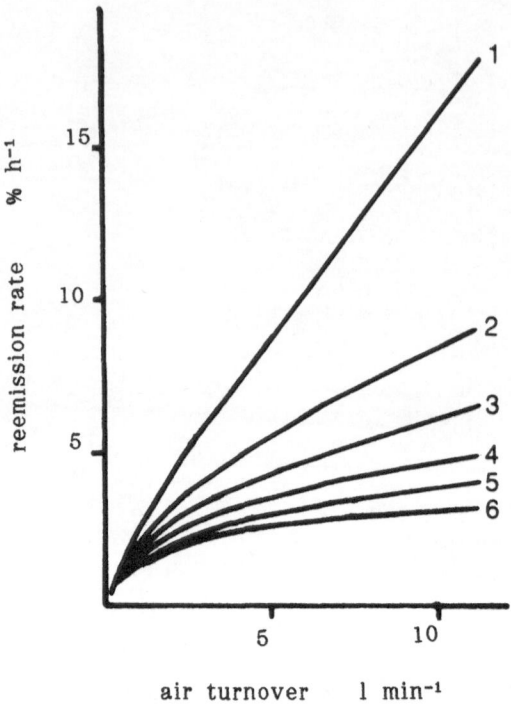

Figure 4   Reemission rate during subsequent hours after HT exposure in
            dependence of the air turnover (1 l min⁻¹ corresponds to a wind
            velocity of about 0.4 m s⁻¹).

The reemission rate at point A was measured under laboratory conditions in
Jülich 2 % h⁻¹ in comparison to the CEA data of 4-6 % h⁻¹. The air turnover in the larger CEA chamber is more intensive than in our experimental
device. Reemission data from the field cannot be taken into consideration
because of a sudden change of the weather conditions after the release.

Finally we propose our method as a rapid sampling procedure which may
enable one to screen quickly large areas. During such a sampling one
should especially take into account different kinds of agricultural use of
land. Especially forest soil is very active at converting into HTO.

ACKNOWLEDGEMENT

The study was supported by the CEC Radioprotection Programme. Travelling
funds were given by the CEC Fusion Programme. The measurements in the
laboratory were carefully done by Mrs. J. Jansen.

LITERATURE

1. ICRP, Committee II on Permissible Dose for Internal Radiation, ICRP Publ., Pergamon, Oxford 2, (1959).
2. McFarlane, J.C., Rogers, R.D. and Bradley, D.V. jr., Tritium oxidation in surface soil. Environm. Sci. Technol., 1978, 12, 590-593.
3. McFarlane, J.C., Rogers, R.D. and Bradley, D.V. jr., Tritium oxidation in surface soils. A survey of soils near five nuclear fuel reprocessing plants. Environm. Sci. Technol., 1979, 13, 607-609.
4. Garland, J.A. and Cox, L.C., The absorption of tritium gas by English soils, plants and the sea. Water Air Soil Pollution, 1980, 14, 103-114.
5. Förstel, H., Uptake of elementary tritium by the soil. Radiation Protection Dosimetry, 1986, 16, 75-81.
6. Djerassi, H. and Lesigne, B. (Eds.), Environmental tritium behaviour - French experiment. CEA/IPSN Report August 1987, 532 p.

UNCERTAINTY ANALYSIS AND MODEL VALIDATION FOR A RETROSPECTIVE ASSESSMENT OF THYROID DOSE RESULTING FROM ATOMIC WEAPONS' TEST FALLOUT

Steven L. Simon†

University of Utah, Salt Lake City, UT USA 84108

ABSTRACT

Modeling internal dose resulting from environmental transport of atomic weapons' test fallout from the Nevada Test Site (NTS) USA, during the period 1951-1962, requires estimates of many important parameters for which little data exists. A retrospective dose assessment study for which model reliability is of great importance is now nearing completion. The study's objectives are to estimate thyroid dose from ingestion of radioiodine in milk and vegetables to a cohort of over 3000 persons. The cohort that had potentially been exposed to NTS fallout were examined for thyroid abnormalities during 1965 and 1985. Model sensitivity analysis and parameter uncertainties were used to help establish priorities for research and uncertainty analysis. Numerous parameters were shown to substantially affect radioiodine concentrations in milk and uncertainty estimates, the most significant being the fraction of fallout intercepted by vegetation and the relative timing of dairy management practices and deposition events. Both parameters were subsequently researched and the results are discussed. Verification of model reliability was attempted by comparing predictions with limited literature data.

† present address: Department of Environmental Sciences and Engineering
University of North Carolina
Chapel Hill, NC USA 27514

INTRODUCTION

Retrospective dose assessment is important because of legal implications and because improved methods are needed for modeling past events and for estimating uncertainties of specific contamination scenarios. This paper is presented not as a description of new methods, but as an example of the use of uncertainty, sensitivity, and validation techniques applied to dose assessment for specific individuals. A nearly completed retrospective study attempts to relate incidence of thyroid disease with exposure to radioiodines from the Nevada Test Site

(NTS) in the United States during the 1950's. A cohort of persons aged 11-18 years who lived near the NTS were examined for thyroid abnormalities from 1965-1968. The subjects were students in southwestern Utah and Nevada and included a control population in Arizona. During 1985-1986, 3,085 subjects were located and re-examined to determine any subsequent occurrence of thyroid disease [1]. Because the cohort members lived in many U.S. locations during the atmospheric nuclear testing (1951-1962), the assessment of thyroid dose from ingestion included residence locations in Utah and the six surrounding states.

The dosimetry development phase is described elsewhere [2]. The level of detail of the modeling effort largely determined the level of detail to be considered for the uncertainty analysis. For example, the assessment model was designed to include site-specific exposure-rate measurements, site-specific dairy management data, individual residence histories, sources of milk and consumption rates for individuals, transfer parameters and dose conversion factors for subjects in-utero, nursing infants, and individuals. Methods to determine uncertainties for selected environmental transport parameters is the subject of this paper.

METHODS

A simple model to predict radioiodine transfer to milk from pasture grass per unit deposition is described with an emphasis on model uncertainties. Although the source term includes the estimates of location and event-specific fallout deposition, the associated uncertainties will not be discussed here. An analytical solution (eq.1) for the integrated concentration in milk is used in the form of a conditional equation, the alternatives applied depending on whether the fallout event occurred before or after the pasture season began. The difference between the date pasture begins (pb) and the deposition or shot date (sd) is computed for the last exponential term. If this is negative indicating that the shot occurred after pasture usage began, the difference is set equal to zero; otherwise, the equation is used as written. Green chop, new hay and ingested soil can also contribute to radioiodine in milk, though, will not be discussed here.

Sensitivity Analysis

Preliminary sensitivity calculations were carried out by programming a stochastic version of the model and using parameter probability distributions derived from the literature and from our own investigations. A sensitivity index [3] (SI) was calculated for each parameter: $SI = 1 - (C_{min})_i/(C_{max})_i$, where C was the minimum and maximum concentration for the ith parameter calculated using the 1st and 99th percentiles respectively. The minimum and maximum for $f_d(p)$ was an exception because of the constraint on the maximum of 1.0. The other parameters for each SI calculation use median values. Tables 1 and 2, respectively, give representative values of the parameter distributions and the calculated sensitivity indices.

$$\frac{C_{int}}{dep} = \frac{(1 - e^{-\alpha * Y})}{Y} * Q_f * f_d(p) * f_m *[\frac{exp(-\lambda_{eff}*(ps-sd))}{-\lambda_{eff}} + \frac{exp(-\lambda_{eff}*(pb-sd))}{\lambda_{eff}}] \quad eq. (1)$$

Definitions for the model:

sd	= shot date or deposition date	α	=	vegetation interception parameter (m^2/kg)
dep	= deposition value (μCi/m^2)			
λ	= radioactive decay constant (1/d)	Y	=	vegetation biomass yield (kg/m^2 dry)
λ_{eff}	= $\lambda_{weathering} + \lambda$ (1/d)	f_m	=	isotope specific feed-to-milk transfer coefficient (d/L)
$f_d(p)$	= fraction of diet from pasture			
Q_f	= total dry matter intake in cow's diet (kg/d)	pb	=	date that pasture usage begins
		ps	=	date that pasture usage stops
C_{int}	= integrated concentration in milk (μCi-d/L)			

TABLE 1

Model Parameters and Distribution Characteristics

Variable	Distribution Type	Std. Dev.(normal) or GSD (log-normal)	1st Percentile	Median	99th Percentile
α (m^2/kg)	lognormal	1.3	0.10	0.19	0.35
Y (kg/m^2)	lognormal	1.8	0.076	0.30	1.19
F_m (d/L)	lognormal	1.7	0.0029	0.01	0.035
λ_{eff} (1/d)	lognormal	1.8	0.143	0.165	0.19
$F_d(p)$ (-)	normal	0.24	0.06	0.62	1.0
Q_f (kg/d)	normal	2.7	7.2	13.5	19.8
pb (d)	normal	4.0	-9.0	0.0	9.0
ps (d)	normal	4.0	-9.0	0.0	9.0

A short period of time before and after cows begin to use pasture in the spring or stop using pasture in the fall will be referred to as the "period of uncertainty." This period represents an interval of time when grazing of pasture grass may begin or stop. Variables were introduced into the stochastic version of the model which describe the probability distributions of pasture begin dates (pb) or pasture stop dates (ps) relative to median dates. These variables are normally distributed, have a mean of zero and a standard deviation of 4 days. The variability was determined from data obtained by a University of Utah survey of dairy managers of the 1950's.

The model was initially seen to be the most sensitive (SI= 0.95) to the interception constant, α; a median value of 0.39 m^2/kg with a geometric standard deviation (GSD) of 1.8 [4] was used. The timing of a deposition event relative to the date of a change in the dairy management operation was also seen to potentially affect the calculated concentrations. Most relevant to this

TABLE 2
Sensitivity Index (SI) Results for Model Variables

Variable	SI				
	Date of Deposition Relative to Pasture Begin Date				
	-8	-4	0	4	8
PB	0.94	0.88	0.77	0.56	0.15
	Date of Deposition Relative to Pasture Stop Date				
	-22	-18	-14	-10	-6
PS	0.11	0.22	0.43	0.84	1.0
$F_d(p)$	0.94				
F_m	0.92				
α	0.70 (reduced from 0.95, see text)				
Q_f	0.64				
λ_{eff}	0.25				
Y	0.10				

discussion is the date when pasture usage begins (pb) or pasture usage stops (ps). The sensitivity of the model to the date that pasture use begins increases as the interval between the deposition date and median pasture begin date increases (Table 2). Conversely, the sensitivity to the stop date increases as the interval decreases. Methods for considering the uncertainties of the two most influential parameters and efforts to perform validation calculations are discussed in the remainder of this paper.

DISCUSSION

Uncertainty Associated with the Interception Fraction

For assessments of chronic or acute fallout, such as from weapons tests, the interception fraction parameter is used to quantitatively describe the fraction of an aerosol intercepted by vegetation. The relationship between the fraction of an aerosol deposited per m^2 on vegetation (F_v) and the biomass yield (Y, kg/m^2) was empirically determined by Chamberlain [5] to be:
$F_v = (1-e^{-\alpha*Y})$ [eq. (2)]. The interception parameter, α, for a number of grass types was originally calculated from interception data for micron-size particles which is less than ideal for modeling close-in weapons fallout which may range up to 100 microns or more in size. The values of α originally reported [5] varied from 2.78 ± 0.14 to 3.33 ± 0.56 m^2/kg.

The efforts of the Utah study to reduce model sensitivity to the interception parameter α were based on suggestions in the literature that α may be particle size dependent [6,7]. It has also been hypothesized that interception is less for large particle sizes characteristic of close-in weapons fallout [4,8,9]. Because close-in fallout particles decrease in size as distance from ground zero (or time of fallout arrival) increases [10,11,12,13], it was suspected that interception may have been less at locations close to the test site.

Data of plant interception by dry deposition following four NTS [14, 15] has recently been reviewed [4] for purposes of modeling interception of fallout particles. A geometric mean value (GSD in parentheses) was determined [4] for α equal to 0.39 m^2/kg (1.8), applicable to close-in NTS fallout on grasses within the distance of 130 to 420 km. The Utah study found the original data useful for predicting values of α at locations of various distances from the NTS where measurements were not available. A statistically significant relationship was determined for α as a function of distance from the test site or as a function of fallout time-of-arrival (TOA) [16]. The functions uses either distance or TOA as a surrogate variable for the particle size distribution which is believed to be a determining factor for interception. The distance relationship was useful for reducing model sensitivity and uncertainty to α.

The functions for α determined by least-squares analysis of the plant interception data are:

$$\alpha \ (\text{m}^2/\text{kg}) = 7.01 * 10^{-4} * (\text{km})^{1.127} \qquad (R = 0.80) \qquad \text{eq. (3)}$$
$$\alpha \ (\text{m}^2/\text{kg}) = 0.0412 * (\text{TOA in hr})^{1.063} \qquad (R = 0.78) \qquad \text{eq. (4)}$$

with the constraint that $\alpha \leq 3.0$ m^2/kg. The uncertainty of the interception constant ideally should be determined from the standard deviation of the predicted value using methods from least-squares analysis. For simplicity, however, the uncertainty in α was approximated as a GSD of 1.3, consistent with the spread of values at most distances. The decreased variability of α at any particular distance contributed to a reduction of model sensitivity.

Predicted values of α were tested by two methods to determine their usefulness for fallout deposition modeling. The first method consisted of comparing concentrations of radioiodine in milk predicted by the model with measured values. This method actually tests the model structure and the choice of all parameter values in their ability to closely predict measured data. The model of eq. (1) modified to predict the time-dependent concentration was tested satisfactorily against a data set of ^{131}I concentrations in milk measured near the NTS following event SMALLBOY [10] (Fig. 1). Although this calculation does not assure model agreement under varied conditions, such calculations are necessary for assessing model agreement at least under the conditions for which the model was intended [3].

The second method was to compare the predicted interception fraction values (via eqs. 2 and 3) with values derived from measured radioiodine concentrations in milk at several different distances from the test site. For this comparison, the interception fraction (F_v) was calculated from reported concentrations in milk following close-in NTS fallout deposition from event SMALLBOY [11]. These values of F_v are denoted for convenience in Table 3 as "observed values" (O) and the interception fraction values determined by the regression equations are denoted as "predicted values" (P).

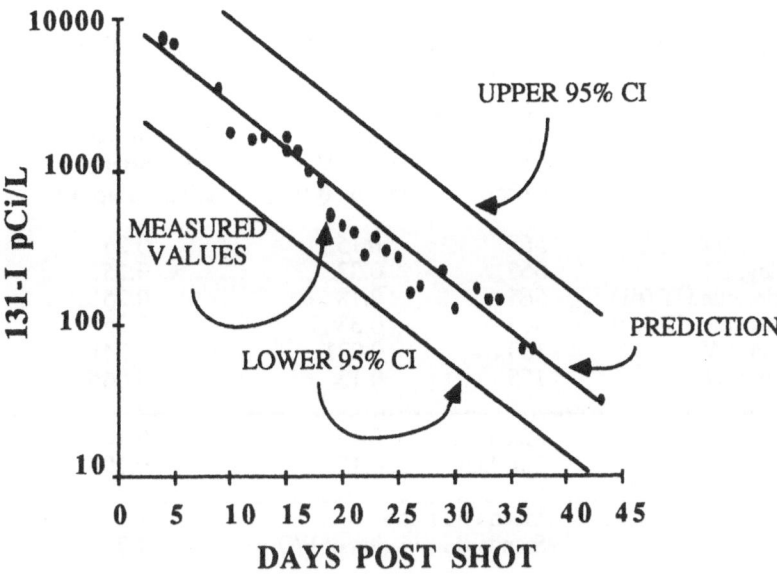

Fig.1 Comparison of predicted 131-I concentration in milk from
model with measurements following event SMALLBOY [10].

The comparison calculations reported here use values of exposure-rate [17], conversion factors
to calculate iodine deposition [18], Q_f [19], $f_d(p)$ [10], Y [20], and f_m [21] from sources
believed to contain relevant data. The peak radioiodine concentrations were used as reference
points for comparison and were determined by decay correcting concentrations back to an
equivalent point in time. By assuming that the ratio of concentrations from the model and the
measurements are equal to unity, eq. (5) can be solved for the interception fraction necessary to
to have resulted in the measured concentration values. Table 3 summarizes the comparison of
the "predicted" and "observed" interception fractions.

$$(F_v)_{model} = \frac{\left[(C_{measurement})_{peak} \right]}{(mr/hr\ @\ H+12) * (\mu Ci/m^2)/(mr/hr\ @\ H+12) * (1/Y) * Q_f * f_d(p) * f_m} \qquad eq.(5)$$

The values of the interception fraction derived from observed concentrations in milk are close to
the values predicted by the distance relationship (Table 3). The mean interception fraction of
0.18 lies between 0.12, derived from α equal to 0.39 m^2/kg, and 0.63, derived from α equal
to 3.0 m^2/kg. The average P/O ratios from the distance and TOA relationships were compared
to unity by a one-sample "t-test," neither of which were significant (i.e., p > 0.05). The results
indicate that the predictive equations for α yield very satisfactory values for modeling purposes
within the distance range of the data and the calculations (i.e., 100 to 565 km).

TABLE 3

Comparison of the Interception Fraction Derived from Measured Radioiodine Concentrations in Milk with Predictions from Distance Relationship

Location	Distance from the NTS (km)	F_v derived from ^{131}I in milk (O)	F_v predicted from distance relationship (P)
Kamas, UT	565	0.15	0.25
Oakley, UT	565	0.33	0.25
Snyderville,UT (#1)	565	0.18	0.25
Snyderville,UT (#2)	565	0.33	0.25
Alamo, NV	100	0.021	0.041
Caliente, NV	155	0.13	0.066
	Mean	0.19	0.18
	Std. Dev.	0.12	0.10
	Mean P/O Ratio		1.05
	Standard Deviation of P/O		1.7

Uncertainty Associated With the Timing of Dairy Management Operations

The Utah dosimetry effort included determining and using a great deal of site-specific information for the environmental transport model. The commitment of resources to acquire this information was necessary in order to make dose estimates for specific individuals. Site-specific dairy information is particularly critical for making realistic person-specific dose estimates because of the variability of dairy management practices.

To obtain site-specific diary management information, the University of Utah conducted interviews during the summers of 1983 and 1985 with 305 individuals who were milk producers in the 1950's in southwestern Utah and Lincoln County, NV. This interview information was used to generate a quantitative description of the annual dairy feeding schedule for each milk producer, referred to in our algorithm design as a "feeding regime." The feeding regime contains estimates of the total dry matter intake (Q_f in kg/d), the proportion (f_d) of Q_f from each of four feed types (i.e., pasture grass, freshly cut alfalfa or "green chop," hay and silage) and the intervals during the year when the proportion of each feed type remained constant. Although values of Q_f and f_d have significant associated uncertainties, only the effect of the timing of feeding changes will be discussed here.

A stochastic calculation that accounts for the uncertainty of the dates when pasture usage began and stopped is a useful approach for estimating C_{int}. This method samples from probability distributions for the pasture begin and stop dates as well as the other model variables. The result of such a calculation is shown in the Fig 2. The spread of the distribution at each point in time as expressed by the GSD is also shown. During the pasture season when there is little uncertainty concerning the use of pasture, the integrated concentration per unit deposition is a near constant value determined by the median values of the model parameters. This is shown by the concentration plateau in Fig. 2. Near the beginning and end of the pasture season, however, the integrated concentration is less than the maximum value during the season. The smaller values near the beginning and end of the season result because a fraction of the activity deposited on the ground and on the vegetation decays during the time before or after the cows are using pasture. The uncertainty (GSD) near the pasture transition times increases due to the additional uncertainty of pasture usage.

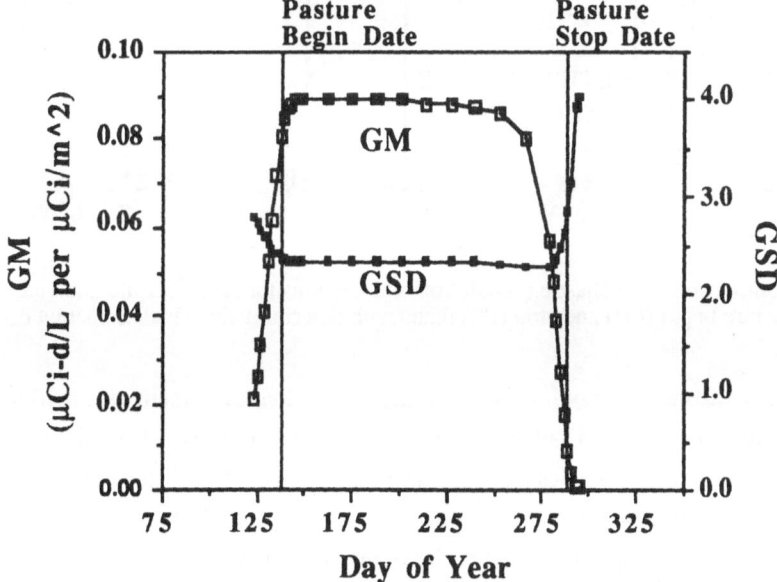

Fig. 2 Results of simulations for the integrated concentration per unit deposition .
The open squares represent the median value (geometric mean or GM) of
1000 iterations for a deposition event on that day, the dark squares
represent the GSD.

A unique feature about the probabilistic calculation that includes pasture begin and stop dates as variables is that the integrated concentration increases more slowly to a plateau value before the season begins than for a deterministic calculation (Fig. 3). The stochastically determined integrated concentration at the beginning of the pasture season increases similarly to a cumulative probability function. In this case, the deterministic concentration estimates have been modified by the cumulative probability of the pasture season having begun before a particular date. At the end of the pasture season, the integrated concentrations decrease slightly more rapidly than for the deterministic calculation, followed by non-zero values after the median pasture stop date (Fig. 3). In this case, the concentrations decrease in a similar manner to an inverse probability function because the deterministic estimates have been modified by the cumulative probability of the pasture season not having ended.

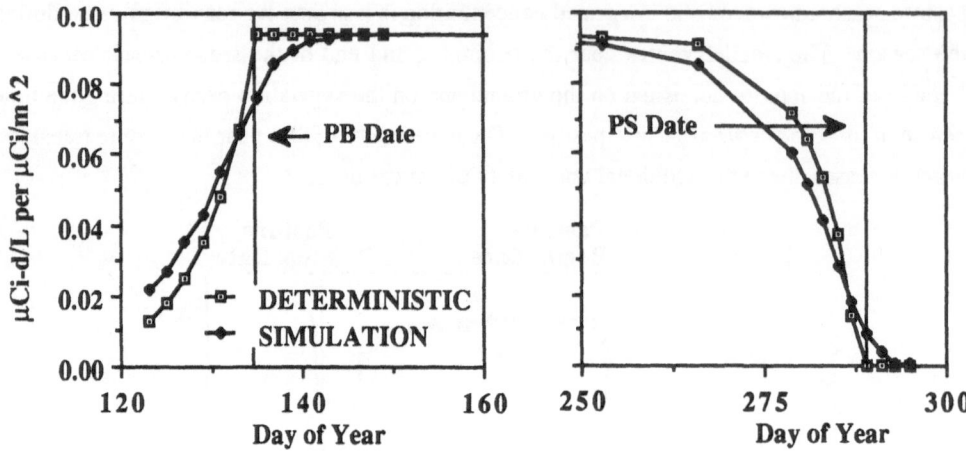

Fig. 3 Comparison of stochastic (simulation) calculations that consider the uncertainty in pasture begin (PB) and stop (PS) dates with deterministic calculations that do not.

In a dose assessment study for a large number of individuals, many of the Monte Carlo simulations for the integrated radioiodine concentration in milk will have the same results because the same parameter values are used. To minimize repetitive calculations, functions can be determined to reproduce the simulation results for any deposition event near the pasture transition time. For example, the fraction (f_b) of the seasonal maximum concentration can be determined for an event near the beginning of the pasture season from the following equation. The predictive equation was determined from a least-squares fit to the simulation data points after their normalization to unity.

$$f_b = \frac{\exp(1.447 + 0.979 * \sigma)}{(1 + \exp(1.447 + 0.979 * \sigma))}$$

The GSD during the "period of uncertainty" at the beginning of the season may be calculated as the product of the GSD during the pasture season (which is a minimum) and the value of the uncertainty factor (UF_b) from the following equation. The equation for the uncertainty factor was also determined by fitting a function to the GSD simulation results after their normalization to unity.

$$UF_b = [0.069 * \exp(-0.406 * \sigma)] + 0.970.$$

At the end or stopping point of the pasture season, the fraction (f_s) of the seasonal maximum concentration and the uncertainty factor UF_s are determined as follows:

$$f_s = \frac{\exp(-1.73 - 1.073 * \sigma)}{(1 + \exp(-1.73 - 1.073 * \sigma))}$$

$UF_s = 0.201 * \exp(0.804 * \sigma) + 0.970$

In these equations σ may be calculated by the standardization procedure: $\sigma = \Delta t/(4\ days)$ where Δt is the number of days (+ or -) from the median pasture begin or pasture stop date to the date that a deposition event takes place. The variable σ is a measure of the time interval between the date of deposition and the median pasture transition date. The units of σ are standard deviations of the probability function for the transition to or from pasture.

CONCLUSIONS

The concentration of fallout radioiodine in milk for the purpose of estimating ingestion dose to the thyroid can be determined by an analytical solution to a simple environmental transport model. This paper describes a model with an emphasis on parameter uncertainties. Sensitivity analysis of the model was used to identify variables that were the most influential to the output. Research efforts were directed toward minimizing the uncertainty, and hence the sensitivity to two important parameters, the vegetation interception constant and the timing of the pasture start and stop dates relative to dates of deposition events.

The interception fraction was determined from analysis of literature data to be a function of distance from the NTS as a surrogate relationship with the unknown particle size distribution of the fallout. The relationship predicted an increase in the interception fraction with increasing distance and was valuable for reducing the uncertainty and, consequently, the sensitivity of the model to the interception parameter.

The variability in pasture begin and stop dates relative to a median date was determined from interviews with managers of dairies during the fallout period. A probabilistic approach for determining the concentration in milk was used that included the uncertainty in the timing of pasture usage relative to the date of deposition events. The method was used to calculate the radioiodine concentration per unit deposition in milk during periods near the beginning and end of the pasture season when there was uncertainty whether cows were using pasture. Consideration of the time-dependent probability of exposure has other applications in dose assessment regardless of whether the exposure is due to the contamination and ingestion of contaminated foods or from external irradiation.

ACKNOWLEDGEMENTS

This study has been funded by the National Cancer Institute, Contract NO1-CO-23917 with the University of Utah, Salt Lake City, UT, USA. Co-investigators include R. D. Lloyd, J. E. Till, H. A. Hawthorne, D. C. Gren, M. Rallison and W. Stevens. Appreciation is extended to

Rita Escher for editorial assistance and to the University of North Carolina, the Greek Atomic Energy Commission and the U.S. DOE for travel support.

REFERENCES

1. Rallison, M. L. and Lotz, T. M. Cohort study of thyroid disease near the Nevada Test Site. Submitted to Health Physics, 1987.

2. Simon S. L. Lloyd, R. D., Till, J. E., Hawthorne, H. A., Grenn, D. C., Rallison, M. and Stevens, W. Development of an algorithm to estimate dose from fallout radioiodine in a thyroid cohort study. Submitted to Health Physics, 1987.

3. Hoffman, F. O. and Gardner, R. Evaluation of uncertainties in radiological assessment models. In Radiological Assessment, A Textbook on Environmental Dose Analysis, ed. J. E. Till, J. E. and H. R. Meyer. NTIS, Springfield, VA. NUREG/CR-3332, ORNL-5968.1983.

4. Whicker, F. W. and Kirchner, T. Pathway: a dynamic foodchain model to predict radionuclide ingestion after fallout deposition. Health Physics, 1987, 52(5):717-737.

5. Chamberlain, A. C. Interception and retention of radioactive aerosols by vegetation. Atmospheric Environment, 1970, 4:57-78.

6. Miller, C. W. Validation of a model to predict aerosol interception by vegetation. In Biological Implications of Radionuclides Released From Nuclear Industries, Proceedings of an International Symposium, 26-30 March, 1979. IAEA, Vienna.

7. Chamberlain, A. C. and Little, P. Transport and capture of particles by vegetation.In Plants and Their Atmospheric Environment. The 21st Symposium of the British Ecological Society, Edinburgh, 1970. eds. Grace J., E. D. Ford and P. G. Jarvis. Blackwell Scientific Publications, Oxford. 1981.

8. Romney, E. M., Lindberg R. G. , Hawthorne H. A., Bystrom, B. G. and Larson K. H. Contamination of plant foliage with radioactive fallout. Ecology, 1963, 4(2):343-349.

9. Anspaugh, L. R.; Koranda, J. J., Ng, Y. C. Internal dose from ingestion in: assessment of radiation dose to sheep wintering in the vicinity of the Nevada Test Site in 1953. ed. Anspaugh and Koranda, U.S. DOE, Las Vegas, NV 89114. DOE-239. 1986.

10. Knapp, H. A. 1963. Iodine-131 in fresh milk and human thyroids following a single deposition of nuclear test fallout. US Atomic Energy Commission, TID-19266. 1963.

11. Storebo, P.B. Prediction of massive wash-out of nuclear bomb debris. Health Physics, 1965, 11:1203-1211.

12. AEC. Operation PLUMBBOB. distribution, characteristics, and biotic availability of fallout, Civil Effects Test Group, US Atomic Energy Commission, WT-1488. 1966.

13. Hicks, H. G. Calculation of the concentration of any radionuclide deposited on the ground by offsite fallout from a nuclear detonation. Health Physics, 1982, 42:585-600.

14. Lindberg, R.G., Romney, E.M, Olafson, J.H. and Larson, K. H. Operation TEAPOT, Nevada test site, factors influencing the biological fate and persistence of radioactive fallout. Civil Effects Test Group,US Atomic Energy Commission. WT-1177. 1959.

16. Simon, S. L. 1987. An analysis of vegetation interception data pertaining to close-in weapons' test fallout. Submitted to Health Physics. 1987.

17. DOE. Town Data Base. Department of Energy, Nevada Operations Office, Las Vegas, NV. 1987.

18. Hicks H. G. Results of calculations of external gamma radiation exposure rates from fallout and the related radionuclide compositions, Parts 1-8. Lawrence Livermore Laboratory, Livermore, CA. UCRL-52152. 1981.

19. Utah. Assessment of leukemia and thyroid disease in relation to fallout in Utah. Quarterly Contract Report. University of Utah, Salt Lake City, UT. 1987.

20. Koranda, J.J. Agricultural factors affecting the daily intake of fresh fallout by dairy cows, Univ. of California, USAEC Rpt. UCRL-12479. 1965.

21. NCRP. Radiological Assessment: Predicting the Transport, Bioaccumulation, and Uptake by Man of Radionuclides Released to the Environment. National Council on Radiation Protection, Bethesda, MD 20814. Report. No. 76. 1984.

AN APPLICATION OF A DIAGNOSTIC WIND MODEL
TO DESCRIBE ATMOSPHERIC TRANSPORT

N. Moussiopoulos and Th. Flassak
Institut für Technische Thermodynamik
Universität Karlsruhe, F.R.G.

ABSTRACT

In this paper a diagnostic model is used to reconstruct the wind field
on the basis of measurements: An initial wind field generated from available
sparse measured data is corrected to satisfy mass conservation by solving a
three-dimensional elliptic differential equation. The actual orography is
taken into account by transforming this equation to a terrain-following co-
ordinate system. For the solution of the transformed equation a fully vectorized fast elliptic solver is applied. By the aid of the obtained wind
fields air parcel trajectories may be calculated to elucidate the prevailing
atmospheric transport mechanisms. As an example, wind fields and air parcel
trajectories are presented for the Athens basin. The results confirm the
features of the sea breeze circulation hinted at by observations and by
previous calculations. The identified air movements are discussed in view of
the elevated pollution levels in Athens.

INTRODUCTION

Atmospheric dispersion is largely affected by topographic features and
by inhomogeneities in the spatial variation of the wind velocity (i.e. by
local circulation systems, shear effects etc.). In such cases simple dispersion models (e.g. of Gaussian type) are inadequate and, therefore, advanced
models should be used. A prerequisite for the application of advanced dispersion models is the knowledge of the three-dimensional wind field.

The most sophisticated models providing the three-dimensional wind
field in complex terrain belong to the class of prognostic models. Basically, such models simulate the dynamics of the planetary boundary layer by
including a large amount of physical information. Unfortunately, this implies both an excessive computational demand and considerable uncertainties
associated with the various parameterizations. Therefore, the utilization of
prognostic models for practical applications is very limited. Contrary to

prognostic models, interpolating techniques are extremely efficient in providing three-dimensional wind fields by a simple interpolation of available
measured data. In spite of the striking simplicity of such techniques, their
use for practical applications is also very restricted because of their
fundamental shortcomings: Firstly, the resulting wind fields in general
violate continuity (i.e. mass conservation); therefore, they would cause
considerable artificial source/sink terms in any dispersion model. Secondly,
these wind fields do not comply with the topographic features and, in particular, they do not satisfy the impermeability boundary condition at ground.

A sound approach for analyses of the air flow in complex terrain consists in applying a diagnostic model: A wind field obtained from an interpolating technique is adjusted to satisfy both continuity and the appropriate boundary conditions. As the computational demand of diagnostic models
is very small, they definitely qualify for on-line performance, especially
in the case of accidental releases. Besides, wind fields obtained may be
used to assess the reliability of prognostic models.

In the first part of this paper the fully vectorized diagnostic wind
model CONDOR [1] is briefly presented. Subsequently, a simple trajectory
model is applied to describe the prevailing atmospheric transport mechanisms
in the Athens basin on the basis of obtained time-varying wind fields. The
results confirm the characteristics of the local air motion in Athens, i.e.
the sea breeze circulation hinted at by observations [2] and by previous
calculations [3].

ANALYSIS

In a first step, an initial wind velocity field $\underline{u}^\circ$ ("observed" field) is
generated by interpolation of sparse measured data (both from surface measurements and upper air soundings). As this "observed" field in general does
not satisfy continuity, in a second step an adjustment is performed: With $\underline{u}$
as the mass-consistent flow field to be calculated ("adjusted" field), the
incompressible form of the continuity equation $\nabla\underline{u} = 0$ yields for the veloc-
ity difference $\underline{c} = \underline{u} - \underline{u}^\circ$ the equation $\nabla\underline{c} = -\nabla\underline{u}^\circ$. For the determination of
$\underline{c}$ it is assumed that the vorticity calculated on the basis of the "observed"
wind field $\underline{u}^\circ$ is equal to the vorticity resulting from the "adjusted" wind
field $\underline{u}$: $\nabla\times\underline{u} = \nabla\times\underline{u}^\circ$. This corresponds to the requirement $\nabla\times\underline{c} = 0$ which is
identically fulfilled, if the velocity difference $\underline{c}$ can be expressed as the
gradient of a scalar field λ (i.e. $\underline{c} = \nabla\lambda$). Therefore, one gets $\nabla^2\lambda = -\nabla\underline{u}^\circ$.
It should be noted that the solution of this equation also minimizes the

variance of the difference between "adjusted" and "observed" field subject
to the strong constraint that the "adjusted" field is non-divergent [1].

The orography is taken into account by transforming the elliptic equation to the terrain-following coordinate system $z = (H - z')/(H - h(x',y'))$;
$x = x'$; $y = y'$, where h is the ground elevation, H the height of the upper
boundary (above sea level) and (x', y', z') the original Cartesian system.
Due to the terrain inhomogeneity, the resulting equation includes several
additional terms and spatially varying coefficients. For its solution a fast
direct elliptic solver is used in the framework of the iterative conjugate
gradient method. The fast direct elliptic solver is based on Fast Fourier
Transformation (FFT) in two directions and the solution of tridiagonal equation systems in the third direction (for more details cf. [4,5]). This algorithm allows a very efficient processing on vector computers. As an example,
the solution of a problem with 64^3 unknowns and Dirichlet boundary conditions takes 217 msec on the two-pipe CYBER 205 computer.

Air parcel trajectories are calculated using the equation $\Delta \underline{r} = \underline{u}\Delta t$ with
$\Delta \underline{r}$ as the displacement of a marker within the time increment Δt (default
value: 60 s). The wind velocity vector $\underline{u}$ at the "old" location of the marker
is obtained using appropriate temporal and spatial interpolation techniques.

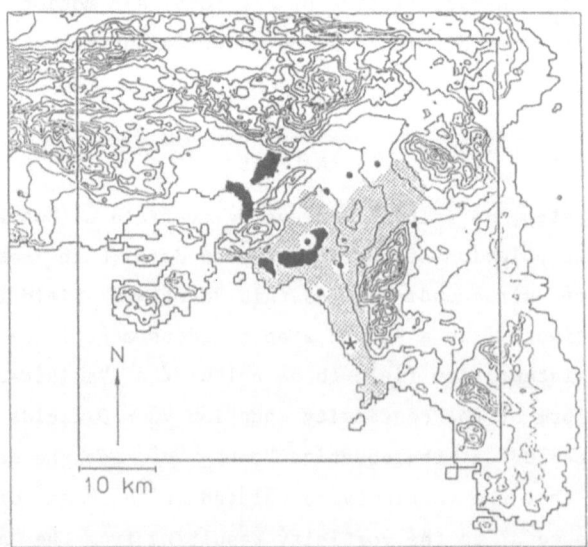

Figure 1. Map of the Attica peninsula showing the location of the computational domain. Contour lines are drawn at intervals of 100 m;
urban areas in the Athens basin are stippled, industrial areas
are inked. Anemometer stations are marked by dots, upper air
soundings are performed at the airport indicated by an asterisk.

69

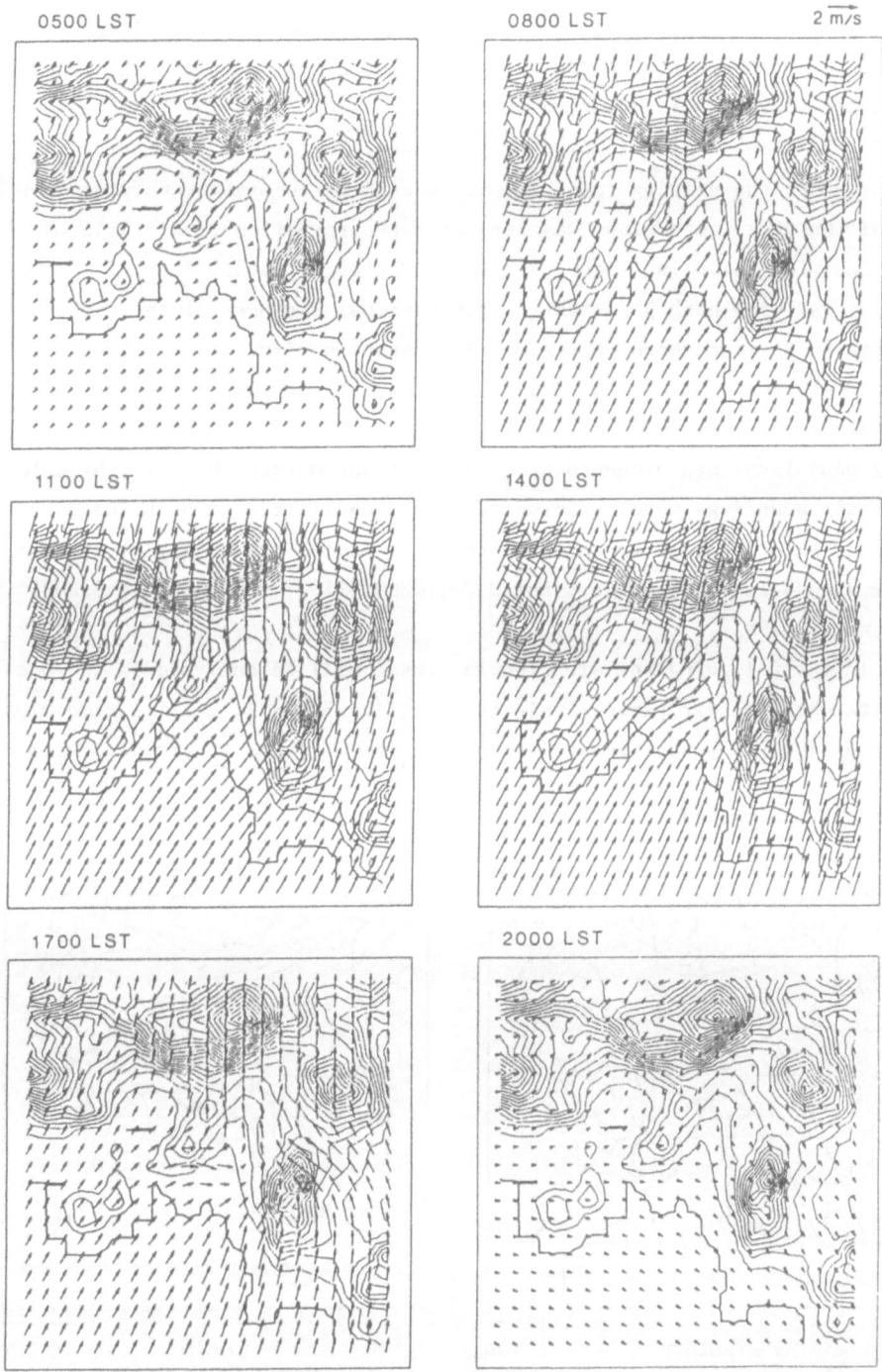

Figure 2. Adjusted wind fields in the Athens basin at 10 m above ground
level for several times of June 26, 1982.

ANALYSIS OF THE AIR FLOW IN ATHENS

Fig. 1 is a map of the Attica peninsula with contour lines drawn at intervals of 100 m. Residential areas are indicated by stippling, industrial areas are inked. Most industrial activity is located in the SW of the basin and in the area of Eleusis west of the basin. The strong insolation in summer provides the driving force for sea breeze circulations. Observations [2] reveal that this phenomenon may lead to a recirculation of air pollutants.

The wind field in Athens was reconstructed for every hour of June 26, 1982, a day with high pollutant concentrations. The base of the computational domain is reproduced in Fig. 1. A constant horizontal meshsize of 2.75 km was selected. In vertical direction nine grid levels were used. The upper boundary was set at 3000 m. "Observed" wind fields were constructed utilizing ground level measurements at eight stations marked in Fig. 1 by dots and upper air soundings were performed near the Hellinicon airport indicated by an asterisk. Fig. 2 shows the adjusted wind fields at the height of 10 m above ground level calculated for 0500, 0800, 1100, 1400, 1700 and 2000 LST of June 26, 1982.

In the following the air flow in Athens will be analysed on the basis of both obtained air parcel trajectories (Figs 3 and 4, initial time 0200 LST and 1400 LST respectively, release height 5 m) and the calculated loca-

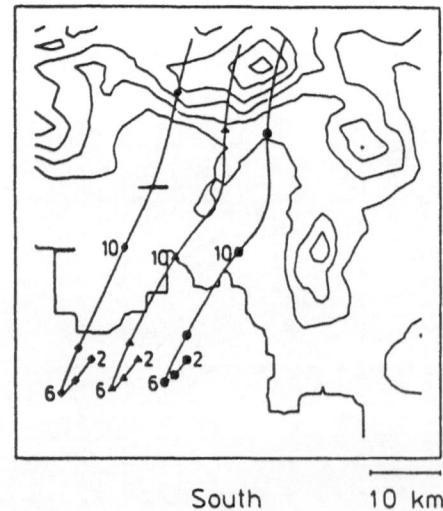

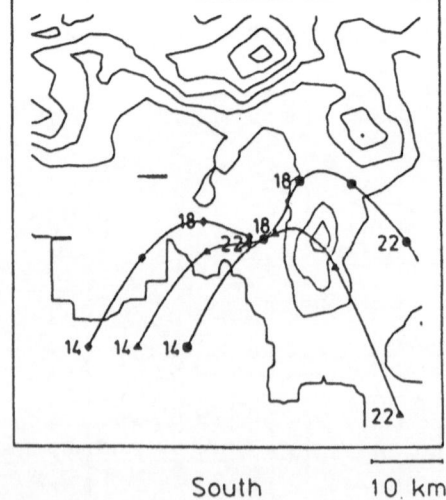

South   10 km   South   10 km

Figure 3. Air parcel trajectories
(start at 0200 LST of
June 26, 1982).

Figure 4. Air parcel trajectories
(start at 1400 LST of
June 26, 1982).

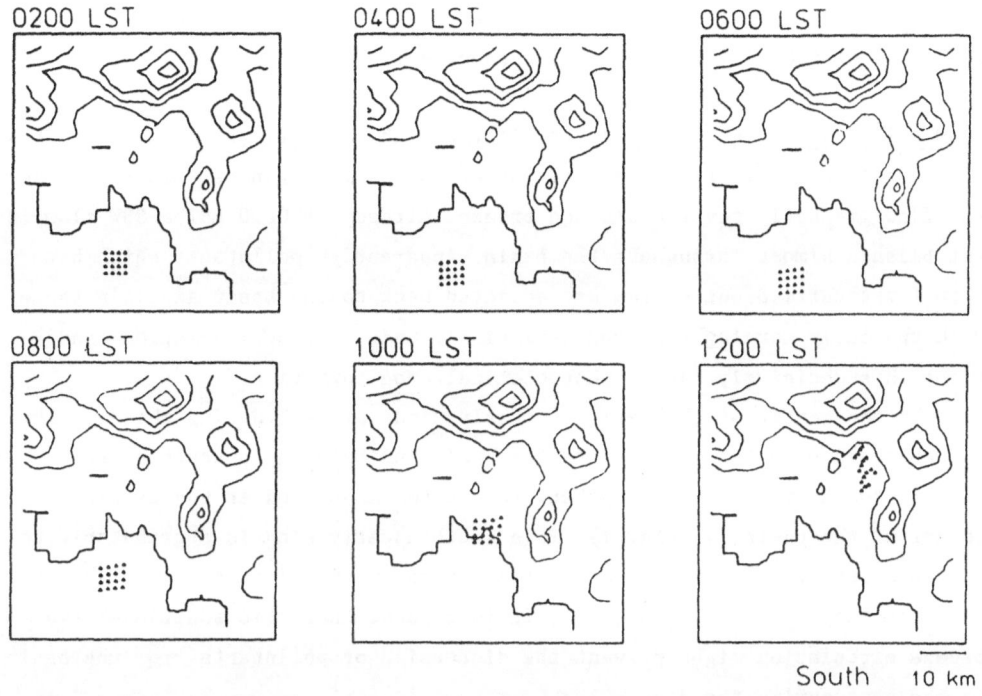

Figure 5. Marker locations (start at 0200 LST of June 26, 1982)

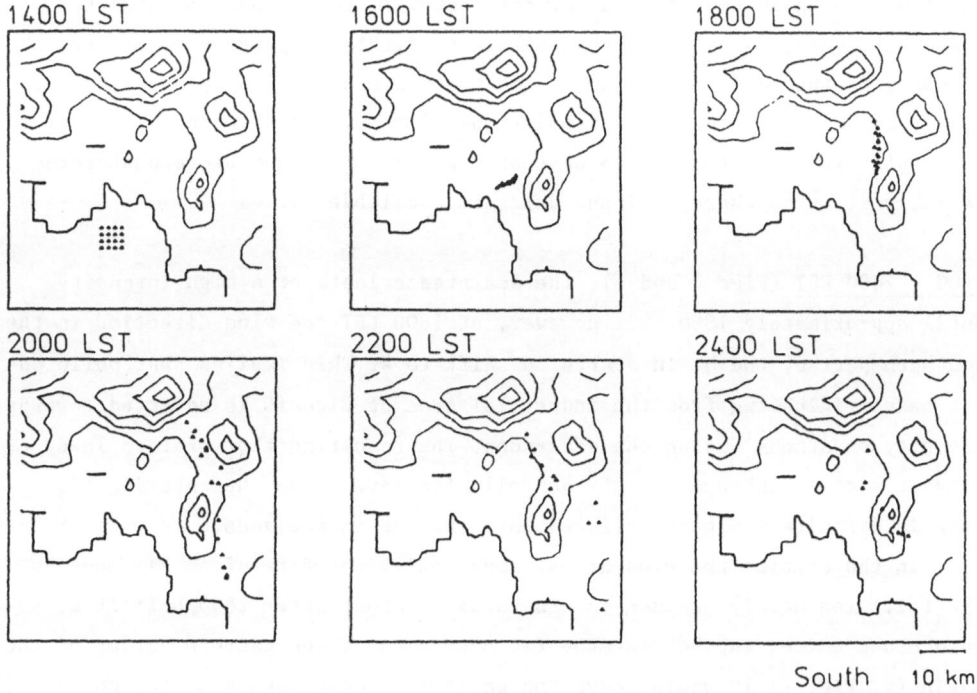

Figure 6. Marker locations (start at 1400 LST of June 26, 1982)

tion of markers at time intervals of 2 h (Figs 5 and 6, released 5 m above
sea level at 0200 LST and 1400 LST respectively).

0200 - 1200 LST (Figs 3 and 5). At night a weak land breeze is blowing. Thus
air mass is advected from the interior of the basin offshore. After sunrise
the flow gradually reverses to sea breeze. Already at 0800 LST a SSW flow is
established almost throughout the basin. Apparently, pollutants which have
been first carried out to sea are advected back to the basin and join there
with the early morning rush hour emissions. Under strong insolation conditions this decisively favors photochemical smog formation.

In the center of the basin southerly winds prevail during the day. This
is connected with the inflow into the basin indicated by a southeasterly
wind at the station in the gap between the two mountains at the eastern
border of the basin (v. Fig. 1). This southeasterly wind is attributable to
a second sea breeze circulation system set up in the plain close to the east
coast of the Attica peninsula [2]. It is evident that this additional sea
breeze circulation might prevent the dispersion of pollutants from the basin
to the east during the day.

According to the obtained results, all regarded markers leave the calculation domain in the early afternoon. However, this cannot be asserted
with confidence, as no measurements were available close to the northern
boundary of the calculation domain and the calculated wind velocities there
merely represent corrected extrapolations of the velocities in the center of
the basin. It is worth mentioning that wind fields cannot be reconstructed
reliably at areas where no input data are available.

1400 - 2400 LST (Figs 4 and 6). The sea breeze lasts at a high intensity
until approximately 1800 LST. However, at 1600 LST the wind direction in the
southern part of the basin starts to shift to W. This implies that polluted
air mass originating from the industrial area of Eleusis is advected towards
the city of Athens during the afternoon. The resulting flow pattern leads
evidently to an enhancement of the pollution levels, as the entering air
mass is definitely higher polluted than the one it replaces.

In the evening the wind drops. Apparently, the fate of an air mass in
the following mainly depends on its location short after 1800 LST: If it
were close to the gap between the two mountains at the eastern border of the
basin (v. Fig. 1) it would leave the basin as a consequence of the north-
westerly wind blowing there during the night; else it would remain in the

basin. It should be noted that the calculated air motion in the plain near the east coast of the Attica peninsula is doubtful, as no measurements were available there (see above).

At night land breeze develops again; with the exception of markers escaping to the SE, all markers remaining in the basin move slowly and uniformly towards the coast. If the synoptic conditions in the following day remain unchanged the whole cycle will be repeated.

The above discussion confirms the features of the sea breeze circulation in Athens hinted at by observational studies [2] and by previous calculations [3].

CONCLUSIONS

Diagnostic wind models represent an adequate basis for analyses of air flow in rough terrain. These models allow an extremely efficient calculation of mass-consistent wind velocity fields which can be used for simulations of atmospheric dispersion. The reliability of the results depends, however, on the availability of representative input wind data for the whole area of interest.

REFERENCES

1. Moussiopoulos, N. and Flassak, Th., Two vectorized algorithms for the effective calculation of mass-consistent flow fields. J. Clim. Appl. Meteorol., 1986, 25, 847-57.

2. Lalas, D.P., Asimakopoulos, D.N., Deligiorgi, D.G. and Helmis, C.G., Sea-breeze circulation and photochemical pollution in Athens Greece, Atmos. Environ., 1983, 17, 1621-32.

3. Moussiopoulos, N., Simulations of the sea- and land-breezes in Athens. Math. Comput. in Simul., 1986, 28, 473-8.

4. Flassak, Th. and Moussiopoulos, N., Modelling the wind field and the dispersion of reactive atmospheric pollutants. In Environmental Meteorology, eds K. Grefen and J. Löbel, Reidel Publishing Company, Dordrecht, Holland, 1987 (in press).

5. Flassak, Th. and Moussiopoulos, N., Direct solution of the Helmholtz equation using Fourier analysis on scalar and vector computers. Submitted to the Environ. Software, 1987.

SEA BREEZE WIND FIELD PREDICTION IN ATMOSPHERIC DISPERSION MODELLING

M. Varvayanni, J.G. Bartzis
NRCPS "Demokritos"
153 10 Aghia Paraskevi Attikis
Athens, Greece

ABSTRACT

The estimation of the dispersion of an airborne radioactive release presupposes the knowledge of the windfield and the level of diffusion in the region under consideration.

The presence of the sea breeze requires particular attention due to the development of circulation patterns and reduced mixing depths within the boundary layer.

The present work constitutes a part of the effort to include the sea breeze capability in the ADREA-I code, a transient three dimensional transport code, under development in NRCPS "Demokritos", and refers specifically to the sea breeze analysis along the Alaskan sea coast [1]. This case has been selected on the way of looking for a well documented sea breeze analytical study in the open literature suitable for comparisons with the model adopted in the present analysis.

It is shown that the generalized three dimensional k-1 model introduced in ADREA-I, canbe also faithfully applied to sea breeze calculations.

NOMENCLATURE

ρ : density

u_i, u_j : velocity components

D_{m_i} : eddy diffusivities

g_i : gravity acceleration components

k : turbulent kinetic energy

ℓ_i : length scales

Θ : potential temperature

Ri : Richardson number

D_{eff} : effective eddy viscosity

T : natural temperature

Z_o : surface roughness parameter

u_* : friction velocity

C_μ : turbulence constant

R : ideal gases constant

u : velocity component along the x axis

v : velocity component along the y axis

w : velocity component along the z axis

1. INTRODUCTION

Dispersion calculations of an airborne radioactive release, at a coastal environment, require the sea breeze features prediction since the phenomenon has been shown to affect significantly the mesoscale concentration patterns.

The present work has been performed on a part of the effort to include sea breeze capability in the ADREA-I code, a transient 3-D code for atmospheric dispersion and other applications, under development in NRCPS "DEMOKRITOS".

The case studied is the sea breeze circulation formation along the Alaskan sea coast [1]. The above case was selected on the way of looking for a well documented sea breeze analytical study, suitable for comparisons with the present model.

2. THE PROBLEM DEFINITION

The present work refers to the analysis of the sea breeze along the Alaskan Beaufort sea coast.

The study area is located at a latitude of approximately 70° N and at a longitude of approximately 148° W. The coastline orientation on the average substends an angle of 115° clockwise from north. The region is characterized by nonexistent topographic complexities.

The synoptic situation is represented by a geostrophic wind at 5 m/s, from 220°. This case was chosen to be representative of large-scale wind velocities [1].

3. DOMAIN

Following Kozo's approach, a two dimensional domain has been selected with the same lateral extension, i.e. 150 km, both landward and seaward.

In the present approach the horizontal distance is divided into 24 mesh points (i=1-24), from land to sea as follows:

$$
\begin{aligned}
DXi &= 30 \text{ km for } i=1,2,23,24 \\
DXi &= 20 \text{ km for } i=3,4,21,22 \\
DXi &= 10 \text{ km for } i=5,6,19,20 \\
DXi &= 5 \text{ km for } i=7-18.
\end{aligned}
\qquad (3.1)
$$

In the vertical direction the grid mesh size is governed by the relation:

$$DZ_{k+1} = 1.2 \, DZ_k, \quad k=1\text{-}18$$

$$(3.2)$$

with $DZ_1 = 10m$.

The domain covers a vertical distance of H = 1280m. The above magnitude of H is well above the expected boundary layer maximum extent.

The domain discretization for i=5-20 and k=1-8 is shown in ref. [2].

4. THE PRESENT MODEL

The present model utilizes mass, momentum and energy balance equations More details about the above equations are discussed elsewhere [2,3,4].

For turbulent stresses the eddy viscosity/diffusivity concept has been adopted [5].

Additionally, for the sake of speed of the calculations the model is applied concerning the Boussinesq approximation, that is the density is set equal to a reference density (ρ_o) and the buoyancy force in the z-momentum equation is set equal to $-(\rho-\rho_o)$, where ρ is given from the equation:

$$\rho = \frac{10^5}{R\Theta}$$

$$(4.1)$$

while the reference density (ρ_o) has the value corresponding to the last top cell.

This approach is consistent with the assumption that there is no vertical motion in neutral conditions (i.e. $\Theta(z)$ = constant).

4.1 Eddy Viscosity/Diffusivity Modelling

In the present report the eddy viscosities were modelled utilizing the turbulent kinetic energy (k) obtained through the corresponding transport equation [5].

The expression of the eddy viscosities Dm_i is the following:

$$Dm_i = \begin{cases} \rho C_\mu k^{\frac{1}{2}} \ell_i : Ri < 0.25 \text{ (unstable and neutral conditions)} \\ \\ D_{eff} = 0.5 \text{ m}^2/s : Ri \geq 0.25 \text{ (stable conditions)} \end{cases} \quad (4.1.1)$$

i=1,2,3

where the generalized Richardson number (R_i) is defined as:

$$Ri = - \frac{g_i Dm_i \frac{1}{\Theta} \frac{\partial \Theta}{\partial x_i}}{\overline{u'_i u'_j} \frac{\partial u_i}{\partial x_i}} \quad (4.1.2)$$

The expression used for the length scales ℓ_i is analogous to the one described in [5]. In the present application the length scales along the x and z axis respectively are given from

$$\ell_x = 0.4h$$
$$\ell_z = \min(z, 0.4h) \quad (4.1.3)$$

where(h) is the boundary layer height, defined as the level where the Richardson number reaches the critical value 0.25.

5. BOUNDARY CONDITIONS

5.1 Lower Boundary

At the ground surface the wind velocity components are set equal to zero: $u_o = u_o = W_o = 0$.

Following Kozo's approach [1] the diurnal temperature variation at the ground surface is a tri-harmonic fit to Kuo's (1968) surface temperature plot [6] appropriately modified for the particular arctic conditions:

T = 273	water and x=0
T = 273+0.5 (25) (ARG)	0 x≤5 km land
T = 273+18 (ARG)	5 km<x≤10 km land
T = 273+20 (ARG)	10 km<x≤20 km "
T = 273+24 (ARG)	20 km<x≤30 km "
T = 273+25 (ARG)	30 km<x "

$$(5.1.1)$$

where ARG = 0.47 sin(13t+265) + 0.15 sin(30t+102)+0.08 sin(45t+306)+0.36.

The surface roughness is approximated by 0.01 m over land [1] while over water z_o is taken to be a function of friction velocity [7].

$$z_o = 0.032 \ u_*^2/g \qquad\qquad (5.1.2)$$

with the requirement $z_o \geq 0.0015$ cm.

5.2 Lateral Boundaries

The boundary conditions at the lateral extremities are:

$$\frac{\partial}{\partial x} (u,v,k,\Theta) = w = 0 \qquad\qquad (5.2.1)$$

5.3 Upper Boundary

The boundary conditions at the upper limit are:

$$u = u_g, \quad v = v_g, \quad w = 0, \quad \Theta = \Theta_s, \quad \frac{\partial k}{\partial z} = 0 \qquad\qquad (5.3.1)$$

where the subscript (g) denotes the geostrophic value and (s) denotes the synoptic value.

6.   INITIAL CONDITIONS

The model starts at 0300 ADST (Arctic daylight standard time) when the large scale velocity, temperature and pressure fields are specified [1].

The geostrophic wind was chosen to be 5 m/s from 220°. The large scale natural temperature field is represented by a linear increase in temperature from the surface (273 K) to 150 m (283 K) where it is isothermal to 550 m. From 550 m on the temperature has a lapse rate of 0.6 K $(100m)^{-1}$.

The large scale pressure field was calculated so that it would balance the given prevailing wind.

The initial horizontal gradients of all variables (except pressure) equal zero.

7. RESULTS

The results are given at 1500 h when the maximum circulation intensity occurs.

Figure (8.1a) shows the velocity field in the u-w plane (where u is in (m/s) and w in (cm/s)). As expected due to the opposing synoptic motion the landward sea breeze penetration is almost non-existent, while the shoreward current over the sea surface does not exceed the horizontal range of 20 km. The vertical extent of the shoreward flow approximates 50m; this low level is reasonable for high latitude sea coasts such as the Alaskan Beaufort sea.

In figure (8.2a) the isotachs of u are shown. The u isotachs also show the shoreward current's small depth and horizontal extent (approximately 50 m and 20 km respectively, see also fig. 8.1a).

In figure (8.3a) the vertical velocity isotachs are depicted. The w contours show that the sea breeze circulation is confined within a narrow range around the coast with the larger portion lying over the sea surface. This result is reasonable due to the large scale opposing motion (see also fig. 8.1a).

In figure (8.3a) the temperature contours T (^{o}C) are also presented. The temperature field shows a surface based inversion layer over water, which is expected due to adiabatic warming in the region of descending motions. The 10, 12 and 14 ^{o}C isotherms show an offshore penetration of approximately 50, 30 and 10 km respectively which is due to horizontal temperature advection by the geostrophic wind.

The above results are in general in a fairly good qualitative and quantitative agreement with the ones obtained by Kozo (1982) (figures 8.1b, 8.2b, 8.3b). This agreement creates confidence that the present model is satisfactorily applied to the sea breeze circulation prediction. On the other hand, it suggests that a simplified model, such as Kozo's, using empirical correlations for exchange coefficients and the hydrostatic approximation, can give a reasonable prediction of the sea breeze circulation features, at least in smooth topography cases.

8. CONCLUSIONS

The main conclusions carried out by the present study are:

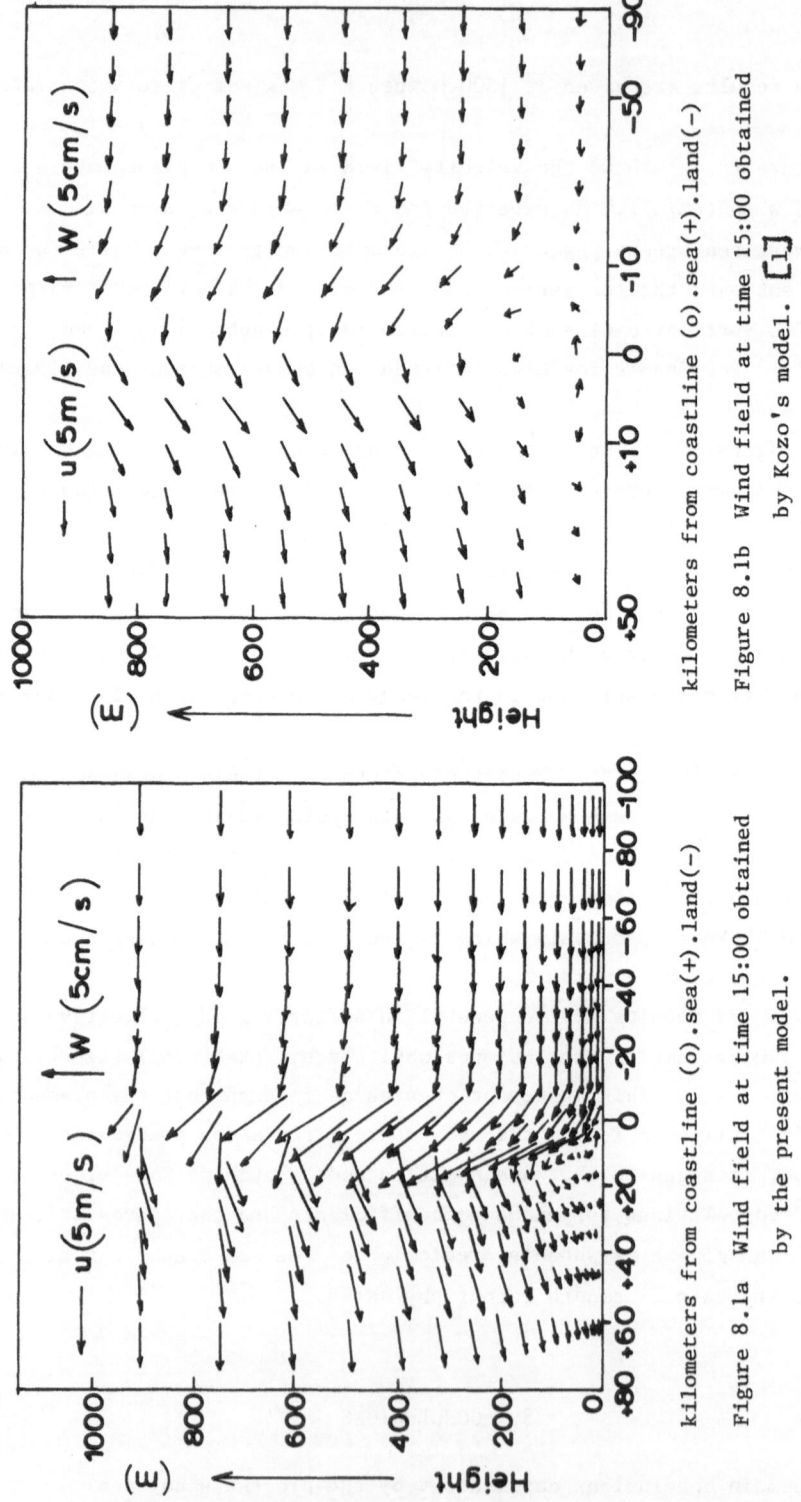

kilometers from coastline (o).sea(+).land(-)

Figure 8.1b Wind field at time 15:00 obtained
by Kozo's model. [1]

kilometers from coastline (o).sea(+).land(-)

Figure 8.1a Wind field at time 15:00 obtained
by the present model.

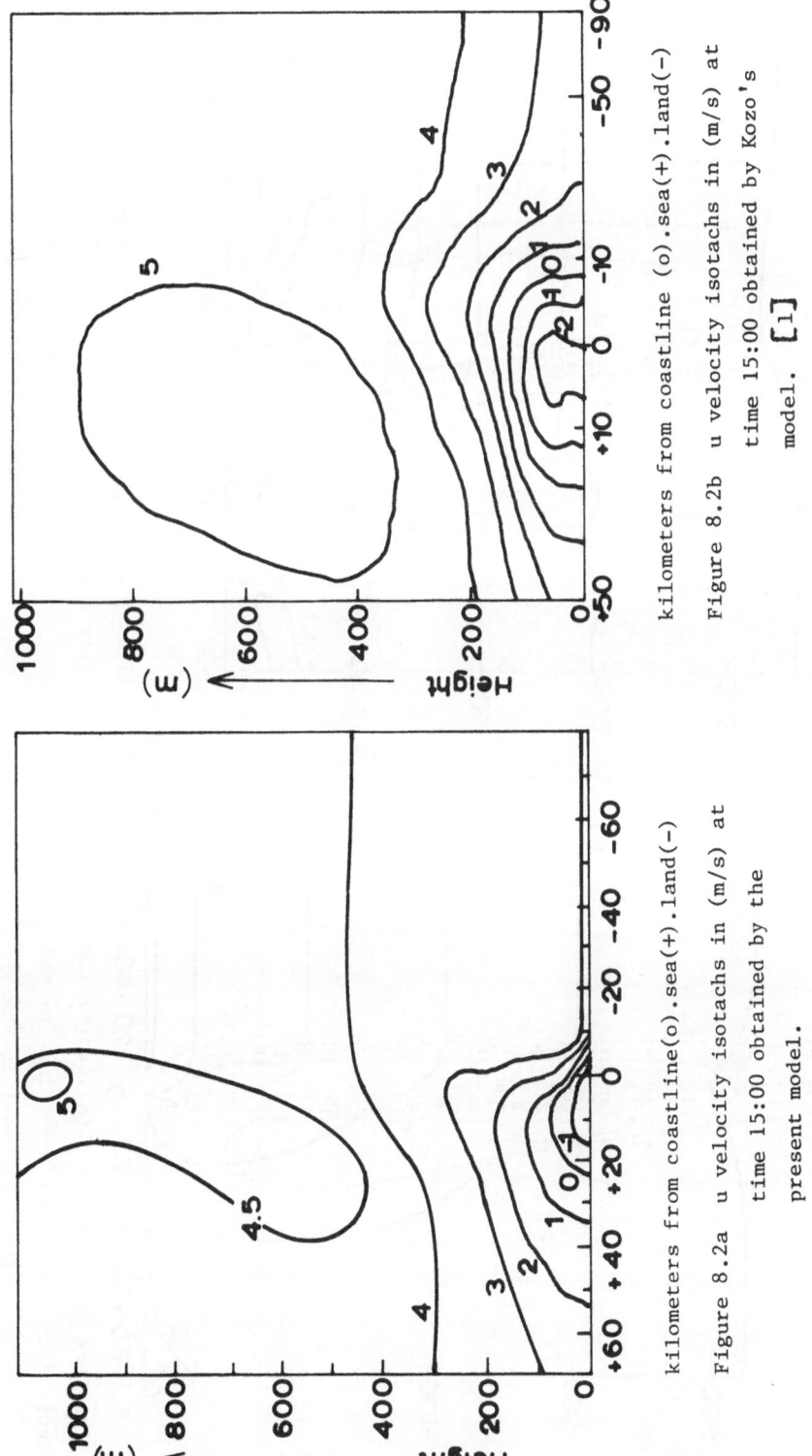

kilometers from coastline (o).sea(+).land(-)

Figure 8.2b u velocity isotachs in (m/s) at
time 15:00 obtained by Kozo's
model. [1]

kilometers from coastline(o).sea(+).land(-)

Figure 8.2a u velocity isotachs in (m/s) at
time 15:00 obtained by the
present model.

82

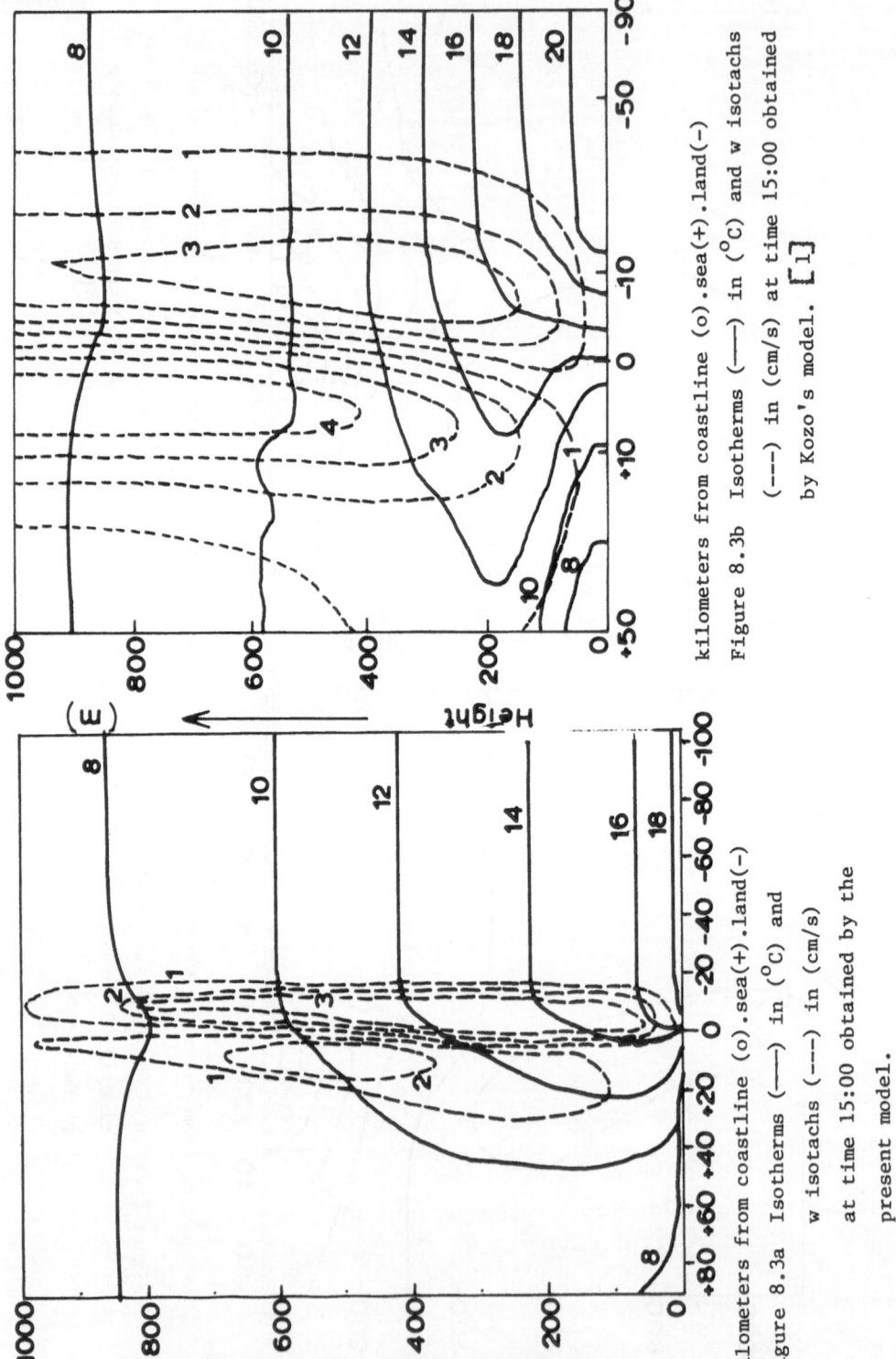

kilometers from coastline (o).sea(+).land(-)

Figure 8.3b Isotherms (———) in (°C) and w isotachs

(– – –) in (cm/s) at time 15:00 obtained

by Kozo's model. [1]

kilometers from coastline (o).sea(+).land(-)

Figure 8.3a Isotherms (———) in (°C) and

w isotachs (– – –) in (cm/s)

at time 15:00 obtained by the

present model.

1. A first application of the ADREA-I code to the boundary layer structure at a coastal environment gave satisfactory results for the sea breeze circulation development in case of an opposing synoptic scale wind, as well as for the temperature field evolution.

2. The present application shows that simplified models, such as Kozo's (1982), can produce satisfactory results for sea breeze theoretical studies, at least in case of smooth topography regions.

ACKNOWLEDGEMENTS

The authors would like to thank European Communities, Directorate General for Science, Research and Development (Directorate for Biology, Radiation Protection and Medical Research) for partial support of the project.

REFERENCES

1. Kozo, T.L., A mathematical model of sea breezes along the Alaskan Beaufort sea coast. Part II. J. Appl. Met., 1982, 21, 900-924.

2. Bartzis, J.G. and Varvayanni, M., ADREA-I code Development: Sea breeze modelling and calculations. DEMO Report 87/8, 1987.

3. Bartzis, J.G., ADREA-I a transient three dimensional transport code for atmospheric and other applications. Some preliminary results. DEMO Report 85/3, 1985.

4. Bartzis, J.G., ADREA-I code development: modelling of the atmospheric stability and verification studies. DEMO Report 86/7, 1986.

5. Bartzis, J.G., Flow modelling in complex terrain. DEMO Report 86/8, 1986.

6. Kuo, H.L., The thermal interaction between the atmosphere and the earth and propagation of diurnal temperature waves, J. Atm. Sci., 25, 682-706.

7. Clark, R.H., Recommended methods for the treatment of the boundary layer in numerical models. Austr. Meteor. Mag., 1970, 18, 51-73.

A COMPARISON OF MODEL PREDICTIONS AND OBSERVATIONS OF THE TRANSFER OF ^{137}Cs THROUGH THE AIR-PASTURE-COW-MILK PATHWAY

Y. C. Ng
Environmental Sciences Division
Lawrence Livermore National Laboratory
Livermore, CA 94550,
USA

and

F. O. Hoffman
Environmental Science Division
Oak Ridge National Laboratory
Oak Ridge, TN 37830,
USA

ABSTRACT

Environmental measurements following the Chernobyl accident for selected locations in the United States and Europe were compared with model predictions of the transfer of ^{137}Cs through the air-pasture-cow-milk pathway. The models include IAEA Safety Series No. 57, AIRDOS/EPA, NRC Regulatory Guides 1.109 and 1.111, the National Council on Radiation Protection and Measurements (NCRP) screening model, and the PATHWAY dynamic food-chain model. Time integrals of the ^{137}Cs concentrations in air, pasture, and milk were estimated, and the predicted and observed grass/air, milk/air, and milk/grass concentration ratios were compared. Predictions of the transfer of ^{137}Cs from air to pasture and from pasture to milk tended to exceed observations. Those of PATHWAY, however, consistently underpredicted the grass/air and milk/air concentration ratios due to the use of parameter values specific for the deposition of large particles and their interception and retention by vegetation. Where possible, parameter values were adjusted to the specific conditions determined for a location, which substantially reduced the discrepancy between predictions and observations.

INTRODUCTION

Initial attempts to use Chernobyl fallout data in models to assess the transfer of Chernobyl ^{137}Cs through the air-pasture-cow-milk pathway overpredicted the amounts transferred from air or grass to milk [1].

Here, we update and extend the earlier comparison of predictions and observations of ^{137}Cs transfer using data sets that are more complete and better matched with respect to location and including predictions by a dynamic food-chain model [2].

METHODS

Data Acquisition and Analysis

Data were selected for the following locations: Aachen, Nordrhein-Westfalen (NRW), Federal Republic of Germany (FRG) [3]; Chester, New Jersey (NJ), United States (USA) [4-7]; Liguria, Italy [8]; Munich, Bavaria, FRG [9]; Netherlands [10]; Northern Italy [11]. The selected data pertain to situations in which cows were grazing when fallout from Chernobyl was deposited. Few locations were chosen because it was difficult to find data that satisfied the requirement that the concentrations in air, grass, and milk represent the same location or region.

The ^{137}Cs data include the total deposition per unit ground area and time-integrated concentrations in air, forage, and fresh farm milk. The concentrations in grass and milk were interpolated or extrapolated as required to estimate the time-integrated concentration.

Selection of Models

The models selected for comparison include IAEA Safety Series No. 57 [12], AIRDOS/EPA [13], NRC Regulatory Guides 1.109 and 1.111 [14,15], the NCRP Screening Model [16], and the PATHWAY food-chain model [2]. All of the models except PATHWAY were developed for predicting equilibrium concentrations as a result of a continuous release of a radionuclide. PATHWAY is a process-level dynamic food-chain transport model that includes numerous processes [2]. Models developed for the assessment of continuous releases were included because the infinite time integral of the concentrations in air, pasture vegetation, and milk from an accidental release is conceptually equal to the equilibrium concentration in these media resulting from the continuous release of the same amount of activity [17]. Therefore, the concentrations predicted by equilibrium models can be compared with time-integrated concentrations from an accidental release.

Differences among the model predictions are expected because the models were developed for different purposes. The IAEA [12] and NCRP [16] models were developed as screening tools. AIRDOS/EPA was developed to assess average exposures to members of the general public [13], and the NRC Regulatory Guides were produced to assess doses to maximally exposed and average individuals [14-15]. PATHWAY was developed to assess terrestrial food-chain contamination and intake by ingestion of radionuclides by humans and livestock after single depositions from the atmosphere [2]. Despite the inherent differences among the models, their predictions should be somewhat similar because of similarities in the data sets from which generic parameter values were developed. Table 1 summarizes the parameter values used for the model predictions.

Because site-specific information was lacking, the following assumptions were made to approximate average environmental conditions at the sampling sites when pastures and grazing cows were exposed to fallout from Chernobyl:

1. A precipitation rate of 2.7 mm/d, which corresponds to an average precipitation of 1 m/year;
2. A 60% supplement of stored feed to the basal diet of fresh pasture vegetation;
3. A dry-matter content of 20% for fresh pasture vegetation to convert concentrations reported on a fresh-weight basis; and
4. A 30-day period of pasture growth prior to harvesting.

TABLE 1
Parameter values for predicting food-chain transport of ^{137}Cs [2,18]

Parameter, symbol, units	Models				
	IAEA	AIRDOS/EPA	NRC	NCRP	PATHWAY
Mass interception factor, r/Y, m²/kg (dry)	2.0	2.0	1.1	2.1	0.39
Deposition velocity, V_d, m/d (dry)	170	160	590		
(wet)	1600	2000	n.a.		
(combined)	1700	2100	n.a.	1000	173
Weathering rate, λ_w, d^{-1}	0.046	0.050	0.050	0.050	0.050
Dry-matter intake, Q, kg/d	16	16	13	16	17
Milk transfer coefficient, Fm, d/L	0.0080	0.0056	0.012	0.0080	0.0071

The assumption of average conditions is not uncommon for generic assessments of accidental releases. Thus, more emphasis is placed on the trends of the comparisons among all locations than on comparisons made for a single location. Actual testing ("validation") of model predictions would require information on the environmental conditions prevailing at each location. Such information was not available at the time of this analysis, but is becoming available through the International Model Validation Study (BIOMOVS, see Haag and Johansson, and Koehler and Nielsen in these proceedings).

RESULTS

Concentration Ratios (CR and R)

Table 2 lists the depositions and time-integrated concentrations of ^{137}Cs in air, grass, and milk. The time-integrated concentrations of ^{137}Cs in grass and milk were normalized by dividing by those in air, and the time-integrated concentrations in milk were normalized by dividing by those in grass. The resulting small differences between concentration ratios (CR) over locations (Table 3) indicate that the processes that affect the transport of ^{137}Cs from the atmosphere to vegetation and milk were similar throughout the northern hemisphere. The time-integrated

TABLE 2
Depositions and time-integrated concentrations of ^{137}Cs in air, grass, and milk

Location	Deposition kBq/m^2	Time-integrated concentration			
		Air Bq-d/m^3	Grass kBq-d/kg DW	Milk kBq-d/L	Ref.
Aachen, NRW, FRG	2.0	3.5	37	0.93	[3]
Chester, NJ, USA	0.029	0.060	0.61	0.020[a]	[4-7]
Liguria, Italy		1.7		1.2[b]	[8]
Munich, Bavaria, FRG	19	13	260	5.8	[9]
Netherlands	1.8	2.1	14[c]	0.77[d]	[10]
Northern Italy		6.8	78		[11]

[a] Extrapolated to 30 June 1986.
[b] The ^{137}Cs activity was assumed to be 2/3 the $^{134,137}Cs$ activity.
[c] Assumes standing crop biomass of 0.2 kg (dry)/m^2.
[d] Includes estimate for the ^{137}Cs that would have been measured if the cows had not been removed from pasture from 4 May to 8 May 1986.

TABLE 3
Ratios (CR) of time-integrated concentrations and ratios (R) of time-integrated concentration and deposition (dep) observed for [137]Cs

	CR			R		
Location	grass/air m^3/kg DW	milk/air m^3/L	milk/grass kg DW/L	dep/air m/d	grass/dep m^2/kg DW	milk/dep m^2/L
Aachen, FRG	11,000	270	0.025	580	18	0.46
Chester, USA	10,000	330	0.033	480	21	0.70
Liguria, Italy		720				
Munich, FRG	21,000	460	0.022	1,500	14	0.30
Netherlands	6,800	370	0.055	870	7.9	0.43
Northern Italy	11,000					

TABLE 4
Ratios (CR) of time-integrated concentrations and ratios (R) of time-integrated concentration and deposition (dep) predicted for [137]Cs

	CR			R		
Model	grass/air m^3/kg DW	milk/air m^3/L	milk/grass kg DW/L	dep/air m/d	grass/dep m^2/kg DW	milk/dep m^2/L
IAEA	56,000	2,900	0.051	1,700	32	1.7
AIRDOS/EPA	67,000	2,300	0.035	2,100	31	1.1
NRC	10,000	620	0.060	590	18	1.1
NCRP	32,000	1,700	0.051	1,000	32	1.7
PATHWAY	3,100	150	0.048	170	18	0.88

concentrations were also normalized to unit deposition (R in Table 3). The normalization effectively removes the processes of wet and dry deposition from the model calculations.

Model Predictions

Concentration ratios CR(grass/air), CR(milk/air), and CR(milk/grass) predicted by the models are summarized in Table 4. The predictions presented for PATHWAY assume a deposition that occurred in late April in arid areas of Utah and Nevada in the United States from a nuclear detonation at the Nevada Test Site [19]. A deposition velocity of 0.2 m/s was used to infer the time-integrated concentration in air from the deposition. The concentration ratios CR(grass/air) and CR(milk/air) predicted by PATHWAY are the lowest, but the CR(milk/grass) predicted by PATHWAY and the other models are nearly the same.

Table 4 also presents predicted ratios (R) of the time-integrated concentration and deposition. The ratio of the deposition and integrated-air concentration, R(dep/air), is the apparent deposition velocity, V_d (combined wet and dry), which is lowest for PATHWAY. The R(grass/dep) and R(milk/dep) predicted by the models vary by less than a factor of two.

Comparison of Predictions and Observations

Table 5 presents a comparison of predicted and observed ^{137}Cs concentration ratios, and Table 6, a comparison of predicted and observed ratios of integrated concentration and deposition. The comparisons are presented in the form of ratios of predictions to observations (P/O). P/O ratios close to unity indicate very good agreement between prediction and observation. P/O ratios greater than unity indicate overpredictions; those less than unity indicate underpredictions.

DISCUSSION

Difference between Predicted and Observed CR Values

AIRDOS/EPA overpredicted CR(grass/air) and CR(milk/air) to the same extent as the IAEA model, and to a greater extent than the NCRP screening model (Table 5). However, one would expect the IAEA and NCRP models, but not AIRDOS/EPA, to overpredict the observed CR values because AIRDOS/EPA was designed to assess average exposures to members of the public and, therefore, should not be conservatively biased.

The deposition velocity, mass interception factor and weathering rate could each influence the ^{137}Cs concentration predicted in vegetation. Lowering the concentrations predicted by the IAEA, AIRDOS/EPA, and NCRP screening models by excluding the wet deposition velocity would not be justified because wet deposition occurred at all the locations. However, an enhanced weathering rate corresponding to a half-life of about 8 d, which was attributed to both washoff by wind and rain and growth dilution, was measured in Munich, FRG [9]. In addition, mass interception factors of 1.0 m^2/kg (dry) or less, which were reduced under the influence of

TABLE 5
Ratios of predictions to observations (P/O) for the grass/air, milk/air,
and milk/grass concentration ratios (CR)

Location	IAEA	AIRDOS/EPA	NRC	NCRP	PATHWAY
		P/O for CR(grass/air)			
Aachen, FRG	5.3	6.3	0.99	3.1	0.30
Chester, USA	5.5	6.6	1.0	3.2	0.31
Munich, FRG	2.7	3.2	0.51	1.6	0.15
Netherlands	8.2	9.8	1.5	4.8	0.46
Northern Italy	4.9	5.9	0.92	2.9	0.28
GM (GSD)	5.0 (1.5)	6.0 (1.5)	0.93 (1.5)	2.9 (1.5)	0.28 (1.5)
		P/O for CR(milk/grass)			
Aachen, FRG	2.0	1.4	2.4	2.0	1.9
Chester, USA	1.6	1.1	1.8	1.6	1.5
Munich, FRG	2.3	1.6	2.7	2.3	2.1
Netherlands	0.94	0.64	1.1	0.94	0.88
GM (GSD)	1.6 (1.5)	1.1 (1.5)	1.9 (1.5)	1.6 (1.5)	1.5 (1.5)
		P/O for CR(milk/air)			
Aachen, FRG	11	8.8	2.3	6.2	0.57
Chester, USA	8.7	7.0	1.9	5.0	0.46
Liguria, Italy	4.0	3.2	0.86	2.3	0.21
Munich, FRG	6.2	5.0	1.4	3.6	0.33
Netherlands	7.7	6.3	1.7	4.5	0.41
GM (GSD)	7.1 (1.5)	5.7 (1.5)	1.5 (1.5)	4.1 (1.5)	0.37 (1.5)

Note: GM denotes geometric mean; GSD, geometric standard deviation.

rain, were measured in Mol, Belgium [20], and can be estimated for Munich,
FRG [9]. The combined effect of reductions in the weathering half-life
and mass interception factor under the influence of rainfall could then
substantially reduce the predicted grass-to-air concentration ratio, which
would improve the agreement between predictions and observations for these
three models.

The PATHWAY predictions are based on a specific deposition velocity
and mass interception factor that are characteristic of large particles
and the absence of rainfall, which would be expected to yield P/O ratios
for CR(grass/air) of less than unity. The apparent agreement between

CR(grass/air) predicted by the NRC model and the observed CR(grass/air) is fortuitous because the NRC model excludes the process of wet deposition.

All of the models overpredicted the grass-to-milk transfer. Parameter adjustments that would lower CR(milk/grass) include a lower intake of dry matter, a higher fraction of uncontaminated stored feed in the intake, and a lower milk-transfer coefficient, F_m. Relatively low F_m values of 2×10^{-3} to 3×10^{-3} d/L have been reported for ^{137}Cs in Chernobyl fallout in the FRG [3,21]. Previously, relatively low F_m values of 2×10^{-3} to 3×10^{-3} d/L were exhibited for ^{137}Cs in worldwide fallout and in diets rich in hay or silage that were high in potassium [22].

The P/O ratio for CR(milk/air) would be expected to equal the product of the P/O ratios of CR(milk/grass) and CR(grass/air). The ^{137}Cs data in Table 5 display this relationship quite well. Indeed, the apparent agreement between the observed CR(milk/air) and that predicted by the NRC models is simply the result of combining underpredictions of CR(grass/air) with overpredictions of CR(milk/grass). The overprediction of CR(milk/grass) by the NRC model is attributable, at least in part, to the relatively high default F_m value of 0.012 d/L.

Differences between Predicted and Observed R Values

The highest P/O ratios for R(dep/air) were from the IAEA and AIRDOS/EPA models, which include wet deposition (Table 6). These models predicted the apparent deposition velocity, R(dep/air), observed in Munich, where precipitation was heavy. The apparent deposition velocity for the PATHWAY predictions was chosen for relatively large particles under dry conditions, which resulted in substantial underpredictions of R(dep/air).

The IAEA, AIRDOS/EPA, and NCRP models overpredicted R(grass/dep). The overprediction may be attributable to some of the factors mentioned previously, i.e., the mass interception factor, the weathering rate, and the influence of rain on these parameters. The geometric mean P/O ratios for R(grass/dep) for the NRC model and PATHWAY are both approximately unity. One might expect PATHWAY to underpredict R(grass/dep) because it includes a relatively low mass interception factor that is characteristic of large particles. However, as indicated by the data in Table 6, the predictions of R(grass/dep) by PATHWAY are in relatively good agreement with observations. PATHWAY is a process-level dynamic model that includes

TABLE 6
Ratios of predictions to observations (P/O) for the ratios (R) of
concentration and deposition

Location	IAEA	AIRDOS/EPA	NRC	NCRP	PATHWAY
			P/O for R(dep/air)		
Aachen, FRG	2.9	3.7	1.0	1.7	0.30
Chester, USA	3.6	4.5	1.2	2.1	0.36
Munich, FRG	1.1	1.4	0.39	0.66	0.11
Netherlands	2.0	2.5	0.68	1.2	0.20
GM (GSD)	2.2 (1.7)	2.8 (1.7)	0.76 (1.7)	1.3 (1.7)	0.22 (1.7)
			P/O for R(grass/dep)		
Aachen, FRG	1.8	1.7	0.98	1.8	1.0
Chester, USA	1.5	1.5	0.83	1.5	0.85
Munich, FRG	2.4	2.3	1.3	2.4	1.3
Netherlands	4.1	4.0	2.2	4.1	2.3
GM (GSD)	2.3 (1.6)	2.2 (1.6)	1.2 (1.6)	2.3 (1.6)	1.3 (1.6)
			P/O for R(milk/dep)		
Aachen, FRG	3.6	2.4	2.3	3.6	1.9
Chester, USA	2.4	1.6	1.5	2.4	1.3
Munich, FRG	5.5	3.6	3.5	5.5	2.9
Netherlands	3.9	2.5	2.5	3.9	2.0
GM (GSD)	3.7 (1.4)	2.4 (1.4)	2.3 (1.4)	3.7 (1.4)	1.9 (1.4)

Note: GM denotes geometric mean; GSD, geometric standard deviation.

plant growth and senescence, soil resuspension, and other processes and considers seasonal changes. Apparently, these complex processes that affect the deposition, interception, retention, and weathering of ^{137}Cs from vegetation compensate for the low value assumed for the mass interception factor.

The P/O ratios for R(milk/dep) are a combination of the P/O for R(grass/dep) and P/O for CR(milk/grass) from Table 5. The P/O ratios for R(milk/dep) exceed those for R(grass/dep) because all the models overpredict grass-to-milk transfer.

CONCLUSIONS

In general, models developed before the Chernobyl accident overpredicted the time-integrated concentrations of ^{137}Cs in pasture grass and milk. With the exception of PATHWAY and the NRC models, the transfer from air to grass was overpredicted to a greater extent than the transfer from grass to milk.

To substantiate the results of this analysis, site-specific data should be employed, replacing the need for the assumption of average conditions. The misprediction of the selected models may be related to the misinterpretation of important processes that differentiate the behavior of Chernobyl fallout from past fallout events. However, many of these processes, such as deposition velocities, vegetation interception and retention factors, and milk transfer coefficients, are not readily determined on a site-specific basis. Nevertheless, site-specific data on the prevailing conditions of daily precipitation and dairy management practices should be more readily determined. As these data become available, the predictive capability of these models can be tested more rigorously. Such testing should lead to further improvement of predictive accuracy and will improve the confidence with which models can be used as tools for decision making.

ACKNOWLEDGMENTS

The authors gratefully acknowledge the assistance of F.W. Whicker and T.B. Kirchner, who provided results of predictions from the PATHWAY model and guidance in their interpretation. The work was performed under the auspices of the U.S. Department of Energy by the Lawrence Livermore National Laboratory under contract number W-74505-Eng-48.

REFERENCES

1. Committee on Model Validation, The Potential Use of Chernobyl Fallout Data to Test and Evaluate the Predictions of Environmental Radiological Assessment Models, Report to the U.S. Department of Energy Office of Health and Environmental Research from the Interlaboratory Task Group on Health and Environmental Aspects of the Soviet Nuclear Accident, Oak Ridge National Laboratory, Oak Ridge, TN, 1987.

2. Whicker, F.W. and Kirchner, T.B., PATHWAY: a dynamic food-chain model to predict radionuclide ingestion after fallout deposition. Health Phys., 1987, 52, 717-737.

3. Bonka, H., Horn, H.G., Küppers, J. and Magua, M., Gemessene radiologische Parameter nach dem Kernkraftwerksunfall in Tschernobyl. Report for the Commission of the European Communities, Brussels, Belgium, June 1986.

4. Juzdan, Z.R., Helfer, I.K., Miller, K.M., Rivera, W., Sanderson, C.G., and Silvestri, S., Deposition of radionuclides in the northern hemisphere following the Chernobyl accident. In A Compendium of the Environmental Measurements Laboratory's Research Projects Related to the Chernobyl Nuclear Accident, ed. H.L. Volchok, USDOE Report EML-460, Environmental Measurements Laboratory, New York, 1986, pp. 105-154.

5. Larson, R.J., Sanderson, C.G., Rivera, W. and Zamichieli, M., The characterization of radionuclides in North American and Hawaiian surface air and deposition following the Chernobyl accident. Ibid., pp. 1-104.

6. Dreicer, M., Helfer, I.K. and Miller, K.M., Measurement of Chernobyl fallout activity in grass and soil at Chester, New Jersey. Ibid., pp. 265-283.

7. Klusek, C.S., Sanderson, C.G. and Rivera, W., Concentrations of ^{131}I, ^{134}Cs, and ^{137}Cs in milk in the New York metropolitan area following the Chernobyl reactor accident. Ibid., pp. 308-326.

8. Corvisiero, P., Salvo, C., Bocacci, P., Ricco, G., Pilot, A., Taccini, G., Scielzo, G., Corso, M., Valerio, F. and Dordo, D., Radioactivity measurements in northwest Italy after fallout from the reactor accident at Chernobyl. Health Phys., 1987, 53, 83-87.

9. Gesellschaft für Strahlen-und Umweltforschung München (GSF), Umweltradioaktivität und Strahlenexposition in Südbayern durch den Tschernobyl-Unfall. Bericht des Instituts für Strahlenschutz GSF-Bericht 16/86, GSF, München-Neuherberg, June 1986.

10. Brofferio, C., Radiological situation in Italy following the Chernobyl accident. Paper presented at GRECA meeting, Paris, 12 June 1986. From ENEA/DISP (Comitato Nazionale per la ricerca e per lo sviluppo dell'Energia Nucleare e delle Energie Alternative/Direzione Securezza Nucleare e Protezione Sanitaria) report, DOC. (DISP/86)1, ENEA/DISP, Rome, 1986.

11. Coordinating Committee for the Monitoring of Radioactive and Xenobiotic Substances (CCRX), Radioactive Contamination in the Netherlands as a Result of the Nuclear Reactor Accident at Chernobyl. CCRX, Leidschendam, October 1986.

12. International Atomic Energy Agency (IAEA), Generic Models and Parameters for Assessing the Environmental Transfer of Radionuclides from Routine Releases, Exposures of Critical Groups. Safety Series No. 57, IAEA, Vienna, 1982.

13. Moore, R.E., Baes, C.F. III, McDowell-Boyer, L.M., Watson, A.P., Hoffman, F.O., Pleasant, J.C. and Miller, C.F., AIRDOS-EPA: A Computerized Methodology for Estimating Environmental Concentrations and Dose to Man from Airborne Releases of Radionuclides. Report ORNL-5532, Oak Ridge National Laboratory, Oak Ridge, Tennessee, 1979.

14. U.S. Nuclear Regulatory Commission (USNRC), _Regulatory Guide 1.109 Rev. 1, Calculation of Annual Doses to Man from Routine Releases of Reactor Effluents for the Purpose of Evaluating Compliance With 10 CFR Part 50, Appendix 1._ Office of Standards Development, USNRC, Washington, D.C., 1977.

15. U.S. Nuclear Regulatory Commission (USNRC), _Regulatory Guide 1.111, Methods for Estimating Atmospheric Transport and Dispersion of Gaseous Effluents in Routine Releases from Light-water-cooled Reactors, Revision 1._ Office of Standards Development, USNRC, Washington, D.C., 1977.

16. National Council on Radiation Protection and Measurements (NCRP), _Screening Techniques for Determining Compliance with Environmental Standards. Release of Radionuclides to the Atmosphere._ NCRP Commentary No. 3, NCRP, Bethesda, Maryland, 1986.

17. Barry, P.J., An introduction to the exposure commitment concept with reference to environmental mercury. Monitoring and Assessment Research Centre Report 12, Chelsea College, University of London, London, 1979.

18. Hoffman, F.O., Bergström, U., Gyllander, C. and Wilkens, A.B., Comparison of predictions from internationally recognized assessment models for the transfer of selected radionuclides through terrestrial food chains. _Nucl. Safety,_ 1984, _25,_ 523-546.

19. Whicker, F.W., Kirchner, T.B., Breshears, D.D. and Otis, M.D., Estimation of radionuclide ingestion: the PATHWAY foodchain model. Submitted to _Health Phys._ (Abstract: _Health Phys.,_ 1987, _52,_ Supp. 1, 578.)

20. Zeevaert, T.H., Private communication. Information obtained during International Biospheric Model Validation Workshop, Vienna, Austria, 26-30 October 1986.

21. Händl, J. and Pfau, A., Feed-milk transfer of fission products following the Chernobyl accident. _Atomkernenergie-Kerntechnik,_ 1987, _49,_ 171-173.

22. Ng, Y.C., Colsher, C.S., Quinn, D.J., and Thompson, S.E., Transfer coefficients for the prediction of the dose to man via the forage-cow-milk pathway from radionuclides released to the biosphere. Report UCRL-51939, Lawrence Livermore National Laboratory, Livermore, California, 1977.

FROM MODEL INTERCOMPARISON TO MODEL VALIDATION
AN EXAMPLE FROM THE BIOMOVS STUDY

H. Koehler
IAEA
Vienna – Austria

S. Nielson
RISØ
Roskilde – Denmark

ABSTRACT

When the BIOMOVS study started in 1985 a scenario (B1) for the transfer of I 131 via the air–pasture–cow–milk pathway was defined for model inter-comparison, as this is one of the most important pathways for radionuclide releases from nuclear power plants and probably the best investigated radioecological problem. The predictions for milk (except from the model PATHWAY) were all within a factor of 20 and generally the modelers were satisfied with their results, as all differences in the predictions could be explained by the individual assumptions or the model intentions, i.e. conservative results or best estimates.

The comprehensive measurements of I 131 in the environment after the Chernobyl accident gave a unique opportunity to receive new and independent datasets for model validation. Since then the necessary site specific data have been assembled for 13 different locations that cover a large range of I 131 air contaminations, and have been offered to participants for calculation (scenario A4). The results will be discussed at the next BIOMOVS workshop in December 1987 in the USA, but the outlines of the study are presented.

INTRODUCTION

For many years mathematical models have been well accepted for the purpose
of radiation dose assessment. Many experimemts have been carried out to
obtain the parameters necessary for these models and big collections of
these experimental data have been assembled (Ng 77, Ng 83, IAEA, IUR 86).
One of the best investigated pathways is the transfer of radioactive
iodine from the air via pasture grass to milk. So when the BIOMOVS study
started there immediately a scenario for this pathway has been defined for
the purpose of model comparison (BIOMOVS 86). It was planned at that time
to look for unpublished data from the nuclear weapons tests fallout, to
also be able to carry out a validation study. But as Chernobyl happened
the plans were changed and this new opportunity was taken to obtain new
and independent datasets that could be used to prove the abilities of the
models in a real situation (BIOMOVS 87).

SCENARIO DEFINITIONS

In the B1 scenario an average long term I 131 concentration of 1 Bq/m^3
in the above ground atmosphere was given, where I isotope shares were
assumed to be 10% elemental, 50% organic and 40% inorganic. Furthermore
an annual precipitation rate of 1000 mm and a growing season of six months
were given (BIOMOVS 86).

As the site specific informations obtained after Chernobyl were more
detailed the input data tables for the A4 scenario contain of 6 pages (see
Appendix 1). Besides daily and time integrated I 131 concentrations in
air, daily precipitation data, information about the location, the pasture
vegetation and the cattle farming were given. This data had been
collected from sites in Sweden (Tranvik), Finland (Loviisa), Denmark
(Roskilde), Netherlands (Petten), Belgium (Geel), Federal Republic of
Germany (Berlin and Munich), Italy (Rome), Hungary (Budapest), India
(Tarapur), Japan (Tokai) and USA (Portland and Oak Ridge). For the
purpose of a blind test these sites were not mentioned on the input data
tables but just have been labelled with arbitrarily chosen letters.

The participants were asked to calculate dry and wet deposition, the
contamination of pasture vegetation and the contamination of milk.

MODELS

The participation in the BIOMOVS study is open to everybody. So B1 has
been performed by 15 modelers, A4 up to now has been run with 20 models,
where from both scenarios have been calculated with 9 models. The models
used for A4 have been used by groups from Finland, Sweden, Belgium,
Netherlands, Federal Republic of Germany (2), Italy, Hungary, Spain,
Canada, USA (4), Japan (3) and the IAEA model from Safety Series 57 (IAEA
82). Calculations according to SS 57 have been performed by Owen Hoffman
and the author and as these have been done independently the results are
quite different due to different assumptions. A list of the models and
their users is attached in Appendix 2.

MATHEMATICAL BACKGROUND

Most of the models were designed for steady state conditions, just in the last years has dynamic modelling become more important. As the Chernobyl accident resulted in a single release of radionuclides one could have the impression that steady state models are not applicable for this situation. But integrating over the whole period from the release of I 131 until its total decay from a single release is equal to time integrated equilibrium concentrations from a continuous release of the same amount of radioactivity (BA79). So the participants were asked to give time integrated results only if their models were created for steady state conditions, and to give time integrated and time dependent results for dynamic models.

RESULTS

The time integrated results from 21 participants for 13 locations for 5 compartments already give 1365 values. As this is far above a number that can be discussed here only the results for the A4 locations N (Munich) and V (Portland) and for B1 are presented. Munich had a rather high air contamination of 102.4 Bq d/m^3 which resulted in a deposition of 76000 Bq I 131/m^2 . Far less iodine of the Chernobyl plume reached Oregon where 0.62 Bq d/m^3 air contamination and a deposition of 480 Bq/m^2 have been measured. The ratios of total depositions to time integrated air concentrations are surprisingly consistent with 742 m/d for Munich and 774 m/d for Portland. From the measurements of pasture grass values of 700000 resp 2510 Bq d/kg d.w. were evaluated. This results in concentration ratios vegetation to total deposition of 9.21 m^2 d/kg d.w. for Munich and 5.23 m^2 d/kg d.w. for Portland, and vegetation to air of 6840 m^3/kg d.w. for N and 4050 m^3/kg d.w. for V. For the milk contamination values of 13100 Bq d/l were amounted for Munich and 57 Bq d/l for Oregon. The concentration ratios of milk to pasture grass follow to 0.019 kg d.w./l for N and 0.023 kg d.w./l for V, while the milk to air concentrations result in 138 m^3/l for Munich and 92m^3/l for the Oregon site. Though the source terms are very different and the composition of the aged plume in Oregon probably differs from what reached Munich, all these concentration ratios are very consistent and do not deviate more than a factor of 2.

The participants' results are given in table 1, where, due to the fact that this table is stuffed with too many numbers already, the values for total deposition, vegetation and milk contaminations were left out and "only" the concentration ratios of total deposition, vegetation and milk to air, vegetation to deposition, milk to vegetation and the predicted to observed (P/O) ratios of total deposition, vegetation and milk are printed.

DISCUSSION

Together with their results each participant was asked to submit a detailed description of the relevant model and give all the equations and parameters used. As yet I have not had a chance to prepare this information, but it will be distributed in time in a working document to the participants of the next BIOMOVS workshop. Therefore, the discussion of the results has to be performed on the submitted values only, disregarding individual assumptions.

Looking at the P/O ratios for milk, the models can be divided into three groups, where SS57, AIRDOS/EPA, FARMLAND, SIRATEC, CRRIS, DOSDIM, CHERBIS, ABG and ENEA PAS show an overprediction of more than a factor of 5, the second group comes rather close to the measured results, while the third group contains only the PATHWAY model, which gives a significant underprediction. RAGTIME, ECOSYS and BILTH give the best estimations.

Going into details shows that many of the good predictions for milk are not satisfying, as the results are compensated from an underprediction of deposition and forage contamination by an overprediction of the transfer into milk. 63% of the results show an underprediction of the total deposition, 24% are below the 50% level. On the other side, only 16% of the models are above the 150% value and the maximum of a factor of 3.75 from AIRDOS/EPA shows less overprediction than PATHWAY gives underprediction.

The relationship between pasture contamination and total deposition was observed to be 9.21 m^2 d/kg d.w. for Munich and 5.23 m^2 d/kg d.w. for Portland. Of the calculations 26% of the results are below the observed values, but only PATHWAY and ECOSYS underestimate this ratio for the Portland site, the highest underprediction made by PATHWAY with shares of 47% and 29% of the observed values. Those models with a more conservative approach show the highest overprediction; 42% are higher than 1.5 times the observed value, 24% are above a factor of 2. This transfer certainly shows acceptable results, as disregarding the extreme values of PATHWAY and SS57 all results scatter around the observations within a factor of 4.5.

The prediction of the contamination of vegetation shows very similar results, 20 of 38 models (53%) underpredict it, where 21% are below the 50% level, with an extreme underprediction by PATHWAY giving only 7% and 11% of the observed values. 29% of the calculations are higher than two times the observations, still 10% are higher than a factor of five and AIRDOS/EPA for Oregon is a factor of 10 bigger. The larger overpredictions now already result from the combination of an overpredicted deposition and an overpredicted interception of I 131 by pasture vegetation.

The underprediction of the pasture contamination that is made by more than half of the models is compensated by a gross overprediction of the transfer to milk. Except BILTH all models assumed a significantly too high transfer from vegetation to milk, only RAGTIME and ECOSYS lie within a factor of 2 above observation, most of the models between factors of 2 and 6, but PRYMA overpredicts by factors of 9 and 10 and ENEA PAS by factors of 32 and 33. This suddenly results in an overprediction of the milk contamination by nearly all models, where only RAGTIME and ECOSYS show negligible deviations, but both these models already have an input of experiences after Chernobyl. Pathway still shows a significant underprediction, while the overpredictions reach factors of 40 above the observations.

The comparison to the results from the B1 scenario shows that the concentration ratios of total deposition do not deviate more than a factor of 2, though different shares of the chemical forms of I have been assumed. The concentration ratios of pasture to total deposition are fairly consistent except BIOPATH, where I assume mistakes in the value.

The concentration ratio milk to forage gives numbers that have to be discussed as low values for A4 always face high values for B1 and vice versa, except SS57.

CONCLUSIONS

This are only preliminary results of an ongoing study and detailed analysis will be made at the next BIOMOVS meeting. At this two locations it can be seen that the predictions of the transfer from air to ground and into plants are underestimated in many cases, but mostly do not deviate from the measurements by an unacceptable factor. The transfer from forage into milk mostly is substantially overpredicted, but compensates errors made in the previous compartments. This should not seduce to rush into overhasted conclusions as there are 11 more sites left for analysis and different situations at other sites result in different predictions and observations.

REFERENCES

BA 79   Barry P.J.,
            An introduction to the exposure commitment concept with
            reference to environmental mercury
            Monitoring and Assessment Research Centre Report 12
            Chelsea College, University of London, 1979

BIOMOVS 86 BIOMOVS Progress Report No 1, January 1986
            National Institute of Radiation Protection
            Stockholm, Sweden

BIOMOVS 87 BIOMOVS Progress Report No 3, January 1987
            National Institute of Radiation Protection
            Stockholm, Sweden

IAEA   Handbook of radionuclide transfer data for the
            terrestrialenvironment.
            Work in progress.

IUR 86   International Union of Radioecologists
            IV, V Report of the Working Group Soil-to-Plant Transfer
            Factors
            Report prepared by RIVM, P. O. Box 1, 3720 BA Bilthoven
            Netherlands 1986

NG 77   Ng Y.C., Colsher C.S., Quinn D.J., Thompson S.E., 1977
             Transfer coefficients for the prediction of the dose to man
             via the forage-cow- milk pathway from radionuclides released
             to the biosphere.
             University of California Lawrence Livemore Laboratory Report
             UCRL-51939, July 1977

NG 83   Ng Y.C., Hoffman F.O., 1983
             Selection of terrestrial transfer factors for radioecological
             assessment models and regulatory guides.
             Seminar on the environmental transfer to man of radionuclides
             released from nuclear installations.
             Vol.2, Oct. 17-21 1983. CEC, Brussels

APPENDIX 2

SS57   Users: Koehler H., IAEA, and Hoffman F.O., ORNL
             Reference:
             Generic models and parameters for assessing the environmental
             transfer of radionuclides from routine releases: exposures
             of critical groups
             Safety Series No. 57, IAEA, Vienna, Austria, 1982

AIRDOS/EPA   User: Hoffman F.O., ORNL
             Reference:
             Moore R.E., Baes III C.F., Mcdowell-Boyer L.M. Watson A.P.,
             Hoffman F.O., Pleasant J.C. and Miller C.W.
             AIRDOS/EPA: a computerized methodology for estimating environ-
             mental concentrations and dose to man from airborne releases
             of radionuclides
             ORNL-5532 , Oak Ridge, TN, USA, 1979

FARMLAND   Users: Brown J., Jones A., NRPB, UK
             Reference:
             Information available from the users.
             National Radiological Protection Board, Chilton Didcot, UK

SIRATEC   User: Kanyar B., NRIRR, Budapest, Hungary
             Reference:
             Information available from the user.
             National Research Inst. for Radiobiology and Radiohygiene
             Budapest 22, Hungary

CRRIS   User: Hoffman F.O., ORNL
      Reference:
      Miller C.W., Sjoeren A.L., Begovich C.L. Hermann O.W.
      ANEMOS: a computer code to estimate air concentrations and
      ground depositions rates for athmospheric nuclides emitted
      from multiple operating sources.
      ORNL-5913, Oak Ridge, TN, USA, 1986

SPADE 2 USERS: Morris C., MAFF, Jones C., ANS, UK
      Reference:
      Information available from the users.
      Associated Nuclear Services
      Epsom, Surrey, UK

DOSDIM Users: Govaerts P., Zeevaerts T., Sohier A., SCK/CEN
      Reference:
      Information available from the users.
      SCK/CEN, Mol, Belgium

ECOSYS Users: Mueller H., Proehl G., GSF, FRG
      Reference:
      Information available from the users.
      GSF, Neuherberg, FRG

Cherbis Users: Barry P., Gentner R., Chalk River, Canada
      Reference:
      Information available from the users.
      CRNL/AECL, Ontario, Canada

ABG   User: Koehler H., IAEA, Vienna, Austria
      Reference:
      Allgemeine Berechnungsgrundlage fuer die Strahlenexposition
      bei radioaktiven Ableitungen mit der Abluft oder in Ober-
      flaechengewaesser, Gemeinsames Ministerialblatt, Nr. 21, 1979

PRYMA Users: Cancio D., Font J.L., CIEMAT, Spain
      Reference:
      Information available from the users.
      Institute for Environmental and Radiation Protection
      Madrid, Spain

CIRCLE   User: Homma T., JAERI, Tokai, Japan
TERFOC   References:
CHRONIC   Information available from the user.
            JAERI, Ibaraki-ken, Japan

BILTH   User: Blaauboer R., RIVM, Bilthoven, Netherlands
            References:
            Information available from the user.
            Laboratory of Radiation Research (RIVM)
            Bilthoven, The Netherlands

RAGTIME   Users: Killough G.G., Hoffman F.O., ORNL, USA
            Reference:
            Pleasant J.C., McDowell-Boyer, Killough G.G.
            RAGTIME: A FORTRAN IV implementation of a time-depen-
            dent model for radionuclides in agricultural systems
            ORNL/NUREG/TM-371, Union Carbide Corp., Nuclear Division,
            ORNL, USA, 1980

ENEA PAS   User: Monte L., ENEA PAS, ROMA, Italy
            Reference:
            Information available from the user.
            ENEA PAS, Casaccia, Italy

BIOPATH   User: Bergstroem U., Studsvik, Sweden
            Reference:
            Information available from the user.
            Studsvik, Nyköping, Sweden

PATHWAY   User: Whicker F.W., Ft. Collins, CO, USA
            Reference:
            Whicker F.W., Kirchner T.B.
            PATHWAY: a dynamic food-chain model to predict radionuclide
            ingestion after fallout deposition
            Health Physics, Vol. 52, No.6, pp 717-737, 1987

DETRA   User: Korhonen R.
            Reference:
            Information available from the user.
            Technical Research Centre of Finland, VTT, Helsinki, Finland

CESIUM TRANSPORT IN FOOD CHAINS - COMPARISON OF MODEL PREDICTIONS AND OBSERVATIONS

H. Müller and G. Pröhl
Institut für Strahlenschutz
Gesellschaft für Strahlen- und Umweltforschung (GSF)
D-8042 Neuherberg, F.R. Germany

ABSTRACT

Measurements of cesium from the Chernobyl accident in different environmental media are compared with respective model predictions. Especially the sub-models for the
- interception of washout by plants
- time-dependence of the contamination of grass
- transfer fodder - milk
- transfer fodder - meat
are considered. Mean values and ranges of several model parameters are derived from our measurements for the post-Chernobyl situation.

These parameter ranges are used as input data for a stochastic version of the radioecological model ECOSYS and a probability function of the activity intake by man is generated. The results are compared with the distribution functions of cesium content in persons actually measured with whole body counters.

INTRODUCTION

A lot of measurements of cesium activity in different environmental media have been carried out in the Munich region where a rather high cesium deposition occured after the Chernobyl accident (about 20 kBq/m² Cs-137 and 10 kBq/m² Cs-134 [1]). These data offer a good opportunity to validate the predictions of the dynamic radioecological model ECOSYS which has been developed at our institute for several years.

INTERCEPTION FACTORS

The interception factor f_w for wet deposition, i.e. that fraction of activity deposited by rain which is retained on the leaves of plants, is of
great importance for the assessment of the initial contamination of human
foodstuffs. The assumption of a constant value for f_w (for example 0.2 in
several models) for all situations is not very satisfactory, because the
interception depends on several parameters as e.g. the plant species, the
stage of its development, the amount of rainfall and the ability of the
radioactive element to be fixed to the leaf. Based on data in the literature [2-5] on the interception and retention of rainwater by plant cano-
pies, an approach has been developed considering all these factors:

$$f_w = k_1 \cdot L \cdot (1- \exp (-k_2 \cdot R)) / R$$

with   L: leaf area index (as estimated e.g. from the herbage density)
       R: amount of precipitation (mm)
       k_1, k_2: element and plant dependent constants

Fig. 1 illustrates the dependency of f_w on the amount of precipitation
for deposition of cesium on grass with different values of the leaf area
index as assumed in our model.

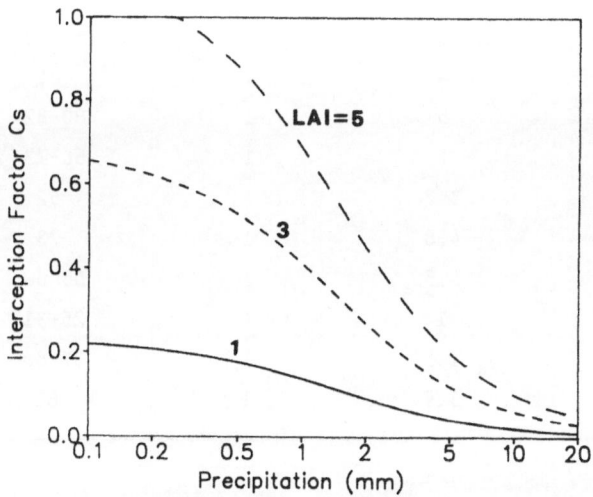

Figure 1: Interception factor for cesium on grass with different leaf area
indices (LAI) as used in ECOSYS.

The measurements of the activity in the precipitation during the Chernobyl fallout and the initial contamination of plants after rainfall have been used to check the model at least at some data points. In addition we have made several experiments in this year where rainwater from the Chernobyl fallout was sprayed onto different plants. This offers some further checks of the model. Table 1 compares the predicted and the measured interception factors for these cases. In general, the agreement between predictions and measurements seems to be fairly good even if there are some cases where the difference between model and measurement is some ten percent. Several other measurements e.g. for other elements in the beginning of May 1986 in the Munich region showed interception factors mostly in the range between 0.05 and 0.2; this fits quite well to the model approach.

TABLE 1

Comparison of predicted and measured interception factors for cesium

Plant species	Herbage density (kg/m²)	Leaf area index	Amount of precipitation (mm)	Interception factor (%) measured	predicted
Grass	0.6-0.8	3-3.5*	5.6	8-10	10-13
Grass	1.4-1.9	5-6*	1.1	39-54	62-71
Wheat		3	1	56-62	40
Wheat		4	1	86	51
Wheat		5	1	80-82	65
Wheat		7	1	66-73	90
Barley		3.2	1	52	40
Barley		4.8	1	73	62
Barley		5.5	1	65-68	70
Potatoes		1	1	25-31	15
Potatoes		1.5	1	37-68	22
Potatoes		1.9	1	61	29

*

  = estimated

PASTURE - COW - MILK - PATHWAY

For this pathway the time dependence of the grass contamination and the transfer from the cow's fodder to the milk are most important. To describe the decrease of the specific activity in grass due to weathering and dilution by biomass growth after the initial deposition we formerly used an exponential decline with a half-life of 14 days until the level of root uptake is reached. In figure 2 some time series of measured activites in grass are shown measured in 1986. The actual decline of the specific activity in May 1986 corresponds to a half-life of about 8 to 14 days.

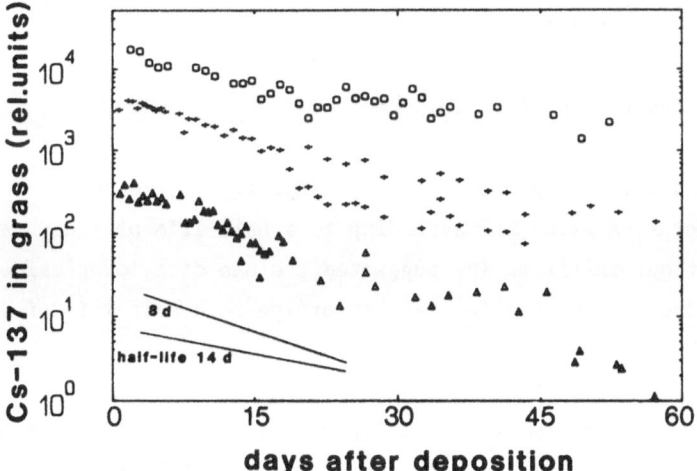

Figure 2: Time-dependency of the specific activity in grass as measured at different sites after the Chernobyl-accident.

After some weeks the decrease becomes slower because weathering effects do not play a role for the second cut of grass. Considering all our measurements of grass and milk we now use the following equation describing the activity in grass:

$$C_g(t) = C_g(0) \cdot \{f \cdot \exp(-t \cdot \ln 2/T_1) + (1-f) \cdot \exp(-t \cdot \ln 2/T_2)\}$$

with $C_g(t)$: specific activity of grass (Bq/kg) at time t

T_1 : fast removal half-life (8... 14 days for deposition in May)

T_2 : slow removal half-life (40...60 days)

f : fraction of fast component (0.9... 0.99)

It should be noted that the fast component (T_1) may be quite different at times in the year with slower growth of biomass.

For the transfer of cesium from the fodder to the milk we use the following approach:

$$C_M(t) = I \cdot T_M \cdot \int_o^T C_g(t) \cdot \{a \cdot \lambda_1 \cdot exp(-\lambda_1 \cdot (T-t)) + (1-a) \cdot \lambda_2 \cdot exp\ (-\lambda_2 \cdot (T-t))\} dt$$

with $C_M(t)$: specific activity of milk (Bq/1) at time t

    I   : daily intake of grass by the cow (kg/d)

    T_M : equilibrium transfer factor (d/1)

    $\lambda_{1,2}$: biological transfer rates (1/d)

    a   : fraction of fast component

For T_M we used before the post-Chernobyl experiences the value of 0.005 d/1; for λ_1 a value corresponding to a half-life of 1 day. A feeding experiment at our institute [6] suggested the use of a lower value of T_M (0.002 d/1) and a main transfer rate according to a half-life of about 3 days.

Besides a lot of measurements of activity in grass and milk of several farms after the Chernobyl accident, we carried out extensive feeding experiments in co-laboration with the Bayerische Landesanstalt für Tierzucht (Bavarian Institute for Animal Breeding), Grub. In Fig. 3 some examples of the measured specific activity in milk are shown as a function of time. In addition, the specific activity is plotted using the formula given above, the indicated parameters T_M, and a = 0.6, $\lambda_1 = 0.46\ d^{-1}$ ($T_{1/2} = 1.5$ d), $\lambda = 0.063\ d^{-1}$ ($T_{1/2} = 11$ d).

This shows that the model permits a fairly good approximation of the measurements if a transfer factor between 0.002 and 0.005 d/1 is used. The latter value seems to be rather high; the only cow in the experiments which reached this value (upper curve in Fig. 3) was pregnant and in the very last stage of the lactation period and had a very low milk yield. A value of 0.003 d/1 seems to be best for mean situations.

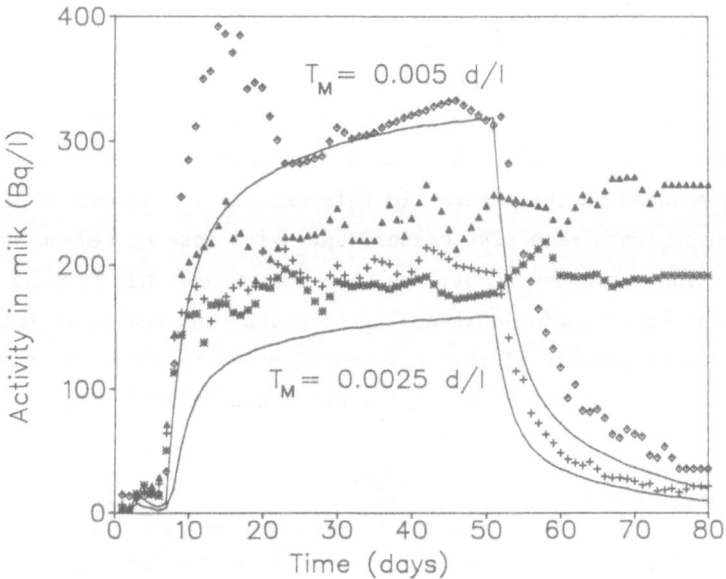

Figure 3: Specific activity in the milk of cows in a feeding experiment and model predictions.

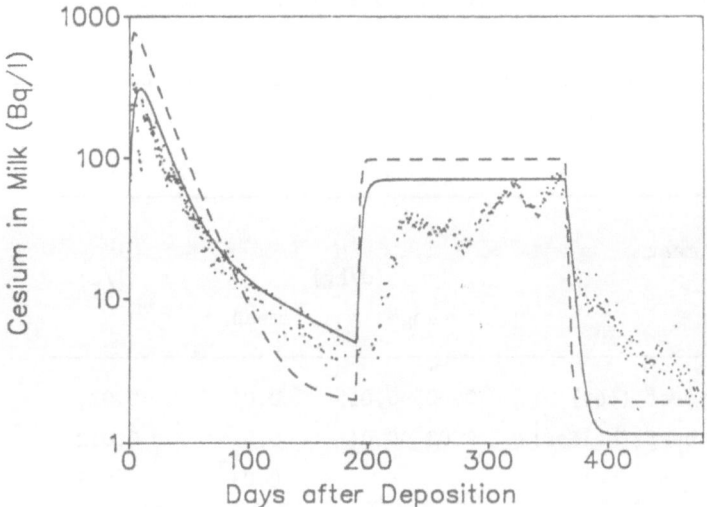

Figure 4: Time-dependency of the specific activity measured in the milk of a dairy-farm in Southern Bavaria. The dashed line is the prediction of the model with the pre-Chernobyl parameters, the solid line that with the present parameters.

In Fig. 4 the measurements of the specific activity of the milk of a dairy-farm in Southern Bavaria are shown for the period from May 1986 until August 1987. The winter peak due to feeding of hay and silage is obvious. The results of our model using the pre-Chernobyl parameters and the revised parameter values (as given above; including the interception approach) are added in the figure. In this example the agreement is very good. Of course there were also measurements with less agreement between model prediction and measurement; the main reason was that several farmers tried to feed the lactating cows as little grass from the first cut in May 1986 as possible; instead of it, less contaminated fodder like grass or hay from the second or third cut, maize etc. were fed.

MEAT

The situation with meat is similar to that with milk: Our feeding experiments in co-operation with Bayerische Landesanstalt für Tierzucht, Grub have shown that the same approach as above for milk (but with the values for T_M and λ_1 from Table 2) fits quite well the measured activities in meat, even in those cases where the animals got contaminated fodder only during a limited period some time before slaughtering.

Table 2: Model parameters for the transfer from fodder to meat (same equation as given above for milk with a = 1)

meat	T_M (d/kg)		λ_1 (1/d)
	range	mean	
beef (cow)	0.005-0.015	0.01	0.023
beef (heifer)	0.03 -0.04	0.035	0.012
beef (bull)		0.03[*]	0.023
veal	0.3 -0.4	0.35	0.027
pork	0.35 -0.45	0.4	0.020

[*]experiment not finished

TOTAL ACTIVITY INTAKE BY MAN

The main aim of a radioecological model is to assess the radiation exposure of man which results from the activity intake via all foodstuffs. The
validation of these results by measurements is much more difficult than
that of the activity of single foodstuffs as done above because it is not
possible to get adequate information on the constitution of daily diets
and their origins.

Since the Chernobyl accident, many whole-body-counter measurements of the
cesium content of members of the public were made in the Munich region
[7]. From these measurements the cesium intake has been assessed (using
certain assumptions about cesium metabolism in man) and a frequency
distribution of the total cesium uptake during the first year (May 1986
till May 1987) has been obtained [8].

These data can be compared with the prediction of the cesium intake by the
ECOSYS-model. For this purpose a stochastic version of the model (which
uses distribution functions for the different model parameters) has been
employed. The ranges of the model parameters were chosen to fit the post
Chernobyl situation (e.g. the above given parameter ranges for milk-,
meat- transfer factors, etc.). The eating habits of the public were chosen
on the basis of mean consumption rates with an adequate variability. The
cesium deposition as measured at our institute near Munich was used for
the calculation. The resulting distribution function of the total Cs-137
intake during the first year after the Chernobyl accident is compared with
the respective function from the whole body counter measurements in
Fig. 5. For the measurements two curves are shown: one corresponds to
those 21 persons with the most reliable assessment of the cesium intake
(these persons have been measured at least once per month starting soon
after the Chernobyl accident); the other curve results from the measurements of 158 persons with fewer measurements per person. The distribution
function of the model prediction is located at cesium intake values about
3 times as high as the measured ones. This results from several differences between model and the real situation:
-    In the model it is assumed that all foodstuffs are produced in areas
     with equal contamination level. In reality the deposition was lower in

many parts of Germany or Europe so that less contaminated foodstuffs
have been imported into the Munich region.

- The model is based on constant mean consumption rates at all times
  while the real consumption of milk and fresh vegetables was reduced at
  least after the accident until the early summer 1986.
- The model assumes normal feeding practices (hay etc.) during winter
  time while in reality many farmers have tried to minimize the cesium
  content of the milk by appropriate feeding with less contaminated
  fodder.
- The beginning of feeding hay and silage at the end of 1986 was some
  weeks later than normal because of a late begin of winter.

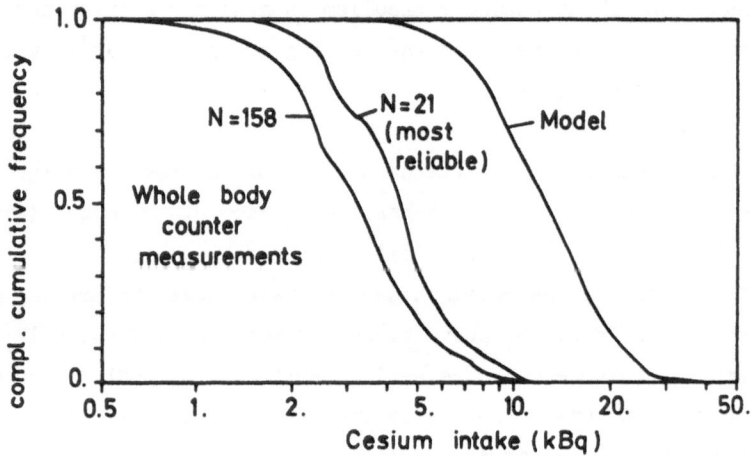

Figure 5: Cumulative frequency distributions of the cesium intake from May
1986 till May 1987: Measurements and model predictions.

While the first point could probably be considered in a model the others
cannot be predicted generally. The behaviour of people is a psychological
problem and depends on the perceived health hazards and on a lot of un-
foreseenable facts.

In summary, the prediction of activity concentrations in food stuffs after
radionuclide depositions can be obtained with reasonable accuracy; it is
easier than the prognosis of the real radiation exposure due to ingestion
of these food stuffs.

REFERENCES

1. Gesellschaft für Strahlen- und Umweltforschung - Institut für Strahlenschutz, Umweltradioaktivität und Strahlenexposition in Südbayern durch den Tschernobyl-Unfall. GSF-Bericht 16/86, Juni 1986

2. Monteith, J.L., Vegetation and Atmosphere, Vol. 1, Academic Press, London, New York, San Francisco 1975

3. Merriam, R.A., Surface Water Storage on Annual Rye-Grass, J. Geophysical Research, 66 (1961), p. 1833-1838

4. Reiniger, P., Levi, E., Coppola, M., Measurement of Surface Storage on Rye-Grass and Clover by Nuclear Methods. Proc. IAEA Symposium Isotope and Radiation Techniques in Soil Physics and Irrigation Studies Vienna 1.-5. Oct. 1973, S. 327-334

5. Angeletti, L., La contamination des pâturages par l'iode - 131, Rapport CEA-R-5056, 1980

6. Voigt, G., Henrichs, K., Pröhl, G., Paretzke, H.G., Experimentelle Bestimmung von Transferfunktionen Futter/Rindfleisch, Futter/ Schweinefleisch und Futter/Milch für Cs-137, Co-60, Mn-54, Na-22, I-131 und Tc-95m. Abschlußbericht Forschungsvorhaben StSch 935, GSF,Mai 1986

7. Bogner, L., Mühle, P., Czempiel, H., Berg, D., Henrichs, K., Ganzkörpermessungen im Raum München nach dem Reaktorunfall von Tschernobyl bis zum Juli 1987. Jahrestagung Deutsche Gesellschaft für medizinische Physik, Innsbruck, September 1987

8. Henrichs, K., private communication, September 1987

COMPARING PREDICTIONS OF TWO RADIOECOLOGICAL MODELS
WITH MEASURED VALUES

H. Maubert - A. Grauby - V. Ponzetto

Commissariat à l'Energie Atomique
Institut de Protection et de Sûreté Nucléaire
Département d'Etudes et de Recherches en Sécurité
Service d'Etudes et de Recherches sur l'Environnement
Centre d'Etudes Nucléaires de Cadarache
13108 - Saint Paul Lez Durance, FRANCE

ABSTRACT

The models most widely used in France for impact studies related to nuclear facilities have been consolidated into computerized reference under the code name "DUNE".
The environmental effects of radionuclides released during the Chernobyl accident allowed observations to be made with applications not only to accidental fallout but also to routine waste discharge.
Two examples are considered which reveal the difficulties encountered in completely verifying a model : one involves the contamination of mutton, and the other concerns radionuclide migration in soils.
Two questions are then raised : "How can a model which produces uncertain results be confronted with field measurements which are variable by nature ?" and "Should existing models be modified in the light of the latest observations ?"

1. <u>INTRODUCTION</u>

The models in widespread use in radioecology can be classified into two broad families, although there is no clear-cut division between them.

The first group includes models developed from field measurements. These are often "black box" constructions for which mathematical analysis of the phenomena involved is not the primary objective, and which often consist of equations yielding realistic or at least conservative sets of values. Such models are subject to statistical verification, in which the accumulated values are assumed to prove that the proposed equations and factors provide sufficiently reliable predictions.

The second group comprises models based more on research, resulting in one or more equations that are often adjusted and verified by experimental findings : i.e. the theoretical model is confronted with a physical model. Although the objective here is to allow implementation under varied conditions, extrapolation to actual conditions is often difficult.

In reality, predictive studies make use of both categories of models depending on the state of knowledge and the extent of adaptation required to suit the problem under consideration. Field studies conducted in the vicinity of French nuclear facilities to verify the performance of impact study models [1] have been unsuccessful because of the low radioactivity levels encountered.

2.   EXAMPLES : MODELS AND VERIFICATION CRITERIA

2.1 Radioactivity in Mutton following the Chernobyl accident

2.1.1 Description of the model used

The measurements acquired following the Chernobyl accident provide data for investigation radioactivity transfers among ecological compartments. In southeastern France, grass was sampled at various locations in the Var River basin and mutton samples were taken from butcher's shops.

With regard to radioactivity fixation in an animal organ following a single ingestion during a single meal, it may be assumed that a fraction of the radioactivity uptake is rapidly transferred to the organ and then eliminated exponentially over a period depending on the physical and biological half-lives of the radionuclide in question.

The following equation is thus applicable in the event of a series of daily ingestion :

$$Ca = \sum_{j} \left[Ra(j) * Ret * EXP -(L_{eff}(t-j)) \right] \tag{1}$$

where : Ca   Concentration in organ on slaughtering (Bq.kg^{-1} fresh weight)

       Ra(j) Daily radionuclide uptake (Bq.d^{-1})

       Ret   Retention factor for the organ (kg^{-1} fresh weight)

       Leff   Time constant for effective radionuclide elimination (d^{-1})

       t   Date of slaughtering (d)

The equilibrium transfer factor then corresponds to the relation :

$$Fm = Ret/L_{eff} \quad (d.kg^{-1} \text{ fresh weight}) \tag{2}$$

which is the factor generally found in predictive models for routine discharges.

2.1.2 Input data and parameter adjustment

*   Calculating the food uptake

A daily ration of 2 kg of dry matter is assumed [2]. The period considered extended from May 1986 to October 1987, and was divided into three intervals :

- Early May 1986 to late October 1986 (0 < t < 180)

During the summer of 1986 it was observed [3] that the ^{137}Cs and ^{134}Cs activity in the grass diminished according to the following relation :

$$Cv(j) = Ds * [0.7 * EXP(- 0.69 * j/28)]$$

where : Cv(j) Concentration in the grass on the dth day after the fallout (Bq.kg^{-1} dry weight)

j   Number of days after fallout (d)

Ds   Surfacic activity (0-5 cm) (Bq.m^{-2})

0.7   Overall initial transfer factor (m^2.kg^{-1})

The mean Ds value in the region was 4235 Bq.m^{-2} for ^{137}Cs and 1360 Bq.m^{-2} for ^{134}Cs.

With a 2 kg (dry weight) daily food ration :

$$Ra(j) = Cv(j) * 2 \quad (Bq.d^{-1})$$

- November 1986 to April 1987 (181 < t < 365)

The sheep were outdoors during this period, and it was assumed that the daily food ration consisted of 1 kg dry weight of field grass and 1 kg dry weight of fodder.

The main radioactivity level in eight fodder samples from the region was 515 Bq.kg^{-1} dry weight for ^{137}Cs and 250 bq.kg^{-1} dry weight for ^{134}Cs.

The radioactivity in the field grass was assumed to be constant at 50 and 20 Bq.kg^{-1} dry weight respectively for the two cesium isotopes ; these values were confirmed by samples taken at the end of 1986 and in 1987.

118

Regardless of the day "j" considered, the winter values were thus :

$$Ra(j) = 515 + 50 \text{ Bq.d}^{-1} \text{ for } ^{137}Cs$$
$$Ra(j) = 250 + 20 \text{ Bq.d}^{-1} \text{ for } ^{134}Cs$$

- Summer 1987 (366 < t < 590)

The daily ration consisted of 2 kg dry weight of field grass :

$$Ra(j) = 50 \times 2 \text{ Bq.d}^{-1} \text{ for } ^{137}Cs$$
$$Ra(j) = 20 \times 2 \text{ Bq.d}^{-1} \text{ for } ^{134}Cs$$

* Adjustement of the Leff and Ret parameters

The calculated results are shown in graphically in Figure 1, based on the cesium isotopes in the flesh. Both parameters were adjusted considering only points 1 and 2.

SHEEP FLESH

Var River Basin (Southeastern France)

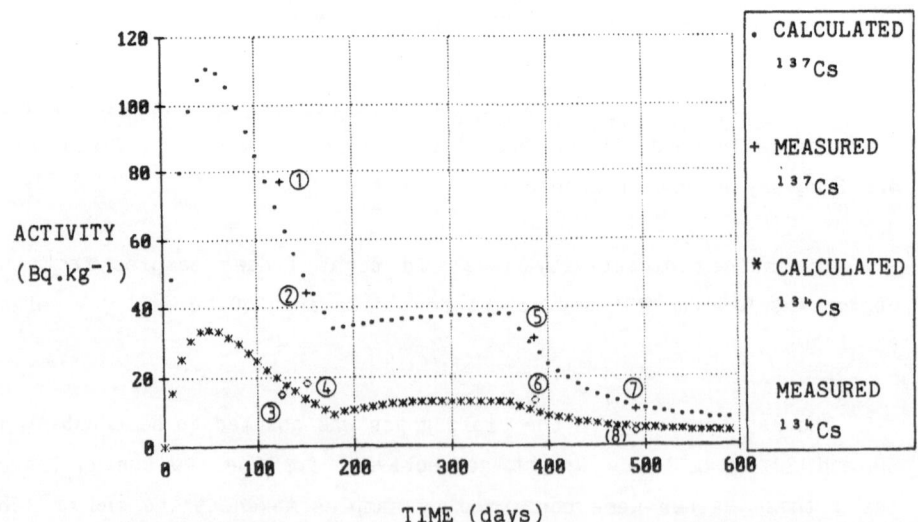

Figure 1. Calculated and measured radioactivity in Mutton

The Leff factor was first selected on the basis of the effective half-life of ^{137}Cs in the human body according to the "Technician's Vade Mecum" [4], i.e. 70 days. The decay rates determined with this value were too low, however, in comparison with the measured findings. The optimum value was thus found to be :

$$Leff = 0.15 \ d^{-1}$$

This value corresponds to an effective half-life of 46 days, or somewhat less than the accepted value in human beings ; this result seems plausible.

The Ret factor was then adjusted to fit the theoretical curve to points 1 and 2 :

$$Ret = 0.001 \ kg^{-1} \ fresh \ weight$$

It may be observed that suitable agreement of the relative positions of the other measurements points (3, 4, 5, 6, 7 and 8) with the curves is obtained simply by modifying the fallout and food uptake values, with no further parameter adjustement.

This is not sufficient to confirm the validity of the model in view of the limited number of measurements, but no significant discrepancies were encountered.

The transfer factor at equilibrium would then be :

$$Fm = 7 \times 10^{-2} \ d.kg^{-1}$$

The "DUNE" code in the Radioecology Manual [2] indicates values of 1.2×10^{-1} for lamb and 2×10^{-2} for beef ; the result here lies within this range.

2.2 Laboratory tests on radionuclide migration in the soil

The second example concerns experiments on ^{137}Cs and ^{85}Sr migration in soil columns under the effect of periodic watering.

2.2.1 Brief description of the experimental setup

The experiment used a set of 12 identical soil columns measuring 15 cm high and 5.5 cm in diameter inside plexiglass tubes. The rainfall/dryout cycle was simulated by dividing the 12 columns into three groups, each submitted to different watering conditions ; 3 x 20 ml, 3 x 30 ml and 3 x 50 ml per week. Each group of four soil columns included two contaminated by ^{137}Cs and two by ^{85}Sr. The tubes were plugged at both ends to allow drainage without desiccation (Figure 2).

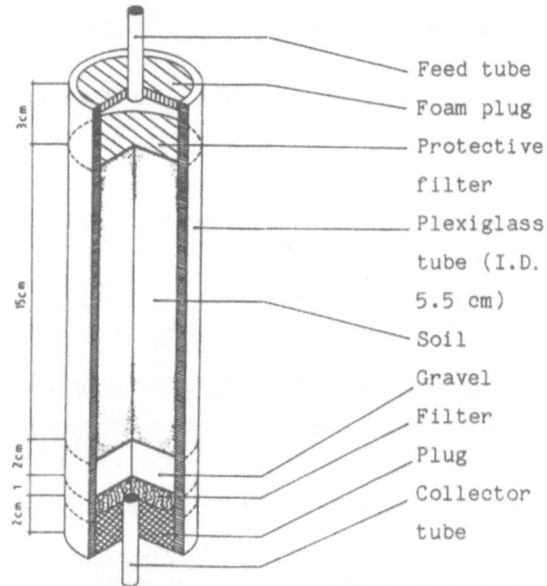

Figure 2. Cutaway schematic of a soil column

The soil sampled from a cultivated field at Cadarache was a calcareous loamy soil with a pH of approximately 8, and with low organic matter (1.2 %) and clay (16.8 %) content. The mean apparent density was 1.2 g.cm^{-3} and the saturation humidity was 47 %. Tap water was used for watering.

The experiment included two steps :

. Water flow behavior was first characterized by percolation with tritium-labeled water.

. The soil samples were contamined by ^{137}Cs and ^{85}Sr during the first watering, then watered regularly for 11 weeks. On completion of the experiment they were cut into 1 cm thick layers using a special saw, and the radioactivity of each layer was determined.

2.2.2 Model and results

The basic equation is the widely used relation governing solute flow in porous media where the solid phase has a sorption capacity :

$$\frac{Da}{H} \frac{\partial S}{\partial t} + \frac{\partial C}{\partial t} = D \cdot \frac{\partial^2 C}{\partial x^2} - \bar{u} \cdot \frac{\partial C}{\partial x} \qquad (3)$$

$$S = Kd \cdot C \text{ at equilibrium} \qquad (4)$$

where :

Da	Apparent density	$(g.cm^{-3})$
H	Mean humidity	
S	Concentration in dry solid phase	$(Bq.g^{-1})$
C	Concentration in liquid phase	$(Bq.cm^{-3})$
$\bar{u}$	Mean flow velocity	$(cm.s^{-1})$
t	Time	(s)
x	Distance	(cm)
Kd	Distribution coefficient defined by the solid/liquid phase concentration ratio	$(cm^3.g^{-1})$
D	Apparent dispersion coefficient	$(cm^2.s^{-1})$

A discrete representation was obtained by the finite differences method using an explicit scheme. The distance step Δx was 1 cm (i.e. the column layers thickness after cutting) and the time step was chosen so that $\Delta t = \Delta x/\bar{u}$.

* Adjustement of the D parameter

The mean flow velocity and the apparent dispersion coefficient were calculated from the measured activity in the percolation tests after tritium spiking. A suitable computer program was used with a discrete version of equation (3) without the sorption term.

Table 1. Water flow properties in the soil columns

Watering rate (ml.week^{-1})	3 x 20	3 x 30	3 x 50
Mean flow velocity (cm.s^{-1}) (a) (b)	5.8 x 10^{-4} 3.3 x 10^{-4}	4.9 x 10^{-4} 5.0 x 10^{-4}	6.7 x 10^{-4} 6.9 x 10^{-4}
Apparent dispersion coefficient (cm^2.s^{-1})	1.3 x 10^{-3}	8.2 x 10^{-4}	7.4 x 10^{-4}

(a) Calculated from tritium activity after percolation tests
(b) Determined using the method defined by Muntz

Variations in the $\bar{u}$ and D values were observed under different watering conditions, but the differences tended to decrease as the amount of water increased, i.e. as the conditions neared saturation.

* Adjustement of hte Kd parameter

After cutting up the soil columns and measuring the actual activity profiles using the parameters already defined, a numerical method was implemented to determine the distribution coefficient Kd that provided the best fit between measured and calculated results in each case. As shown by the typical calculated and measured migration profiles in Figure 3, the agreement is very satisfactory.

The Kd values ranged from 48 cm³.g⁻¹ for ⁸⁵Sr depending on the amount of water supplied, and were 1820 cm³.g⁻¹ in all cases for ¹³⁷Cs. These results show remarkable uniformity.

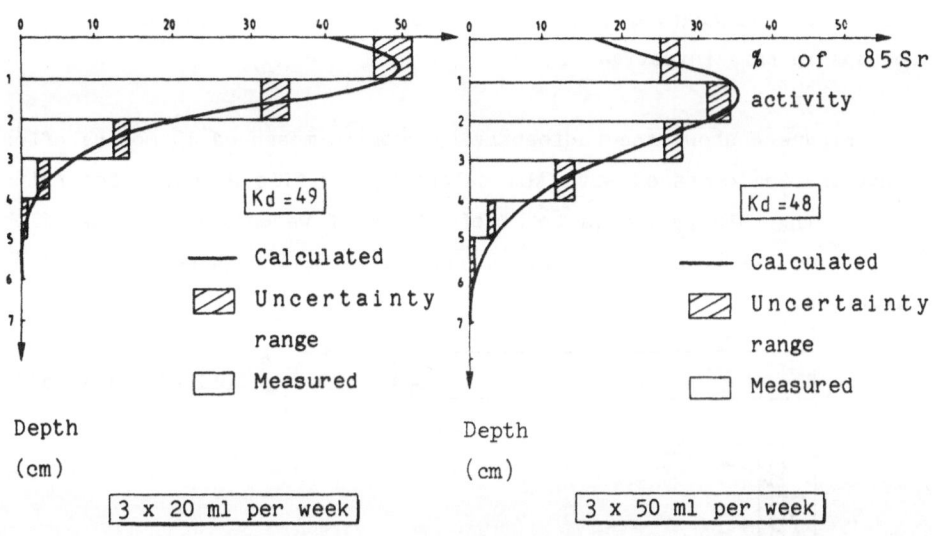

Figure 3. Experimental and calculated migration profiles for ⁸⁵Sr

2.2.3 Extrapolation in the field

This work is of interest only if it can be extrapolated to natural conditions. This was attempted by applying the model to an experimental plot at Cadarache labeled with ¹³⁴Cs fallout from Chernobyl.

The following specific parameters were measured for the test plot :

- The percolation rate as measured by the double ring method was 1.33×10^{-3} cm.s⁻¹
- The effective macroporosity was 35 %
- The apparent density was 1.2 g.cm⁻³

The experimental ^{134} Cs distribution coefficient (1820 cm^3.g^{-1}) was used.

The apparent dispersion coefficient was 8 x 10^{-4} cm^2.s^{-1} as reported in the literature [5] [6] and as measured in most of the soil columns in this investigation.

Figure 4 shows the radioactivity profile measured 10 months after contamination compared with the calculated profile allowing for rainfall during this period (a total of 189 mm of water). The two profiles are in satisfactory agreement, although this will require further confirmation.

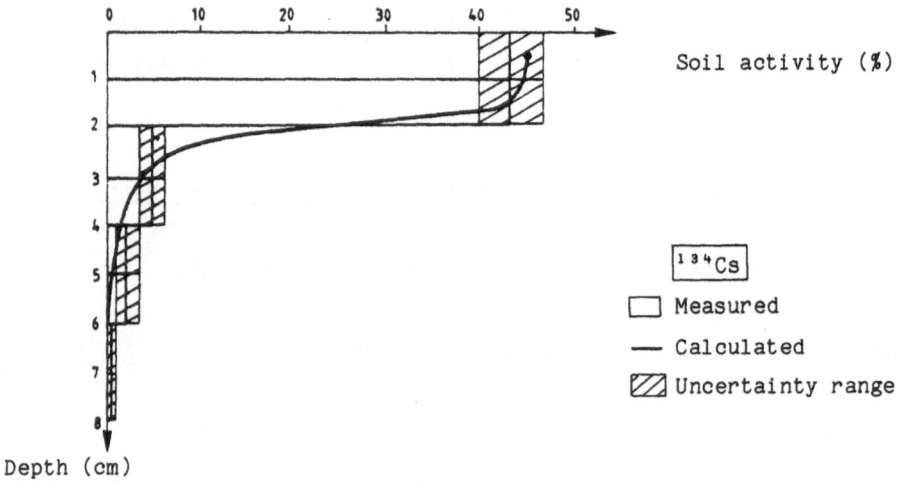

Depth (cm)

Figure 4. In situ ^{134} Cs migration profile 10 months after deposition
Field measurements versus calculated values

2.2.4 Consequences on Kd distribution coefficient measurements

Some authors [5] have shown that the validity of the proposed model depends primarily on the measured Kd coefficient. The soil column method provides good results, but is long and relatively difficult to carry out. The batch method is much quicker, but generally gives much higher values that are not compatible with the model.

Another method is therefore proposed here that is both fast and representative. Based on the numeric nature of the model, the method consists in isolating a soil layer and contaminating it after obtaining the desired moisture content. After equilibrium conditions are reached, the mobile water fraction is drawn off and the Kd coefficient is determined by measuring the activity of the percolated water. The mean values for five measurements performed in this way were compared with the values calculated on the basis of soil column experiments :

Radionuclide	Measured Kd	Soil column Kd
^{137}Cs	2020	1820
^{85}Sr	58	49

The results are of the same order of magnitude, even though the method can no doubt be improved.

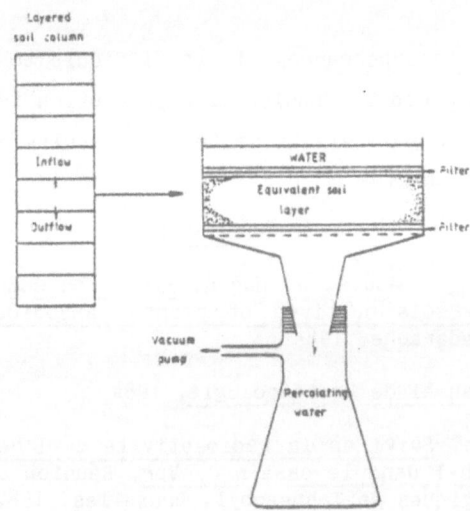

Figure 5. Experimental setup for quick "column" Kd measurement

3.   CONCLUSION

Both examples discussed here seem to indicate that the proposed models provide realistic results. Nevertheless, some uncertainties remain, but for different reasons in each case :

- The small number of meat samples is not sufficient for a statistical analysis that would raise any remaining doubts. The input data supplied to the model are subject to error, especially for the fodder. Thus while the mean ^{137}Cs activity is 515 Bq.kg^{-1} dry weight, the extreme values ranged from 25 to 1920 Bq.kg^{-1} dry weight depending on the time and place the samples were taken : this is the natural variability range. If the model appeared to work properly, the reason is probably that the animals were not fed from a single stock, and that the mathematical mean value approximated the "physical" mean value resulting from random commercial practices.

- The basic equation used for the migration study applies to a constantly saturated porous medium, and not for actual soil. The mathematical analysis is thus questionable, even if many authors have emphasized the good results obtained with the model : this is no doubt attribution to the fact that migration is very slow in a desaturated sorbant medium for lack of motive force.

Thus, despite appearances, it is difficult to provide irrefutable verification of a model. Considerable precaution is therefore required before modifying any models known to give results even with unnecessarily wide safety margins.

REFERENCES

1. S. Roussel, A. Jouve, J. Hugon. Evolution comparée de la radioactivité des sols cultivés et plantes associées. Programme mixte EDF/CEA. Cadarache, 1984

2. CEA/EDF. Manuel de Radioécologie. 1984

3. H. Maubert. Suivi de la radioactivité de l'herbe après 'accident de Tchernobyl dans le bassin du Var. Réunion sur les conséquences radioécologiques de Tchernobyl. Bruxelles, 1987

4. R. Pannetier. Vade Mecum du technicien. 1965

5. H. Bachuber, K. Bunzl, W. Schimmack. Migration of ^{137}Cs and ^{85}Sr in multilayered soils. Results from batch, column and fallout experiments. Nuclear technology, vol 59, Nov 1982, pp. 291-301.

6. M.J. Frissel, P. Reiniger. Simulation of accumulation and leaching in soils. International symposition on migration in the terrestrial of long lived radionuclides. N. Nuclear fuel cycle. USA, 1978

VALIDATION OF THE FARMLAND MODELS FOR RADIONUCLIDE TRANSFER THROUGH TERRESTRIAL FOODCHAINS

J BROWN, S M HAYWOOD AND B T WILKINS
National Radiological Protection Board
Chilton, Didcot, Oxon, UK

ABSTRACT

The NRPB dynamic foodchain models, FARMLAND, are general models to simulate the transfer of radionuclides through terrestrial foodchains.

In order to assess the ability of the models to represent general conditions, predictions have been compared with two sets of data. The first set are measurements made after the Chernobyl reactor accident of activity concentrations in milk, lamb and green vegetables. The suitability of the data for model validation, the problems encountered and any significant differences between predictions and measurements are identified and discussed. The second set is of UK average activity concentrations in milk from fallout due to weapons testing over the last thirty years.

Model predictions have also been compared with site-specific data of levels in milk resulting from the Chernobyl accident, for two different cattle management regimes.

INTRODUCTION

It is desirable to compare predictions of environmental transfer models such as FARMLAND with environmental data which have not been used in the development of the model: this may identify deficiencies in model structure, mechanisms which have been overlooked, and the use of inappropriate parameter values. Comparison with appropriate sets of data will assess the ability of FARMLAND to represent the general conditions for which it was developed, and also test its flexibility to simulate site-specific conditions.

In this paper the predictions of the FARMLAND models are compared with measurements (i) following the Chernobyl accident and (ii) due to fallout from weapons testing. The applicability and completeness of the sets of measurement data and the problems encountered in model validation studies are discussed, with the identification of any differences seen between model predictions and the measured data. The studies described in this paper will be discussed in more detail in a future report [1].

REQUIREMENTS FOR MODEL VALIDATION

FARMLAND comprises a set of dynamic sub-models, each of which independently simulate radionuclide transfer through a different part of the foodchain. Details of the models used here have been described elsewhere [2,3]. Figure 1 shows the main features of the FARMLAND models for intake by grazing animals. In the model, activity enters through deposition onto plant and soil surfaces, and subsequently transfers between interlinked compartments.

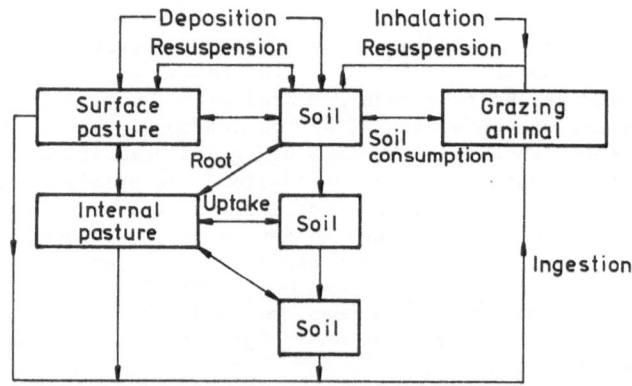

Figure 1. FARMLAND models for intake by grazing animals

For reliable model validation, compatibility between data on deposition and food concentration is required. Ideally, both sets of measurements should have been made at the same location to eliminate the variation due to different deposition patterns, agricultural practices and weather conditions which may occur even over quite small localised areas. To enable the time dependent features of the model to be studied, measurements are required at regular time intervals over a relatively long period. Monitoring programmes do not always provide information suitable for validation purposes, because the activity concentrations are often below the detection limit of the measurement technique.

COMPARISON OF FARMLAND MODEL RESULTS WITH MEASUREMENT DATA APPLICABLE FOR LARGE REGIONS OF THE UK

Comparison with monitoring data made after the Chernobyl accident

The emphasis in the UK following the Chernobyl accident was on the monitoring of activity levels in food and other environmental materials for radiological protection purposes. In most cases the data obtained are insufficient for model validation; measurement of deposition and food concentrations were seldom obtained at the same location, and were not followed for a sufficient length of time.

Several sets of measurements of activity levels in various foods on a regional basis are, however, available, together with some indication of deposition onto grass in these regions [1], and these sets of data may be used to compare model predictions with measured concentrations averaged across large regions of the UK. Model predictions based on average deposition levels in these regions have been compared with the average measured concentrations in food as a function of time. The radionuclides considered are ^{131}I, ^{137}Cs and ^{134}Cs, which were the major radionuclides deposited. The foods considered are milk, green vegetables and lamb.

The areas considered are necessarily large to enable the model predictions to be compared with sufficient data. There are comparatively few measurements of deposition and it is known that there was considerable variation within an area. This factor, combined with the likely incompatibility between deposition and food concentration, results in considerable uncertainty in the deposition data which must be considered when interpreting the results of the comparison. As the data are in the form of activity on grass, it has been necessary to assume an interception factor for grass to convert the measurements to total deposit. An interception factor of 50% was assumed but it is recognised that this is uncertain as interception is influenced by factors such as foliar density and rainfall.

Figure 2 shows measured ^{131}I concentrations in cows' milk and the FARMLAND model predictions, with the range of values associated with them. The measurements are of concentrations in England, but exclude

Cumbria which had much higher levels of deposition. The model predicts time-dependence well, but apparently overestimates by about a factor of two; however, as the figure shows, the variability in both the deposition data, as reflected in the range of predicted values, and in the measured concentrations in milk is of greater significance.

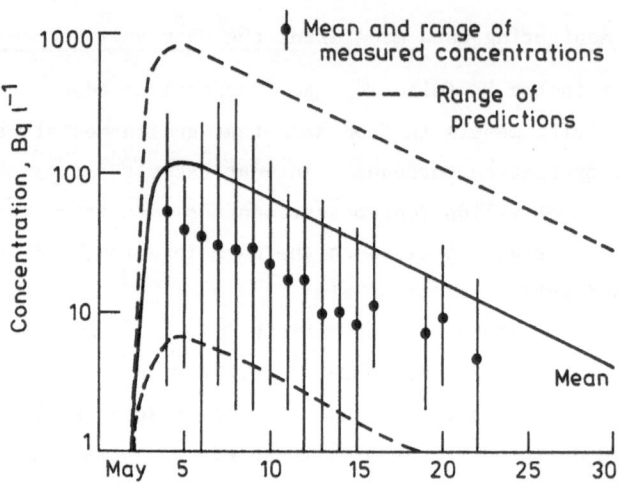

Figure 2. Comparison of FARMLAND model predictions with measured levels of ^{131}I in milk in England, excluding Cumbria.

Comparisons between predictions and measurements for ^{137}Cs and ^{131}I in cows' milk and green vegetables in other regions show a similar pattern - in general, the time dependence is predicted well but the total transfer is overestimated.

Measurement data for lamb are very limited and mainly for upland areas where restrictions have been imposed. These data are only representative of these small areas and levels are therefore high. Model predictions of levels of caesium in lamb are significantly lower than measurements due to differences between the lowland pasture grass modelled in FARMLAND and the upland pasture which the sheep graze in those areas where restrictions have been imposed. However, for sheep grazing a mixture of lowland and upland areas in Northern Ireland the model, based on average deposition in the region, predicts the time-dependence well over the period during which data are available.

Comparison with measurements of UK fallout from weapons testing

Another set of measured levels of radioactivity are the quarterly deposition rates of ^{137}Cs and ^{90}Sr and the resultant concentrations of activity in milk due to fallout from nuclear weapons testing, from the mid 1950s to the present.

Average UK annual deposition rates can be used as input to the pasture-cow-milk model, and the resulting predictions of concentrations in milk compared with measured values. In an earlier study [4], concentrations of ^{137}Cs and ^{90}Sr in milk were predicted for the years 1957-1977, using a simple model for transfer in the cow. The predictions were compared with measured levels and found to generally overestimate by less than a factor of two. A number of revisions were made to both the structure and input parameters of the pasture-soil model following a review of UK transfer data; this gave a close fit to the measurements.

The study has been repeated [1] to investigate the effect of incorporating more complex cow models [3] that have been developed for both radionuclides, and which give improved time dependence. In particular the long term process of recycling of strontium from bone to circulating fluids is included. The main difference for caesium is a more detailed treatment of the soft tissues in the cow.

These more complex cow models were used to predict the levels in milk resulting from the measured deposition rates from 1957 to 1983. For ^{137}Cs the use of the new model was found to have no effect on the predictions, which agree well with the measurements. The inclusion of recycling of

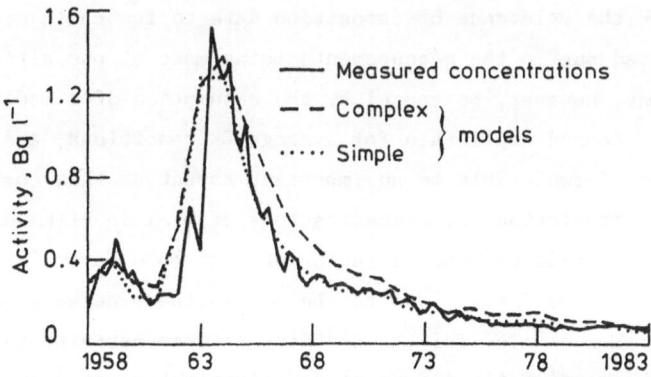

Figure 3. FARMLAND model predictions and measured levels of ^{90}Sr in milk from nuclear weapons fallout.

strontium from bone results in a small overestimation of activity levels
in milk in the mid 1960s (Figure 3). This effect could be compensated for
by small changes in other model parameters; e.g. it could be removed by
the adoption of a more rapid strontium migration rate from the surface
soil, which would be more consistent with average values obtained from the
literature. However, despite the small overestimation, the model
continues to predict levels which are in good agreement with the
measurements.

Comparison of FARMLAND models with site specific measurement data

Detailed post Chernobyl measurement data are available for two dairy
farms in the UK [5]. The farms are located in Berkshire, in the south of
England, and Cumbria in the north. Deposition data, and activity
concentrations in animal fodder and milk are available for ^{131}I, ^{137}Cs and
^{134}Cs throughout May 1986. The Berkshire farm operates a block grazing
feeding regime whereby each morning cattle are given access to an area
that has not been grazed recently. At the Cumbrian farm cattle are
allowed to graze relatively freely and continuously over a large area of
pasture, which is the assumption made in the FARMLAND model. Deposition
at the farm in Berkshire was low, whereas at the farm in Cumbria rainfall
occurred resulting in deposition about an order of magnitude higher.

In this study the measured concentrations of activity in milk have
been compared with model predictions based on the cows' daily intake and
also on deposition on pasture. It is thus possible to see if any
differences are due to the pasture or cow modules of the model. The
uncertainty in the relevance of deposition data to foodchain measurements
has been removed due to the measurements being made at one site.
Differences may, however, be caused by the assumption of a model structure
and parameters deemed applicable for average UK conditions, and inappropriate to these farms. This is an important aspect of this comparison.

The model predictions of concentrations of ^{131}I in milk were found to
exceed measured levels by about a factor of 4 at both farms (Figure 4).
This is thought to be largely due to the use in the general model of the
literature consensus value for the quotient between activity in milk and
activity in the cows' daily intake at equilibrium (Fm) of $9 \ 10^{-3} \ d \ 1^{-1}$
whereas the value implied by the measurements at both farms was around

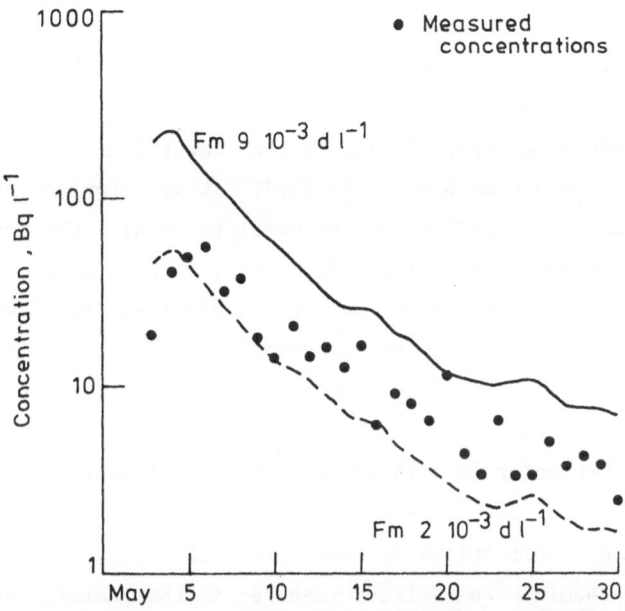

Figure 4. FARMLAND model predictions and measured levels
of ^{131}I in milk at a farm in Berkshire.

$2 \ 10^{-3}$ d 1^{-1}. Using this value of $2 \ 10^{-3}$ d 1^{-1} the model gives predictions much closer to the measurements (Figure 4). The accurate temporal
variation of the predictions for the Berkshire farm implies that the model
is flexible enough to simulate the block grazing feeding at this farm.

For ^{137}Cs, the time variation of the predictions for both farms
agreed well with measurements, and for the Cumbrian farm the overall
magnitude was also in good agreement. For the Berkshire farm, however, the
overall magnitude was a factor of about 4-5 too high. This is again
thought to be related to the quotient between measured activity in milk
and cows' intake used by the model. The value implied by the Cumbrian
data was near the consensus figure used in the model, while the value
implied by the Berkshire data is about five times less. This suggests
that ^{137}Cs deposited in dry conditions may be less available for transfer
than that deposited during rain, at least in the first few weeks. Study
of the transfer of both iodine and caesium radionuclides is in progress
and the implications of this work for general UK modelling will then be
considered.

CONCLUSIONS

Comparisons of FARMLAND model results with measurement data applicable for large regions of the UK have shown that the models predict the time dependence well. In using the post-Chernobyl monitoring data there are considerable uncertainties in the compatibility of the deposition measurements with those in food. It is often not possible to draw conclusions on the ability of the model to predict the scale of transfer to a particular food from these data, because many of the differences can be attributed to these uncertainties. From this point of view these data are disappointing for detailed model validation. Comparisons using site-specific data, where the uncertainties in interpretation are reduced, have shown that there may be factors influencing the transfer to milk which might be relevant in applying the model in more general situations.

Further model validations are not worthwhile until experiments studying the mechanisms that affect transfer in the pasture-cow-milk and upland pasture-sheep pathways are completed.

REFERENCES

1. Brown, J and Haywood, S M. Validation of the FARMLAND models for radionuclide transfer through terrestrial foodchains. Chilton, NRPB report (to be published).

2. Simmonds, J R, The influence of season of the year on the transfer of radionuclides to terrestrial foods following an accidental release to atmosphere. Chilton, NRPB-M121 (1985).

3. Brown, J, and Crick, M J. The development of dynamic models for the transfer of strontium, caesium and iodine in sheep for use in radiological protection (to be published).

4. Haywood, S M, Simmonds, J R and Linsley, G S, The development of models for the transfer of ^{137}Cs and ^{90}Sr in the pasture-cow-milk pathway using fallout data. Chilton, NRPB-R110 (1980) (London, HMSO).

5. Wilkins, B T, Bradley, E J and Fulker, M J, The influence of different agricultural practices on the transfer of radionuclides from pasture to milk after the Chernobyl accident. Sci. Tot. Environ. (in press).

MODELLING THE RADIOCESIUM CONTENT IN MILK AND COMPARISON
WITH THE MEASURED DATA AFTER THE CHERNOBYL ACCIDENT

B.Kanyar , N.Fulop , A.Kerekes and L.Kovacs

National Research Institute of Radiobiology and Radiohygiene
Budapest, H-1221
HUNGARY

ABSTRACT

After the .first peak of the radioactive cesium in the milk at the
middle of May 1986 there has been occurred a second one during the
winterseason. To describe these two peaks in the milk we have taken into
consideration that the radioactivity of the fresh animal feed decreases
exponentially and of the dry one is constant, exept the first year when
it increases due to the storage of the most contaminated dry feed. The
consumption rate of the freshes is derived as a proportional amount of
the agricultural productivity that has been given by a Fourier series.
The total consumption rate has been kept constant during the whole year.
By use of the model it can be calculated the concentrations in milk for
several years even by taking into consideration the root uptake of the
pasture forage. The uncertatinty analysis has given confidence ranges of
about one magnitude of the cesium content in the milk.

INTRODUCTION

Following the Chernobyl accident on 26 April 1986 the first peak
of the radioactively contaminated air was detected in Hungary on the
30th April. Somewhat less contamination occurred on the 3rd May, and more
significant third one on 6-8 May [1,2]. Our institute was rather involved
in the measurements of samples around of the capital, Budapest. This
region was one of the most contaminated part of the country.
    Meanwhile the monitoring program we have made efforts to use and
develope models to describe the dynamic processes of the radioactivity
transfer in the environmental components [3]. Our activities in the
modelling and in the uncertainty analysis were rather initiated by taking
part in the BIOMOVS-investigations [4]. In this presentation we are mainly
focused on the modelling of the seasonal and periodical behaviours of the
radiocesium concentration in milk after a single contamination of the
animal feed.

MODEL EQUATIONS

The model takes into consideration that the animal feed is shared into fresh and dry parts [5]. Namely the total intakes of the radio-activity is

$$U^*_{tot}(t) = C_{fr}(t) * U_{fr}(t) + C_{dry}(t) * U_{dry}(t) * exp(-\alpha * \tau)$$

where U^*_{tot} : is the total intakes (Bq/d),
C_{fr}, C_{dry}: are the radioactive concentrations (Bq/kg) in feed,
U_{fr}, U_{dry}: are the consumption rates of the animal (kg/d),
α : is the radioactive decay constant,
τ : is the duration of the store the dry feed and
t   : is the time variable.

Following a single deposition onto the pasture the radioactive concentrations can be approximated by a single exponential function

$$C_{fr}(t) = C_{\circleddash} * exp[-(k + \alpha) * (t - t_{\circleddash})],$$

where $C_{\circleddash}$: is the radioactive concentration of the animal
feed at the date of deposition and
k   : is the weathering rate constant.

For general case the $C_{fr}(t)$ can be given by a sum of two exponentials or even by a more complex function. The consumption rate of the freshes is derived as

$$U_{fr}(t) = S * H(t),$$

where $H(t)$: is the agricultural productivity of the pasture, normalized to the dry matter and one animal (kg/d) and
S   : is the fraction of the agricultural yield $H(t)$ that the
animal consumes by freshes (dimensionless).

Because of the U_{tot} is kept constant, the dry consumption rate is

$$U_{dry}(t) = U_{tot} - U_{fr}(t).$$

To formulate the intake of the radioactivity by dry feed, in general an average concentration is used during a period of one year. This value is derived from the time integrated concentration of the pasture by

$$\bar{C}_{dry}(j) = \frac{\int_{t_1}^{t_2} C_{fr}(t)\, dt}{t_e - t_b} * M_j,$$

where $\bar{C}_{dry}(j)$: is the average concentration in the comsumption
cycle of j after the fall-out,

t_b, t_e : are the dates of beginning and ending the vegetation in
every year, respectively,

$t_1 = j * T + t_b,$

$t_2 = j * T + t_e \quad (j=0,1,2,\ldots),$

$j = int [(t - t_o) / T],$

T   : is the cycle period (T=365 days) and
M_j : is a modifying factor, the mass ratio of the contaminated
     and uncontaminated dry feed.

  In general the value of M_j is 1, but by mixing the contaminated
hay with uncontaminated one from another agricultural site the M_j may
be different from 1.
  Because of the cyclic behaviours of the agricultural yield the function H(t) has been determined by a Fourier series like

$$H(t) = \frac{a_o}{2} + \sum_{i=1}^{N} a_i * \cos[i * \Omega * (t + \Phi)],$$

where Φ is a time delay and

$\Phi = (t_b + t_e) / 2,$
$\Omega = 2 * \pi / T, \quad (T=1 \text{ year}),$
$a_o = 2 * U_{tot},$
$a_i = a_o * sinc(i * T_1 / T) * sinc[i * (T_o + T_1) / T],$

$sinc(x) = \sin(\pi * x)/(\pi * x)$

  The shape of the function H(t), together with the fresh consumption
rate are given in Figure 1. The parameter T_1 introduced in the form of
a_i means the time period of the increasing and decreasing parts, and

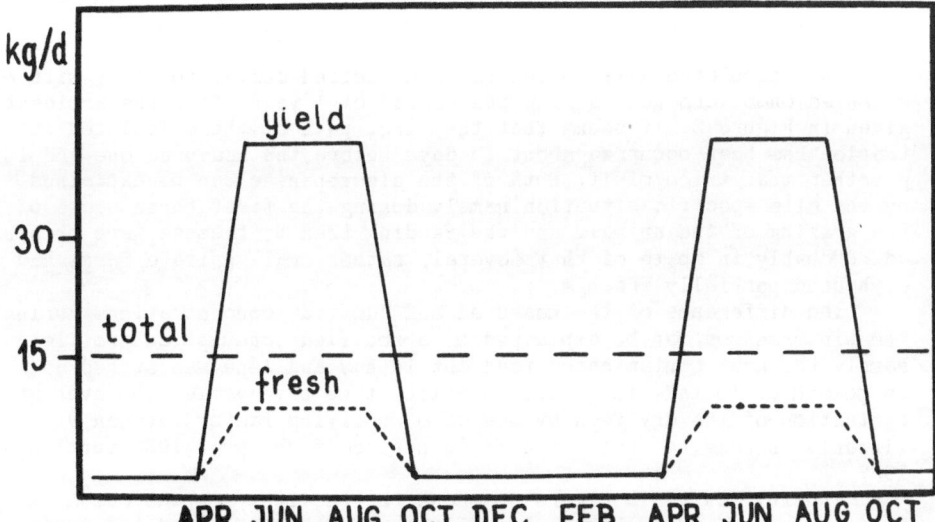

Figure 1. The agricultural yield, the fresh and the total
consumption rates as periodical functions.

$T_\emptyset$ is the duration kept the H(t) at the maximal value of

$$H_\emptyset = U_{tot} * T / (T_\emptyset + T_1) .$$

The advantages of the given mathematical form of H(t) are that the function in periodical and continous in time. The ratio of the vegetation interval ($T_\emptyset/T$) and the increasing section (T_1) can be varied easily depending on the climatic conditions.

COMPARISON OF THE CALCULATED AND THE MEASURED VALUES

The radioactivity content in the animal milk can be derived by the product of the pasture-milk transfer factor (F_m) and the intake of the radioactivity (U^*_{tot}) [5]. As an example let us have data of the Cs-137 transfer parameters given in Table 1. These values concern rather on the site specific data from Hungary after the Chernobyl accident. The F_m value was taken from the SS. No. 57 [5].

Table 1. The parameter values used for computer simulation of the Cs-137 concentration in cow milk.

U_{tot}(kg/d)	S	α(1/d)	k(1/d)	F_m(d/1)	τ_m(d)
15	0.25	8E-5	0.07	0.008	90

$T_\emptyset$(d)	T_1(d)	t_b(d)	t_e(d)	$t_\emptyset$(d)	$C_\emptyset$(kBq)
120	20	110	270	125	1.0

The calculated milk concentrations (dotted curve) together with our measured ones (crosses) during the period of 1 year after the accident are given in Figure 2. It seems that the first peak of the calculated concentration has been occurred about 15 days before the measured one and is greather than twice of it. Both of the discrepanses can be explained by the site specific situation, namely during the first three weeks of May the grazing of the animals and the feeding them by freshes were prohibited. Probably in spite of that several, rather small private farms had even used partially freshes.

The difference of the measured and "dotted" concentrations during the winterseason can be explained by a modified consumption profile. Namely the most contaminated feed cut in May and June was stored as long as possible. To take into consideration it we have varied the average concentration of the dry feed by use of a modifying factor between 0 - 2 and linearly increasing during the first period, from April 1986 until April 1987.

The full curve in Figure 2. shows the results from the cases that the cows were feeding by freshes only from 20th of May and the modifying

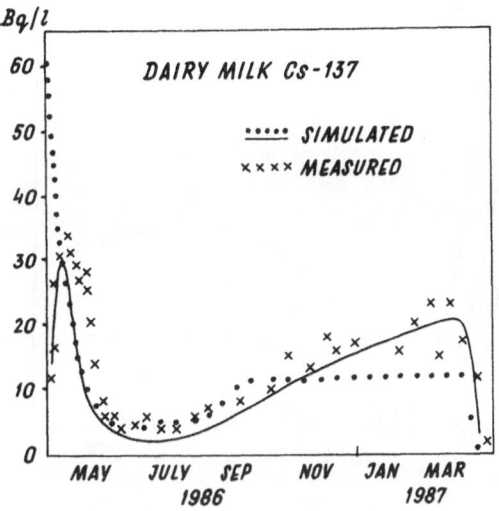

Figure 2. The Cs-content in the dairy milk.

factor is linearly varied in range of 0-2. In this case the differences between the calculated and measured concentrations are in an acceptable range.

UNCERTAINTY ANALYSIS

Uncertainty analysis has been performed to investigate the confidence intervals of the simulated concentrations either due to our lack of knowledge of the parameters or due to the stochastic variability of them [6,7].

Table 2. The parameter intervals for uncertainty analysis.

U_{tot}	S	k	F_m	C_o
12-18	0.2-0.3	0.04-0.10	0.004-0.012	0.6-1.4

T_{1o}	T_1	t_b	t_e	-
100-140	15-25	100-130	240-300	-

The intervals of the simple random varied parameters are given in Table 2. The mean values of the parameters were the same as defined in Table 1. for the improved simulation and the distribution was a normal type for every parameter. Because for the most parameters we could not made a difference between the parameter variation due to the lack of knowledge and stochastic variability our results refer to an somehow joint parameter variation distribution.

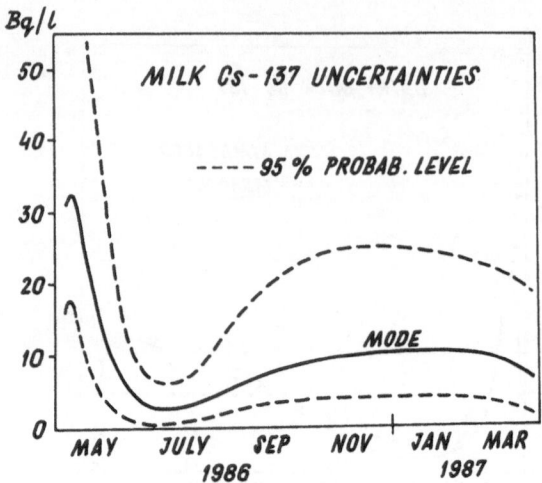

Figure 3. The uncertainties of the simulated Cs-137 in milk.

Figure 3. shows the mode of the distribution of the predicted concentration values got from 100 simulations at each point. The upper and lower curves give the confidence intervals belong to the 95 % probability level. According to the results the uncertaintes are about one of the magnitude meanwhile the greatest variation of the parameters is between a factor of 3.

REFERENCES

1. Andrasi,A., Beleznay,E., Deme,S., Feher,I., Monitoring the radiation consequences due to the disaster at the Chernobyl nuclear facility from 28 April to 12 June, 1986. KFKI-Reports, Budapest, 1986/49/K.
2. Biro,T., Feher,I. and Sztanyik,L.B., Radiation consequences in Hungary of the Chernobyl accident. Special issue of the Hungarian Atomic Energy Commission, Budapest, July 1986.
3. Fulop,N., Kanyar,B. and Koblinger-Bokori,E., Modelling the radionuclide transport in the terrestrial food chain after Chernobyl. Proc. of the 14th Regional Congress of IRPA, Kupari (Yugoslavia), Sept. 29 - Oct 2, 1987 (to be published).
4. BIOMOVS (Biospheric Model Validation Study) Progress Reports, No. 4., Natl. Inst. of Rad. Prot., Stockholm, June 1987.
5. Generic Models and Parameters for Assessing Environmental Transfer of Radionuclides from Routine Releases. Safety Series, No. 57, IAEA, Vienna, 1982.
6. Hoffman,F.O., Bergstrom,U., Gillander,C. and Wilkens A.-B., Comparison of predictions from internationally recognized assessment models for the transfer of selected radionuclides trough terrestrial food chains. Nuclear Safety, 1984, 25, 533-546.
7. Procedures for Evaluating the Reliability of Predictions Made by Environmental Transfer Models. Draft, to be published as an IAEA Safety Series Report.

PARTICULATE TRANSPORT PROCESSES IN AGROECOSYSTEMS: VALIDATION OF PREDICTIVE MODELS

J. E. Pinder III, K. W. McLeod and D. C. Adriano
Savannah River Ecology Laboratory
Drawer E
Aiken, South Carolina 29801
USA

ABSTRACT

The accuracy of three radionuclide transfer models for predicting surficial radionuclide contamination levels in agricultural crops was investigated using Pu-labelled aerosols and soil particles. Transport processes included the interception and retention of atmospheric deposition and the resuspension of soil particles to plant surfaces. Crops included wheat, soybeans and corn. The models evaluated were: 1) FOOD, a relatively general model; 2) AGNS, a model developed for the southeastern United States; and 3) SRP, an empirical model derived from studies on the U. S. Department of Energy's Savannah River Plant in South Carolina. Predicted concentrations were compared to independent field data. The three models generally predicted the surficial Pu concentrations due to the interception of deposition to within a factor of 3, but showed differential abilities to predict concentrations due to the resuspension of Pu-bearing soil particles to plant surfaces.

INTRODUCTION

Because of questions concerning the environmental behavior and potential dose-to-man of plutonium and other transuranics released to the atmosphere from fuel fabrication and fuel reprocessing facilities [1, 2], studies were initiated on the fate and behavior of [238]Pu-bearing particles released to the atmosphere from the H-Area nuclear fuel chemical separations facility on the U. S. Department of Energy's Savannah River Plant in Barnwell County, South Carolina. These studies resulted in empirical models that predicted the movement of Pu from atmospheric deposition and the resuspension of soil in agroecosystems [3, 4]. Resuspension refers to the movement of particles from the soil surface to plant surfaces due to the effects of wind and raindrop splash. To test the validity of this model, hereafter refered to as the SRP model, studies were

performed at the F-Area nuclear fuel chemical separations facility, also on
the Savannah River Plant. These studies were designed to obtain independent
field data to compare to the predictions of the SRP model.

The accuracy of two other models was also tested. The first of these
models, FOOD [5], is a simple model using the same transfer parameters for
a variety of crops to predict the interception, retention and transfer to
grains of atmospherically deposited radionuclides. The second model, AGNS
[6], was developed as an extension of the U. S. Nuclear Regulatory
Commission's Regulatory Guide 1.109 methodology for predicting the
radionuclide concentrations of crops [8]. The model contains site
specific parameter estimates for South Carolina. FOOD and AGNS were
selected from a variety of possible models because they represented the
range from a simple, general model to a model specifically designed for use
with fuel reprocessing facilities in this region of the United States.

Plutonium moves in terrestrial ecosystems due to the transport of
small (< 100 μm) Pu particles or as Pu adhered to small soil particles
[7]. Thus, the ability of a model to accurately predict plutonium
movement is a reflection of its ability to predict the transport of these
small particles. The movement of these particles is likely to determine
the behavior of other insoluble radionuclides and heavy metals and
hydrophobic organics.

DESCRIPTION OF RADIONUCLIDE MODELS

Descriptions, including the equations and parameters used to predict
concentrations (Bq / kg dry mass) in whole-plant vegetation (C_v) at
harvest and grain (C_g) for radionuclide deposition at constant rates
are given by Baker et al. [5] for FOOD and by Boone et al. [6] for
AGNS. Baker et al. [5] give water contents of 0.8 g water per g wet
vegetation and 0.12 g water per g grain, and these water contents have been
used to convert the concentrations for wet masses predicted from FOOD to
concentrations per dry masses. We have used crop yields of 2.5 kg wet mass
/ m^2 and growing seasons of 90 days in estimating concentrations from
the FOOD model.

The FOOD and AGNS models have similar structures and the parameter
estimates are derived from similar data bases. For example, both models
use 0.0495 d^{-1} as a loss rate of deposition particles from plant surfaces.
Both models make use of interception fractions to determine the proportion
of the deposition initially intercepted and retained by the vegetation. In

FOOD, a constant interception fraction of 0.25 is assumed. In AGNS, interception fractions vary during the growing season and are estimated from plant biomass using the filtration model [9].

The SRP model is an empirical model where the transfer coefficients have been estimated from measurements on crops grown adjacent to the H-Area nuclear fuel chemical separations facility. The model predicts surface concentrations for whole-plant vegetation and for grain for two modes of surface contamination. These are: (1) the interception and retention of dry deposition; and (2) the resuspension of Pu-bearing particles from the soil surface to the surface of the plant tissue.

The concentration for vegetation due to deposition (C_{vd} in Bq / g dry mass) is given by equation (1):

$$C_{vd} = D_y * Fr / Y_v ,$$

where D_y is the deposition rate in Bq / m^2 per year, Fr = the fraction of a year's deposition retained on the crop at harvest, and Y_v = the crop yield in g dry mass / m^2. The concentration (Bq / g dry mass) for grain due to deposition (C_{gd}) is given by equation (2):

$$C_{gd} = D_y * Fr * Fc / Y_g ,$$

where Fc is the fraction of the surface contamination of the vegetation that is transferred to the grain from the vegetation during combining, and Y_g is the grain yield in g dry mass / m^2.

The equations for predicting concentrations due to deposition are different from those in FOOD or AGNS. Moreover, the SRP model is based on measurements conducted after the development of FOOD and independent of the development of AGNS. Estimates of Fr, Y_v, Fc and Y_g are listed in Appendix 1.

The concentration (Bq / g dry mass) on the vegetation due to soil resuspension (C_{vr}) is given by equation (3):

$$C_{vr} = Rv * I / Y_v ,$$

where I = the radionuclide inventory in the soil (Bq / m^2 in the top 15 cm), and Rv = the fraction of the soil inventory on the the surface of the vegetation. The concentration (Bq / g dry) for the grain (C_{gr}) due to resuspension is given by equation (4):

$$C_{gr} = [(Fc * Rv * I) + (Rg * I)] / Y_g ,$$

where Rg is the fraction of the soil inventory resuspended to the grain during mechanical combining.

Estimates of the parameters Rv and Rg are given in Appendix 1 are for Pu that has bee well mixed into the soil profile. Where a contaminant is recently deposited on the soil surface or preferentially sorbed to smaller and more readily resuspended particles [10], resuspension to vegetation and grain surfaces may occur at greater rates.

AGRICULTURAL AND SAMPLING METHODS FOR FIELD STUDIES

Wheat, soybeans and corn were grown in a 0.5 ha field adjacent to the F-Area nuclear fuels chemical separations facility. The crops were grown using standard agricultural practices. Edible grains were obtained by harvesting the crop with the appropriate commercial combining equipment. Wheat (*Triticum aestivum* var. Coker 68-19) was planted 15 December 1979 and harvested 1 August 1979. Soybeans (*Glycine max* var Coker 488) were planted 23 July 1981 and harvested 9 December 1981. Corn (*Zea mays* var DeKalb XL82) was planted 30 March 1982 and harvested 28 September 1982.

Samples were obtained from each crop to estimate vegetative biomass, grain biomass, and the concentrations of ^{239,240}Pu in vegetation and combined grains. The relative contributions of atmospheric deposition and soil resuspension to the ^{239,240}Pu concentrations of the vegetation were determined by growing wheat, soybeans and corn in enclosures designed to permit atmospheric deposition onto the crops but prevent soil resuspension. The contribution due to resuspension for the field crop was estimated as the concentration of the field crops minus the concentrations of the enclosure plants. The ^{239,240}Pu inventory of the soil was estimated from 10 randomly located soil cores. The deposition rate of ^{239,240}Pu was measured using 0.25 X 0.33 m tacky papers suspended 7.6 m above the center of the field.

COMPARISONS OF PREDICTIONS OF THE FOOD, AGNS AND SRP MODELS

Table 1 compares predictions from FOOD, AGNS and the SRP model for wheat, soybeans and corn grown under a constant deposition rate of 1 Bq ^{239,240}Pu / m^2 per d. Predicted concentrations for whole-plant vegetation and for grains generally differ by less than a factor of 3. The similarity in predictions from FOOD and AGNS are understandable because the structure of the models is similar. The similarity of the predictions of the FOOD and AGNS models with those of the SRP model cannot be explained by similar model structures. The similarity occurs because predictions from the FOOD and AGNS models show general agreement to the results for atmospheric deposition obtained at H-Area that were used to construct the SRP model.

TABLE 1

Comparison of the predicted concentrations of ^{239,240}Pu in vegetation and grain for the models FOOD, AGNS and SRP. Predictions are for a site receiving a constant deposition rate of 1 Bq / m^2 per d.

| MODELS | Predicted Concentrations (mBq/g dry mass) | | |
	Wheat	Soybeans	Corn
	----- Vegetation -----		
FOOD	10	10	10
AGNS	27	28	17
SRP	31	23	26
	----- Grains -----		
FOOD	0.23	0.23	0.23
AGNS	0.27	0.28	0.17
SRP	0.39	0.15	0.08

Figure 1 shows the comparisons of observed ^{238}Pu concentrations in wheat, soybeans and corn grains from the H-Area studies and predicted concentrations from FOOD and AGNS for two fields of wheat [3], two fields of soybeans [4] and four fields of corn [Pinder, unpublished data]. The predicted and observed values show close correspondence. The maximum ratio of observed to predicted is 1.9. The errors (i.e., error = observed concentration minus predicted concentrations) averaged -0.55 (Standard Error = 1.1) μBq / g for FOOD and -0.60 (S. E. = 1.0) μBq / g for AGNS There appears to be greater error for predicted concentrations greater than 5 μBq / g, but this reflects a disturbed and atypical soil that produced poor yields.

COMPARISON OF MODEL PREDICTIONS WITH F-AREA FIELD DATA

Table 2 compares the predicted concentrations of Pu from the FOOD, AGNS and SRP models to those measured in the field studies at F-Area. The ^{239,240}Pu concentrations. Table 2 also contains predictions of resuspended Pu for the SRP model. Studies of root uptake of Pu from soils on the

Savannah River Plant have indicated that the contributions of root uptake to the Pu concentrations in Table 2 are negligible. The mean deposition rate for ^{239,240}Pu was 0.53 (S. E. – 0.018) mBq / m^2 per d. This rate is based on n – 6 analyses. The mean soil inventory, as estimated from 10 soil cores, was 427 (S. E. – 71) Bq / m^2.

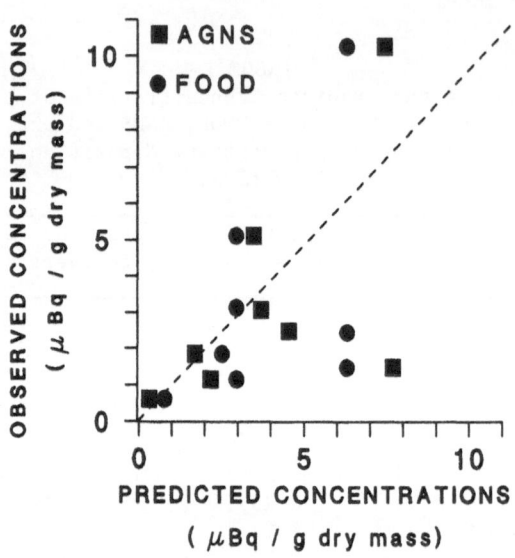

Figure 1. Comparison of ^{238}Pu concentrations due to deposition predicted from the FOOD and AGNS models with those observed at H-Area.

The predicted concentrations for whole-plant vegetation at F-Area show general agreement with the observed concentrations, except for soybeans. The total ^{239,240}Pu concentrations for soybeans are as much as 20 X that predicted. The concentration due to deposition as estimated from the polyethylene enclosures is 100-200 X the predicted, but shows considerable variation among samples. The tacky paper sample for the month preceding the harvesting of the soybeans indicated a deposition rate approximately 3 times the mean. This observation and the large concentrations in the enclosures suggest that a period of unusually large deposition occurred just prior to harvesting.

For wheat and corn, the predicted whole-plant concentrations for deposition are within a factor of 3 of the observed concentrations due to deposition as estimated from the plants in the polyethylene enclosures. An exception is the predicted value from FOOD for corn which is a factor of 5 lower than observed.

TABLE 2

Comparison of the predicted concentrations of ^{239,240}Pu in vegetation from the models FOOD, AGNS and SRP with those observed for crops grown at F-Area. The predictions are for a constant deposition rate of 0.53 mBq / m^2 per d and a soil inventory of 427 Bq / m^2.

| | Concentrations (μBq/g dry mass) | | |
	Wheat	Soybeans	Corn
	Predicted Concentrations for Vegetation		
FOOD	5	5	5
AGNS	14	15	9
SRP			
Deposition	17	12	14
Resuspension	18	24	29
Total	35	36	33
	Observed Concentrations: Mean $\pm$ Standard Error		
Deposition	8$\pm$2	1800$\pm$1040	26$\pm$4
	n=5	n=3	n=5
Resuspension	29		15
Total	37$\pm$11	105$\pm$70	41$\pm$12
	n=5	n=5	n=10

The observed concentrations for field grown wheat and corn were greater than those due to deposition and suggested that approximately one-third to one-half the the Pu content of the vegetation was due to resuspension of soil to plant surfaces. For wheat, the field concentration of 37 (S. E. = 11) μBq / g was significantly greater than the concentration in the enclosure of 8 (S. E. = 2) μBq /g (t = 2.69; df = 8; P < 0.05). The difference for corn was not statistically significant.

The predicted concentrations due to resuspension from the SRP model are within a factor of 2 of the estimated concentration due to resuspension obtained for wheat and corn as the difference in concentration between field and enclosure plants. The total predicted concentrations from the SRP model are closer to the observed concentrations for field grown crops than either the FOOD or AGNS predictions and differ from the observed concentrations by less than a factor of 1.5.

Resuspension may be expected to make a significant contribution to the radionuclide contamination of plant surfaces where the inventory of the contaminant in the soil has been elevated due to previous deposition rates being greater than current rates or where large, accidental releases have occurred in the past.

Although most radionuclide transfer models do not contain terms that account for resuspension, they do contain terms that predict radionuclide concentrations in grain due to root uptake and translocation to grain [11]. Root uptake is usually predicted using a concentration ratio which expresses the ratio of radionuclide concentration in vegetation to that in soil. The relative importance of resuspension and root uptake in moving radionuclides from soil to grain can be determined by recomputing the parameters R_g, R_v, F_c and Y_g into a concentration ratio due to resuspension (CR_{Resusp}) using equation (5):

$$CR_{Resusp} = [(R_v * F_c + R_g) / Y_g] / (1 / S),$$

where S = the soil mass in the top 0.15 m of 1 m^2. For S = 195000 g / m^2, the values for wheat, soybean and corn grains are 1.7×10^{-4}, 1.9×10^{-4} and 2.6×10^{-5}. Most of the concentration ratios for soluble radionuclides are greater than these values. For insoluble radionuclides, whose concentration ratios for grains may be $< 10^{-5}$, the effects of resuspension can be incorporated into predictions by using ratios $\geq 2 \times 10^{-4}$. The concentration ratios for Pu in FOOD and AGNS are 2.2×10^{-4} and 1.5×10^{-3}, respectively.

Unfortunately, no comparisons of predicted and observed concentrations could be made for wheat or corn grain. For wheat, mechanical problems with the combine caused the grain to be contaminated with soil. For corn, the predicted concentrations were less than the detection limits for Pu.

CONCLUSIONS

The intercomparisons of the predictions of the FOOD, AGNS and SRP models, the comparisons of the predictions of the FOOD and AGNS models with the observed [238]Pu data from H-Area, and the comparisons of the predictions of the FOOD, AGNS and SRP models with the observed data for wheat and corn at F-Area all support the accuracy of the models in predicting the Pu contents of vegetation due to atmospheric depositon. Predicted and observed concentrations usually agreed within a factor of 3. The models FOOD and AGNS do not contain terms which predict radionuclide concentratios for resuspension of soil particles to vegetation or grain surfaces. Resuspension can be an important pathway for insoluble elements such as Pu.

ACKNOWLEDGMENTS

This research was supported by Contract DE-AC09-76SR00819 between the U. S. Department of Energy and the Savannah River Ecology Laboratory. We thank T. G. Ciravolo and K. C. Sherrod for their assistance.

REFERENCES

1. Barr, N. F., The radiological significance of transuranic radioisotopes released to the environment during the operation of the LMFBR nuclear fuel cycle. In Transuranium Nuclides in the Environment. International Atomic Energy Agency, Vienna, 1976, pp. 649-656.

2. Vaughn, B. E., Wildung, R. E., and Fuquay, J. J., Transport of airborne effluents to man via the food chain. In Controlling Airborne Effluents from Fuel Cycle Plants. American Nuclear Society, Hinsdale, Illinois, 1976, pp. 8-1 to 8-13.

3. McLeod, K. W., Adriano, D. C., Boni, A. L., Corey, J. C., Horton, J. H., Paine, D. and Pinder, J. E., III, Influence of a nuclear fuel chemical separations facility on the plutonium contents of a wheat crop. J. Environ. Qual., 1980, 9, 306-315.

4. Adriano, D. C., Pinder, J. E., III, McLeod, K. W., Corey, J. C. and Boni, A. L., Plutonium contents and fluxes in a soybean crop ecosystem near a nuclear fuel chemical separations facility. J. Environ. Qual., 1982, 11, 506-511.

5. Baker, D. A., Hoenes, G. R. and Soldat, J. K., FOOD - An interactive code to calculate internal radiation doses from contaminated food products. In Environmental Modeling and Simulation, ed. W. R. Ott, U. S. Environmental Protection Agency, NTIS, Springfield, Virginia, 1976, pp. 204-208.

6. Boone, F. W., Ng, Y. C. and Palms, J. M., Terrestrial Pathways of Radionuclide Particulates. Hlth. Phys., 1981, 41, 735-747.

7. Watters, R. L., Edgington, D. N., Hakonson, T. E., Hanson, W. C., Smith, M. H., Whicker, F. W. and Wildung, R. E., Synthesis of the research literature. In Transuranic Elements in the Environment, ed. W. C. Hanson, U. S. Department of Energy, NTIS, Springfield, Virginia, 1980, pp. 1-44.

8. U. S. Nuclear Regulatory Commission, Regulatory Guide 1.109, Calculation of Annual Doses to Man from Routine Releases of Reactor Effluents for the Purpose of Compliance with 10 CFR Part 50. Revision 1, 1977.

9. Chamberlain, A. C., Interception and retention of radioactive aerosols by vegetation. Atmos. Environ., 1970, 4, 57-78.

10. Anspaugh, L. R., Shinn, J. H., Phelps, P. L. and Kennedy, N. C., Resuspension and redistribution of plutonium in soils. Hlth. Phys., 1975, 29, 571-582.

11. Hoffman, F. O., Miller, C. W., Shaeffer, D. L. and Garten, C. T., Jr., Computer codes for the assessment of radionuclides released to the environment. *Nucl. Safety*, 1977, **3**, 343-354.

APPENDIX 1

Parameter estimates and their standard errors for radionuclide transfers in the SRP model for Pu transport in wheat, soybean and corn agroecosystems. Standard errors are computed from estimates from different field sites and are based on sample sizes of 2, 2 and and 3 fields for wheat, soybeans and corn, respectively.

TRANSFER	SYMBOL	PARAMETER ESTIMATES FOR CROPS		
		Wheat	Soybeans	Corn
Fraction of deposition retained	F_r	0.030 ±0.001	0.020 ±0.012	0.032 ±0.009
Fraction transferred to grain by combining	F_c	0.0043 ±0.0004	0.0021 ±0.0001	0.0010 ±0.0006
Fraction of soil inventory resuspended to vegetation	R_v	1.5E-05 ±4.9E-06	1.8E-05 ±3.7E-06	2.0E-05 ±3.0E-06
Fraction of soil inventory resuspended to grain during combining	R_g	4.2E-08 ±3.3E-08	6.1E-08 ±2.1E-08	0.0E-00 ±0.0E-00
Vegetation biomass (g dry mass / m^2)	Y_v	350	320	450
Grain biomass (g dry mass / m^2)	Y_g	120	100	150

.THE IUR PROJECT ON SOIL-TO-PLANT TRANSFER FACTORS OF RADIONUCLIDES
EXPECTED VALUES AND UNCERTAINTIES

M.J. Frissel and J. Koster
National Institute of Public Health and Environmental Hygiene (RIVM)
3720 BA Bilthoven
The Netherlands

ABSTRACT

The large uncertainties which are associated with the values of the
soil-to-plant transfer factors prompted the IUR (International Union of
Radioecologists) some years ago to initiate a soil-to-plant transfer factor working group. This group is carrying out a joint project in which participate 20 research institutions. The aim of the project is to provide reliable estimates of the soil-to-plant transfer factors of different crops
for different soils, environmental conditions, etc. Only data of experiments
which met predefined criteria are included. Data of five years of experiments are now available.

Expected values for transfer values have been derived statistically for
a large number of combinations of different factors as radionuclides, crops
and soils. The 95 percent confidence interval was derived for averaged estimates based on the observations present in the data set as well as the 95
percent confidence interval of single predicted values for specific combinations of factors. The latter confidence interval is much larger, due to
the relatively limited number of data for many of the possible combinations
of factors.

INTRODUCTION

This manucript gives an overview of the results of statistical calcul-

ations on transfer data of Am, Co, Cs, Pu, Sr, Tc, Mn, Np and Zn which are

stored in the data bank of the IUR Soil-to-Plant Transfer Factor Working

Group. Three types of data are presented:

- Standardized values for expected transfer factors with upper and lower

  limits for 95 percent confidence intervals.

- Conversion data which can be used to convert values from standardized conditions to actual conditions, e.g. a conversion of the value for pH = 6 (standardized value) to the value of the actual pH.
- Uncertainty factors caused by various conditions and which elucidate the 95 percent confidence intervals.

THE IUR SOIL-TO-PLANT TRANSFER FACTOR PROJECT

The project is a joint project of about 20, mostly Western European research institutions. The aim of the project is to provide reliable estimates of the soil-to-plant transfer factors of different crops for different soils, environmental conditions, etc. Only data of experiments which met predefined criteria are included. The criteria are choosen in such a way as to minimize artefacts. They include recommendations for minimum plot size, fertilization, period to take into account between the addition of a radionuclide to a soil and the time of harvest, etc.

All transfer factors are stored into a data bank together with experimental conditions, soil type, pH, organic matter content, depth of the soil-layer to which the radionuclide is added, period that the radionuclide is present in the soil, irrigation, etc. The data are published annually, together wich results of statistical analysis as expected values, confidence intervals, etc. [1]. For participants see the last section of this manuscript.

STANDARDIZED VALUES FOR EXPECTED TRANSFER FACTORS

Crop Groups

The number of crops for which predictions have to be made is very large, so it seems an impossible task to handle each crop separately. Therefore crops are divided into crop groups which are used in all calculations, tables and discussion. The groups are:

Cereals	= grain of barley, wheat, oat, rye, corn
Fodder	= maize (without grain), clover, alfalfa
Grass	= grass, vegetation of pastures and grass land
Pods	= pods and seeds of peas and beans
Roots	= roots of root crops as beet and radish
Tubers	= potatoes, swedes
Vegetables	= green vegetables as lettuce cabbage, spinach, leek

Soil Type, pH and OM (Organic Matter)

The impact of the type of soil on the uptake is large; this effect is, however, for a considerable part caused by differences in pH and OM (organic matter). Therefore the impacts of pH and OM are considered separately,

while three types of soils are used: sand, loam and clay soils. A peat soil
is considered as a sandy soil with a high OM percentage. The standardized
value for OM is 4 percent for all crops, with the exception of grass for
which an OM of 10 percent is used. For other OM percentages corrections have
to be made with the eqns (1) or (2). The standardized pH (KCl) value is 6.

Radionuclide Availability, Contamination Depth, Irrigation

The availability of some radionuclides decreases with time. The period
elapsed since the moment that a radionuclide reached the soil is therefore
considered separately. The standardized value is 2 year. Also the position
of the radioactivity in the soil is taken into account. For the statistical
analysis it is assumed that the radionuclide is homogeneously mixed within a
layer which reaches from the soil surface to a certain depth. The standardized value of this depths is for all crops 20 cm, with the exception of
grass for which 10 cm is used. For other contamination depths a correction
has to be made. From the data it appeared that irrigation enhances the uptake, it is therefore considered as a separate variable. The standardized
value is 0 mm.

Units

The transfer factors are expressed as the ratio between the amount of
radionuclides in the edible parts of the crop and the amount of radionuclides in the dry soil. As well as for food as for animal feed the dry
weight is used; its use avoids the inaccuracy associated with different
moisture contents in fresh products.

EXPECTED VALUES

At present data are available for Am, Co, Cs, Pu, Sr, Tc, Mn, Np and
Zn. Because of lack of space only data for Cs are presented here. For the
other radionuclides see [1]. The expected standardized values, listed in the
first column of table 1 are based on the eqns (1) and (2). The 2nd and 3rd
columns show the lower and upper limits of the 95 percent confidence interval for average estimates, based on the stored observed values. The 4th and 5th
column show the expected lower and upper limits for a single new observation, also
based on the 95 percent confidence interval. The difference between these two
types of confidence intervals is large. The explanation is that the observed data reflect normal agricultural practice. Numerous data are available.
The confidence intervals for the observed data are therefore rather narrow.
In fact these confidence intervals reflect the uncertainty of well known
environmental conditions, as e.g. the surroundings near a particular power

plant. The confidence interval for single predicted values takes into account all factors which may have an impact on the uncertainty. Only limited data are available for combinations of such conditions. In fact these confidence intervals reflect possible combinations of extreme environmental conditions. They are hardly applicable to normal agricultural conditions, but may have to be used for very long term impact studies as e.g. studies to the effects of the disposal of nuclear waste in geological systems.

TABLE 1

Transfer factors for the uptake of Cs from soils to food and animal feed for standardized conditions ($Bq.kg^{-1}$ dry weight crop per $Bq.kg^{-1}$ dry soil)

Soil	Product	Transfer Factors				
			Confidence interval			
			Average value		Single value	
		Expected value	lower limit	upper limit	lower limit	upper limit
Sand	Cereals	$1.8 \ 10^{-2}$	$1.4 \ 10^{-2}$	$2.2 \ 10^{-2}$	$1.4 \ 10^{-3}$	$2.2 \ 10^{-1}$
	Fodder	$6.0 \ 10^{-2}$	$4.4 \ 10^{-2}$	$8.1 \ 10^{-2}$	$4.9 \ 10^{-3}$	$7.4 \ 10^{-1}$
	Grass	$1.5 \ 10^{-1}$	$1.2 \ 10^{-1}$	$1.9 \ 10^{-1}$	$6.2 \ 10^{-3}$	$9.2 \ 10^{-1}$
	Pods	$1.0 \ 10^{-1}$	$6.8 \ 10^{-2}$	$1.5 \ 10^{-1}$	$8.0 \ 10^{-3}$	1.3
	Root crops	$6.3 \ 10^{-2}$	$4.0 \ 10^{-2}$	$9.8 \ 10^{-2}$	$5.0 \ 10^{-3}$	$7.9 \ 10^{-1}$
	Tubers (potato)	$8.9 \ 10^{-2}$	$6.6 \ 10^{-2}$	$1.2 \ 10^{-1}$	$7.2 \ 10^{-3}$	1.1
	Vegetables	$2.2 \ 10^{-1}$	$1.7 \ 10^{-1}$	$2.9 \ 10^{-1}$	$1.8 \ 10^{-2}$	2.7
Clay	Cereals	$1.5 \ 10^{-2}$	$1.1 \ 10^{-2}$	$2.0 \ 10^{-2}$	$1.2 \ 10^{-3}$	$1.8 \ 10^{-1}$
	Fodder	$5.0 \ 10^{-2}$	$3.4 \ 10^{-2}$	$7.4 \ 10^{-2}$	$4.0 \ 10^{-3}$	$6.2 \ 10^{-1}$
	Grass	$1.3 \ 10^{-1}$	$9.1 \ 10^{-2}$	$1.7 \ 10^{-1}$	$5.1 \ 10^{-3}$	$7.8 \ 10^{-1}$
	Pods	$8.4 \ 10^{-2}$	$5.3 \ 10^{-2}$	$1.3 \ 10^{-1}$	$6.7 \ 10^{-3}$	1.1
	Root crops	$5.3 \ 10^{-2}$	$3.1 \ 10^{-2}$	$8.9 \ 10^{-2}$	$4.1 \ 10^{-3}$	$6.7 \ 10^{-1}$
	Tubers (potato)	$7.4 \ 10^{-2}$	$5.1 \ 10^{-2}$	$1.1 \ 10^{-1}$	$6.0 \ 10^{-3}$	$9.3 \ 10^{-1}$
	Vegetables	$1.9 \ 10^{-1}$	$1.3 \ 10^{-1}$	$2.7 \ 10^{-1}$	$1.5 \ 10^{-2}$	2.3
Loam	Cereals	$1.4 \ 10^{-2}$	$1.1 \ 10^{-2}$	$1.8 \ 10^{-2}$	$1.2 \ 10^{-3}$	$1.8 \ 10^{-1}$
	Fodder	$4.8 \ 10^{-2}$	$3.5 \ 10^{-2}$	$6.6 \ 10^{-2}$	$3.9 \ 10^{-3}$	$6.0 \ 10^{-1}$
	Grass	$1.2 \ 10^{-1}$	$1.1 \ 10^{-1}$	$1.5 \ 10^{-1}$	$5.0 \ 10^{-3}$	$7.5 \ 10^{-1}$
	Pods	$8.1 \ 10^{-2}$	$5.4 \ 10^{-2}$	$1.2 \ 10^{-1}$	$6.5 \ 10^{-3}$	1.0
	Root crops	$5.1 \ 10^{-2}$	$3.2 \ 10^{-2}$	$8.0 \ 10^{-2}$	$4.0 \ 10^{-3}$	$6.4 \ 10^{-1}$
	Tubers (potato)	$7.2 \ 10^{-2}$	$5.3 \ 10^{-2}$	$9.6 \ 10^{-2}$	$5.8 \ 10^{-3}$	$8.8 \ 10^{-1}$
	Vegetables	$1.8 \ 10^{-1}$	$1.4 \ 10^{-1}$	$2.4 \ 10^{-1}$	$1.5 \ 10^{-2}$	2.2

Conversion to non-standardized conditions

For the statistical analysis of the data the eqns (1) and (2) were used. Eqn (1) applies to all crops except grass, and eqn (2) applies to grass. The difference between the 2 equations is that eqn (1) refers to a standardized depth of the radionuclide of 20 cm and an OM of 4 percent, while these values for eqn (2) are respectively 10 cm and 10 percent. The other standard values are pH = 6, the period elapsed since the radionuclide reached the soil = 2 years and the amount of irrigation water irr = 0 mm.

The maximum value to be applied for OM = 10 percent.

$$TF = TF_{stand} \cdot e^{(pH-6)a} \cdot e^{(OM-4)b} \cdot e^{(Period-2)c} \cdot e^{(Depth-20)d} \cdot e^{irr.g} \quad (1)$$

$$TF_{grass} = TF_{stand} \cdot e^{(pH-6)a} \cdot e^{(OM-10)b} \cdot e^{(Period-2)c} \cdot e^{(Depth-19)d} \cdot e^{irr.g} \quad (2)$$

It will be clear that the derived transfer factors are geometric means. The values for the coefficients a-g are: a = -0.64, b = 0.011, c = -0.082, d = -0.016 and g = -0.008.

Identification of specific uncertainties

Due to the statistical analysis the expected values are averaged for space, time and variation of crops within one crop type. Because data of different investigators are used they are in fact also averaged for differences in local conditions, which may be described as differences in way of management, climate and chemical form of the radionuclide at the time it reached the soil.

The impact of these factors can be described with the Uncertainty Factor UF. For the estimation of UF's populations of transfer factors again are assumed to be log normally distributed. After logarithmic transformation of the data the arithmetic rules of the normal distribution hold, i.e. if the original data are TF_i the following equations apply:

Mean of the log transformed data: $\mu = \sum_{i=1}^{n} \dfrac{\ln TF_i}{n}$

Geometric mean of the original date: $e^{\mu} = \exp \left(\sum_{i=1}^{n} \dfrac{\ln TF_i}{n} \right)$

Variance of the log transformed data: $var = \dfrac{\sum_{i=1}^{n} (\ln TF_i)^2 - \dfrac{(\sum \ln TF_i)^2}{n}}{n-1}$

Standard deviation of the log transformed data: $\sigma = \sqrt{var}$

For the original data this gives a spread factor: $S = \exp(\sigma)$

A property of the normal distribution is: 95 percent of the observations are between $\mu - 2\sigma$ and $\mu = 2\sigma$.

For the original data this means that 95 percent of the observations are between e^{μ}/UF and $e^{\mu} \times UF$.

The Uncertainty Factor UF is thus defined as $UF = S^2$.

To estimate the UF's from particular data sets selected series have to
be used. The uncertainty associated with a particular crop requires e.g.
transfer values from one plot and one crop for a sufficiently long period.
A problem is that the availability of some radionuclides decreases with
time. So there should be sufficient data available within a rather short
period of time, say a few years. Very short term experiments have to be excluded also, in that case there exists unsufficient equilibrium between the
solute and adsorbed phase of the radionuclide. Other variables require other
series. Although such series of data are rather scarce, sufficient data were
available to derive the following UF values for Cs:

cereals UF = 4.1 - 12

grass, fodder UF = 2.3 - 8.8

green vegetables UF = 8.9 - 22

Impact of local conditions

When glancing over all available data it seems that the uncertainty
which is related with the investigator, i.e. local conditions as type of
management, climate and chemical form of radionuclide, is of the same order
of magnitude as the forementioned uncertainties.

For a statistical evaluation sufficient data should be available to
calculate standardized TF values per investigator. Data of 4 investigators
appeared to be suitable. The calculated standardized geometric mean values
for each author are shown in table 2 together with the standardized general
geometric mean values.

It is remarkable that the general mean value is sometimes lower than
each of the other 4 values, in one case the general mean value equals the
highest of the investigators' values. Probably the number of observations
is too low for a reliable estimation of the range R. Moroney [2] provides
a relation between R and standard deviation. From this relation it can be
estimated that for large ranges of observations the value of R is 1.5 times
the value from a range of 4 values. With four authors, for many combinations of crop and soil a value of R = 6 seems to be a reasonable estimate,
fodder with an R of 35 forms an exception. It is therefore expected that
for a larger number of authors R will vary between 10 and 50.

TABLE 2
Standardized TF values for Cs. All values have to be multiplied by 10^{-2}

Soil		General geometric mean	Heine	Stout-jes-dijk	Haak	Cawse	R*	UF
Sand	Cereals	1.8	2.3	3.8	11	6.8	5	2.5
	Fodder	6.0	4.6	8.8	150	6.8	33	7.5
	Grass	7.1	15	13	34	7.3	5	2.4
	Tubers	8.9	2.5	14		8.6	6	2.7
	Vegetables	22		63		15	4	2.3
Clay	Cereals	1.5		0.57	3.6	8.7	15	4.8
	Fodder	5.0		1.3	47	8.7	36	7.9
	Grass	5.9		2.0	11	9.3	6	2.7
	Tubers	7.4		2.1		11	5	2.6
	Vegetables	19		9.4		19	2	1.5
Loam	Cereals	1.4	2.7	1.2	10	6.2	8	3.4
	Fodder	4.8	5.4	2.7	140	6.2	52	9.8
	Grass	5.7	17	4.1	31	6.6	8	3.2
	Tubers	7.2	2.9	4.4		7.8	3	1.8
	Vegetables	18		20		13	2	1.3

* R = maximum value/minimum value

Therefore, the uncertainty UF associated with different investigators, which can be calculated using equation (3) will range from 3.8 to 10.

$$\sigma_{ln}^2 = (lnR)^2/12 \qquad \text{eqn (3)}$$

DISCUSSION

Within this manuscript the spread of data has been described by uncertainties. There is not distinguised between 'variability' which can be considered as a spread which can hardly be decreased by further research and 'uncertainty' which can be considered as a spread of data which can be reduced by further studies. Unsufficient data are available to make such a separation.

CONTRIBUTORS IUR SOIL-TO-PLANT TRANSFER FACTOR PROJECT

Boikat, U., University Bremen, FRG; Cawse, P.A., AERE, Harwell, UK; Dehut, J.P., Université Catholique, Louvain-la-Neuve, Belgium; Eriksson, A, Swedisch University of Agriculture, Uppsala, Sweden; Fagniart, E., CEN/SCK, Mol, Belgium; Frissel, M.J., RIVM, Bilthoven, The Netherlands, Ginkel, J.H.van, RIVM, Bilthoven, The Netherlands; Grogan, H., Imperial College, Ascot, UK and Fed.Inst. Reactor Res., Würenlingen, Switzerland; Haak, E., Swedish University of Agriculture, Uppsala, Sweden; Haisch, A., Bayerischer Landesanst.Bodenk. u. Pflanzenbau, München, FRG; Ham, G.J., NRPB, Chilton Didcot, UK; Handl, J., GSF, Hannover, FRG; Heine, K., Bundesanst. Milchforschung, Kiel, FRG; Kirchmann, R., CEN/SCK, Mol, Belgium; Kühn, W., GSF, Hannover, FRG; Linsalata, P., New York University, Tuxedo, USA; Pimpl, M., Kernforschungsanlage Karlsruhe, FRG; Popplewell, D.S., NRPB, Chilton Didcot, UK; Roussel, S., CEN, Cadarache, France; Saas, A., CEN, Cadarache, France; Steffens, W., Inst. Radioagronomy, Jülich, FRG; Stoutjesdijk, J.F., RIVM, Bilthoven, The Netherlands

ACKNOWLEDGEMENT

The authors acknowledge with pleasure the helpful comments they received from Dr.H.W.Köster.

REFERENCES

1. Vth Report of the Workgroup Soil-to-Plant Transfer Factors of the IUR, 1987. Contact person, M.J.Frissel, RIVM, P.O.Box 1, 3720 BA Bilthoven, The Netherlands

2. Moroney, M.J., Facts from figures, Penguin Books Ltd, Harmondsworth, UK, 1951

VALIDATION OF DYNAMIC FOODCHAIN MODELS -
RESULTS OF AN EXPERIMENTAL PROGRAMME

S Nair
Berkeley Nuclear Laboratories
Central Electricity Generating Board
Berkeley
Gloucestershire
GL13 9PB
United Kingdom

ABSTRACT

This paper summarises the current status of a programme of work
involving the production of appropriate dynamic foodchain models and the
acquisition of time-dependent data to validate and improve these. The
models are used to derive equilibrium transfer factors for the assessment
of ingestion doses from routine atmospheric releases and to derive
time-dependent transfer factors for use in similar assessments for discrete
releases arising both from normal operation and from hypothetical
accidental atmospheric releases.

The structure of two new dynamic foodchain models recently developed
by CEGB to study the uptake of S-35 and C-14 by crops is described. These
models are additional to the generic model for studying actinide and
fission product transport in foodchains that has been published elsewhere
[1]. The results of a recent Field Study carried out in the environs of an
operating CEGB nuclear power station are summarised, this study having
successfully resulted in the measurement of S-35 and C-14 equilibrium
transfer factors to a number of UK crops. These measured transfer factors
are compared with the model predictions.

The use of dynamic models to derive equilibrium transfer factors
requires validation against experiment of the time-dependent model
predictions. This is also needed before the models can be used for
discrete release studies. Thus details are provided of two further
experiments, as follows:

(i)   a study aimed at measuring crop interception factors, and

(ii)   a study of the dynamics of radionuclide transport in soils and in
        plants using a system of lysimeters.

Both the above studies utilise contracted experimental facilities and
expertise in the Department of Pure and Applied Biology and the Reactor
Centre at Imperial College. CEGB input has been in defining the
objectives, in participating in the planning of the experiments and in the
subsequent data analysis, particularly with regard to the implications for
foodchain models.

Finally, details will be provided of the objectives of a project,
jointly funded by CEGB and the UK Ministry of Agriculture, Fisheries and
Food (MAFF), which will study the uptake by lamb of a selection of
radionuclides relevant to both routine and accidental releases. Details

will also be provided of the objectives of a joint project with Studsvik
aimed at studying the behaviour of Chernobyl deposition in soils.

INTRODUCTION

One of the key objectives of the controls exercised by the authorising
ministries on radioactive discharges during normal nuclear power station
operation in the UK is to ensure compliance with the ALARA principle. In
practice, the discharges are sufficiently low to ensure that the resulting
enhanced concentrations in environmental materials are small compared to
background and are thus difficult to measure accurately. As a result,
assessments of individual and population doses have to be made using
computer models.

The key nuclides and pathways contributing to individual and
population doses from atmospheric releases arising during normal operation
of CEGB reactors are as follows:

(i)   Magnox: External exposure from Ar-41, ingestion of C-14, ingestion
        (and, to a lesser extent, inhalation) of S-35 and of tritium, in
        that order.

(ii)   AGR: Ingestion of C-14, external exposure to Ar-41, ingestion
        (and, to a lesser extent, inhalation) of S-35 and of tritium, in
        that order.

(iii)   PWR: Ingestion of C-14, ingestion (and to a lesser extent
        inhalation) of I-131 and 133, external exposure to a range of
        Kr/Xe noble gases and external exposure to Ar-41, in that order.

For the case of Magnox and AGR, the releases are associated with the
unavoidable minor discharges of the carbon dioxide coolant which occur
during reactor operation. In addition, spike discharges of S-35 and C-14
occur during blowdowns, when the entire coolant is expelled prior to a
shutdown, say, for maintenance. The predominant chemical form of the C-14
is as carbon dioxide for Magnox and for AGR. For PWRs, much seems to
depend on the detailed design with C-14 being released to varying degrees
as carbon dioxide, methane and ethane. However, for radiological
assessment purposes, the pessimistic assumption is made that all the C-14
is emitted as carbon dioxide. S-35 is emitted from Magnox and AGRs as
carbonyl sulphide during normal operation and also as hydrogen sulphide
during blowdowns. This nuclide is not discharged from PWRs.

SUMMARY OF CEGB'S DYNAMIC FOODCHAIN MODELS

Dynamic S-35 Models for Crop Uptake

Dynamic foodchain models for S-35 uptake into (i) leafy vegetables and
(ii) leguminous and root vegetables and cereals are shown in Figure 1.
They account for the following processes: (i) stomatal uptake of the
gaseous pollutant and its subsequent translocation to roots and, where
appropriate, from inedible to edible tissue, (ii) transfer from tissue to
soil arising from the dying back of tissue, (iii) root exudation to soil,
(iv) root uptake from soil and transfer to both edible and inedible tissue,

and (v) direct uptake of the gaseous pollutant by soil and its subsequent transfer to the crop.

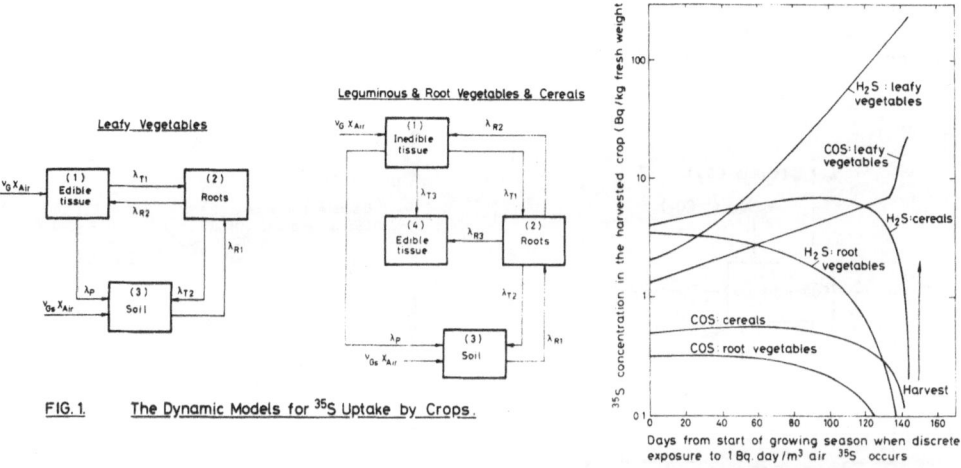

FIG. 1.   The Dynamic Models for ^{35}S Uptake by Crops.

FIG. 2: The Dependence of the ^{35}S Concentration in the Harvested Crop on its Chemical Form and on the Time of Occurrence of the Discrete Exposure.

Studies carried out to date (e.g. [2]) have established that the dry deposition velocity to soil of carbonyl sulphide is approximately three orders of magnitude smaller than that to the crop; a large difference also exists for hydrogen sulphide. Direct stomatal uptake is therefore always more important than uptake from soil. For this reason, uptake and transport within the crop is modelled in greater detail than chemical behaviour and speciation in soil.

The dependence of S-35 concentrations in the harvested crop on the time within the growing season when a discrete exposure of 1 Bq.day/m**3 occurs is shown in Figure 2 as a typical example of the type of seasonality calculations that can be performed using the model.

Dynamic C-14 Model for Crop Uptake

Normal operation of Magnox and AGR can lead to ambient atmospheric C-14 specific activity concentrations which are upto 50% greater than background levels close to the reactor, progressively reducing to background levels with distance from the reactor. In addition, discrete releases of coolant occurring during blowdowns can lead to higher localised concentrations lasting for short periods. The dynamic C-14 model for uptake into crops can be used to study both these cases. It is shown in Figure 3. The model considers uptake and subsequent transfer of both C-14 and stable carbon as carbon dioxide and accounts for the following processes: (i) nett stomatal uptake into tissues, (ii) root uptake of soil bicarbonate, (iii) transfer between the bicarbonate and organic soil pools, (iv) transfer from edible tissue to soil arising from the dying back of tissue and (v) return transfer to atmosphere arising from humic decomposition. A default data base for the model is currently being created, from literature surveys.

For routine releases entailing a constant enhanced C-14 atmospheric specific activity concentration, the model predicts results consistent with an earlier specific activity model [3]; however, for a discrete release of both C-14 and stable carbon at a higher than background level, the model can be used to predict the temporal variation of the C-14 specific activity

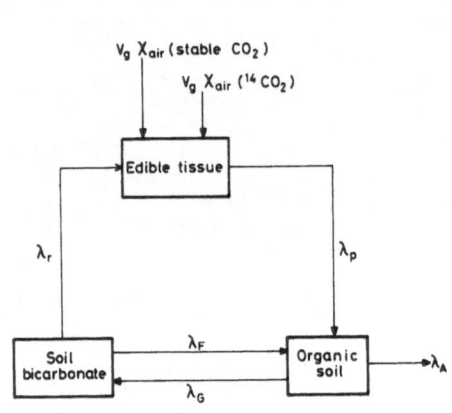

FIG 3 The Dynamic Model for ^{14}C Uptake by Crops.

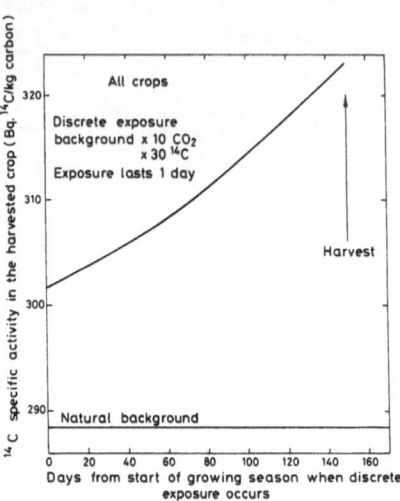

FIG. 4. The Dependence of ^{14}C Specific Activity in the Harvested Crop on the Time of Occurrence of the Discrete Exposure.

in the crop at harvest. A typical result, using a preliminary model data base, for a hypothetical discrete release occurring at different times in the growing season, entailing a tenfold increase in the atmospheric stable carbon dioxide concentration and a thirty-fold increase in the background atmospheric C-14 concentration for a day, is shown in Figure 4.

Generic Dynamic Model for Actinide and Fission Product Releases

The CEGB's generic dynamic model for studying the foodchain transport of actinides and fission products has been presented at a previous CEC seminar and is published elsewhere [1]. This model is principally applicable to accidental releases but can also be used to derive equilibrium transfer factors for routine releases of certain fission products and structurally activated nuclides that may occur during routine operation of PWRs (I-131 and 133, Co-58 and 60, Cs-134 and 137, Sr-89 and 90).

Work presently in hand is aimed at finalising the default data library for the C-14 crop model. It is proposed also to develop parallel models to study the dynamics of S-35 and C-14 uptake into animal products (milk and meat).

THE HINKLEY FIELD STUDY

A Field Study was carried out over a period of about a year, jointly with Drs Kluczewski and Bell of Imperial College under the terms of a CEGB

contract, at the CEGB's Hinkley Point site. The aim of the study was to
measure air-to-crop transfer factors for S-35 and C-14. The Hinkley Point
site consists of the Hinkley Point 'A' Magnox station and the 'B' AGR
station. The study was carried out using a plot about 150 m from the two
'B' station reactors. An experimental programme was devised in which a
number of different crops were sown in or transplanted to the plot at the
appropriate times of the year, and allowed to grow to maturity when they
were harvested for radiochemical analysis of their edible portions. The
crops studied included leafy vegetables such as cabbage and lettuce,
leguminous vegetables such as beans and "root" vegetables such as potatoes,
onions and carrots.

Sequential sampling of the ambient S-35 and C-14 in the air alongside
the plot was also carried out over this period. The S-35 air
concentrations were measured by pumping air through a sample of activated
charcoal, the trapping efficiency of which had been pre-determined to be in
excess of 99.996%. This charcoal was then processed using procedures
outlined in [4] to determine the S-35 content of the charcoal. The average
S-35 air concentration over the period in question was then readily
determined knowing the air pumping rate (ca. 4-5 litres/minute), the time
duration and the overall efficiency.

C-14 air concentrations were determined by sucking air through a
solution of decarbonised sodium hydroxide. At the end of the sampling
period, the sodium hydroxide was analysed for its total carbon and C-14
content by AERE, Harwell, under contract to CEGB. Again, knowledge of the
pumping rate (ca. 2 litres/minute) and the sampling period enabled the
stable carbon and C-14 air concentrations to be determined, in addition to
the atmospheric specific activity. The C-14 content and specific activity
in the crops were also measured by AERE Harwell under contract while the
S-35 concentrations in the crop samples were measured using procedures
summarised in [4].

The ambient S-35 and C-14 air concentrations above the plot were
reasonably constant during the course of the study, varying by no more than
a factor of two. The C-14 specific activity was more constant, varying by
a maximum of about 60%. For the case of S-35, this relative constancy
allowed the derivation of air-to-crop transfer factors directly from the
experimental data. They are plotted in Figure 5 as a complete group. The
data may be sub-divided into the categories "green vegetables" and "root
vegetables" and mean and extreme value transfer factors derived, the latter
corresponding to the "mean + 3 standard deviations" of the log-transformed
normal distribution. The values obtained, expressed in units of Bq/kg
fresh weight per Bq/m**3 ambient air concentration, are as follows:

    Green vegetables - Mean = 150 Extreme = 280
    Root vegetables - Mean = 58 Extreme = 190.

These values compare favourably with the values of 220 and 120 Bq/kg
fresh weight per Bq/m**3 air concentration derived using the dynamic S-35
model, when it is noted that the model data base is based on experimental
data from studies that exclude this particular study.

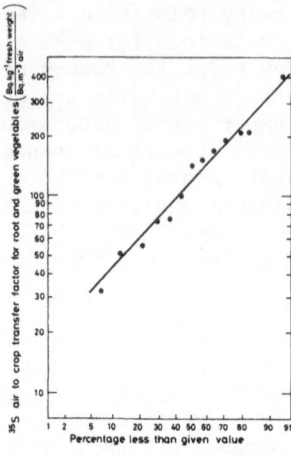

FIG. 5. Log – Probability Plot of Experimental
³⁵S Transfer Factors for Vegetables.

The effect on the measured S-35 concentrations of boiling the crop was also investigated. A reduction factor in the range 0.33 to 0.73 was observed, this factor being defined as the ratio of the concentration in the boiled part to that in the unboiled part referenced to a constant original fresh weight of material.

For the case of C-14, the average specific activity over the period was about 50% in excess of the background specific activity. However, the available data were found, as would be expected, to validate the specific activity model.

ON-GOING AND FUTURE STUDIES

Following on from the Hinkley Field Study, the CEGB have initiated a number of projects which are aimed at acquiring data for validating the various models that CEGB have developed. These studies are being carried out via a programme of contracts involving joint technical collaborations with selected external organisations. Two relevant continuing studies are described below.

Crop Interception Study - (Contractor - Imperial College - Drs Glauert, Bell and Miss Minski)

This study aims to measure interception factors on the crop following a particulate release external to the canopy. The study involves the production of non-radioactive respirable particles containing a tracer that has a very low leachability (and hence is unlikely to translocate within the crop following deposition thereon). The particles are released upwind of the crop or above the crop canopy and the concentrations per unit area of crop and on filter papers located at different heights within the canopy are measured using activation analysis. These data can then be used to estimate the crop interception factor.

The release method was tested by spraying an aerosol consisting of an aqueous solution containing gold chloride over a crop of winter wheat in early 1987. The aerosol used was released under low windspeed, dry conditions upwind of the plot using a system of nebulisers. Interception factors of about 12% were found, this factor being defined as the ratio of the amount deposited on the crop to the total amount deposited. The corresponding crop density was about 23 grams dry weight per square metre, consisting entirely of leaves given the early growth stage. This measured value of interception factor compares with a value of 6% predicted by the dynamic model.

The above experiment was followed, in August 1987, by a release over the same wheatfield of a respirable aerosol of silica particles labelled with dysprosium. The wheat was divided into the ears and the remainder of the crop (straw + stem); activation analysis is currently in progress and is expected to provide separate interception factors for these plant parts which may be compared directly with model predictions.

Dynamics of Radionuclide Transport in Soils and Plants (Contractor - Imperial College - Mr Shaw, Dr Bell and Miss Minski).

This study commenced in March 1987. The objectives are twofold: (i) to study the time-dependent behaviour in a number of soils of a selection of radionuclides and of the influence of initial chemical form on radionuclide speciation and migration, and (ii) to use the Imperial College wind tunnel to study the time-dependence of translocation from inedible to edible parts of radionuclides initially applied to the former. Significant progress has been maintained to date in preparing insoluble chemical forms of Ru, Cs, Co, and Mn using commercially available soluble radionuclide chemical solutions as a basis. Thus, a very low leachability has been demonstrated for Ru precipitated by a combined iron/aluminium hydroxide matrix and for Cs, Co and Mn bound to silica deriving from a silica gel preparation. However, the relative solubility of amorphous silica preparations in comparison to metal hydroxide may preclude its use in favour of the iron/aluminium hydroxide matrix.

The first part of the study will utilise a system of eight outdoor lysimeters five of which contain a sandy loam soil while the other three contain a silty loam, a loam and a silty clay soil respectively. All the lysimeters except one containing the sandy loam were contaminated with Cs-137 (as caesium chloride) in 1983. Soluble and insoluble preparations of the radionuclides will be applied separately to two pairs of the sandy loam lysimeters with just [Cs-137]-chloride being applied to the fifth (uncontaminated) sandy loam lysimeter. [Cs-134]-chloride and the other soluble radionuclide chemical forms will be applied to the lysimeters containing the other three soil types. All the lysimeters will be divided into halves separated by a plastic sheet. One half will remain fallow while wheat and cabbage will be grown in rotation on the other. The objective is to investigate the effect of vegetative cover on soil radionuclide transport. This will be studied by extracting soil cores and studying the physico-chemical speciation of the radionuclides at different layers. The grain and straw of the mature wheat crop and the cabbage will also be analysed for their radionuclide content.

In addition to the above, the CEGB have initiated the following studies, which have just commenced:

(i)   The Dynamics of Radionuclide Uptake by Lamb (Contractor - UK Institute of Terrestrial Ecology and the Hill Farming Research Organisation)

This study is jointly funded by CEGB and by the UK Ministry of Agriculture, Fisheries and Food (MAFF). The objective is to obtain time-dependent metabolic data for a selection of radionuclides (including C-14 and S-35). They will be used to support improved animal product models that will be developed by CEGB.

(ii)   The Behaviour of Chernobyl Fallout Radionuclides on Urban Surfaces and in Soils (Contractor - Studsvik)

The objectives are to study the weathering of Chernobyl deposition on a number of different urban surfaces and in a selection of Swedish soils that are also commonly found in the UK. The latter part of the study will consider both pastures and ploughed soils and will involve studies of the time variation of the vertical soil profile and of the chemical speciation therein, for a number of Chernobyl deposited radionuclides. The results will be used to validate the CEGB's generic dynamic foodchain model.

SUMMARY OF CONCLUSIONS

This paper describes the current status of a programme of research involving the development of dynamic foodchain models by CEGB and the acquisition of data to test these via external contracts. The emphasis of the paper is on routine operational atmospheric releases rather than on accidental releases. However, even in this case, it is important to understand the dynamics of radionuclide transfer since both continuous and discrete releases can occur during normal operation of CEGB reactors. Thus, the essential aspects of two models recently developed to study the dynamics of S-35 and C-14 uptake by crops are described and typical results obtained using these models are shown. Details are also provided of an experimental programme aimed at acquiring time-dependent data to test both the models specific to S-35 and C-14 as well as the generic dynamic foodchain model for actinides and fission products that has previously been published [1].

REFERENCES

1. Nair, S., Models for the evaluation of ingestion doses from the consumption of terrestrial foods following an atmospheric radioactive release. CEGB report RD/B/5200/N84, 1984.

2. Kluczewski, S.M., Bell, J.N.B., Brown, K.A., Minski, M.J., The uptake of S-35 - carbonyl sulphide by plants and soils. CEGB report TPRD/L/2382/N82, 1983.

3. Nair, S., A method for assessing the annual dose to the most exposed individual from tritium and C-14 reactor discharges to atmosphere. CEGB report RD/B/N4668, 1979.

4. Kluczewski, S.M., Nair, S., Bell, J.N.B., A field study of the uptake of S-35 and C-14 into crops characteristic of the UK diet. CEGB report TPRD/B/0735/R86, 1986.

TRANSFER OF RADIONUCLIDES TO FOOD PLANTS:
ROOT VERSUS FOLIAR UPTAKE

P. KOPP, O. OESTLING and W. BURKART

Swiss Federal Institute for Reactor Research
Biology and Environment 81/SU
CH-5303 Würenlingen, Switzerland

Abstract

The amounts of cesium-134 and -137 were measured in vegetables which were contaminated by the fall-out due to the Chernobyl reactor accident. In the case where the plants were sown or planted before this incident occurred they were contaminated almost entirely by deposition of the radionuclides on their exposed parts (stalks and foliage), plants which were sown or planted after 1st May, contamination occurred mainly by transfer from soil to plant via the roots.

The activities found were in most cases too small to be either measured accurately or for real differences to be found in various parts of the plants. Some of the values were within the fluctuations of the background. A normal consumption of vegetables grown in Northern Switzerland could therefore not lead to a significant increase of ingested radioactivity. Some nuts, cereals and berries on the other hand contained considerably higher activities. This was surprising since the migration of cesium in soil is too slow to have reached the root areas. We assumed that the cesium must have been incorporated into the fruit via the leaves. To prove this, we applied aqueous solutions of the radionuclides cesium, strontium and iodine to the leaves of some vegetable plants and measured the resulting activity in the edible parts. In some cases we found a transfer of more than 40% of the total cesium applied to the leaf in the edible part of the plant.

Further experiments, such as washing the leaves at certain intervals after the application, competition with other non-radioactive elements, influence of the experimental conditions as well as the structure of the leaf are under investigation.

1. Introduction

Of the many pathways possible, for example external irradiation, inhalation and ingestion, the latter one poses a big health hazard. Since the reactor accident at Chernobyl we are faced with the fact that large amounts of radioactivity were released and distributed through the atmosphere to the ground contaminating our food. In this case the contamination must have occurred by two quite different pathways, first by direct deposition of the fall-out from the air on leaves and other parts of plants above ground, and second the transfer from soil to plant by uptake through the roots.

The fall-out reached Switzerland at the beginning of May 1986. For plants that were planted before this time we would expect to find some contamination from deposition on the above ground parts, but little or no activity in the roots because cesium is known to migrate very slowly in soil, as was shown in Würenlingen (1), where the cesium deposited between 1956 and 1962 from atmospheric weapons' tests migrated only about 15 cm in twenty years in undisturbed ground (2, 3); furthermore cesium is bound very strongly to clay minerals. From these facts one would not expect cesium to reach the depth of roots before the fields have been ploughed and tilled and therefore considerable transfer from soil to plant through the roots cannot be expected before next year.

For plants which were planted or sown since the Chernobyl fall-out reached the ground the situation is different;

the ground will have been ploughed and tilled and the radionuclides will have been well mixed into the cultivation layer before sowing the seeds or planting the seedlings. In this case we expect to find actvity in the soil around the roots, in the roots as well as in the plants.

A different situation prevails with the fruit and nuts on trees and bushes which in some cases contained higher activities. The radionuclides would have hardly had time to reach these through the roots, in the amounts that they were present. In this case the uptake must have taken place through the leaves with a subsequent transport to the fruit. In Spring 1986 there was a prolonged cold period and plant development as well as formation of fruit and nuts were delayed. The activity they contained could not have been caused by direct contamination and since root uptake can also be excluded, the cesium must have been taken up through the leaves and transported to the developing fruits and nuts within the plants.

Our experiments to determine the concentrations in various parts of food plants were designed to investigate transport pathways and possible places of deposition on the one hand and on the other hand to measure the concentrations of cesium-134 and -137 in the edible parts of commercially grown plants.

2. Methods

The sampling, at nearby farms, where vegetables were grown mainly to supply canning firms, or at a local market gardener's who supplies shops with his produce, was carried out as follows: all the plants that were growing in one of 5 1 m^2 patches were collected. These 5 patches were situated at each corner and in the centre of a square of 5 x 5 m.

The plants were segmented as follows: roots, stems and leaves, fruit or seed, peeled or unpealed, and the soil around the roots was also collected. The parts were washed to remove as much of the deposited radioactive material as possible, bulked and the gamma activity was measured spectrometrically using a Ge-Li detector.

Six categories of food plants were investigated up to the present:

Leafy vegetables:	cicorino rosso salad
	frisee salad
	chard
Root vegetables:	potatoes
	beetroot
Legumes:	peas
	beans
Cereals:	winter wheat
	corn (maize)
Fruit:	tomatoes
	apples
Nuts:	walnuts

Of these one leafy vegetable, cicorino rosso, both the root vegetables, one of the legumes, peas, tomatoes and winter wheat were sown or planted before Chernobyl, the others afterwards. The apples and walnuts grew on old-established trees.

The transfer of any radionuclide is expressed by the transfer factor (T.F.) which we defined as follows:

$$T.F. = \frac{\text{concentration of radiouclide in 1 kg fresh vegetable}}{\text{concentration of radionuclide in 1 kg dry soil}}$$

For the studies of foliar uptake three plant species were selected where the edible parts were not identical with the leaf to which the activity was applied: radishes, beans and potatoes. A mixture of cesium-134, strontium-85 and iodine-131 in a nearly carrier free solution was applied in small droplets to the upper surfaces of the leaves by means of an Eppendorf pipette during the late development of the plant. The applied volume was between 100 and 2000 ul per leaf and the concentration of each nuclide was about 100 Bq/ ul.

3. Results and discussion

Table I shows the amounts of cesium-134 and -137 contained in the various parts of the plants which were either sown or planted in the fields before the Chernobyl fall-out occurred. The activities are expressed in Bq/kg fresh plant material and in Bq/kg dry soil. The time between the planting and the incident and the period of time between the fall-out and the harvest is also listed.

Due to the low activities the results have large uncertainty factors and there is a considerable discrepancy. Overall errors of the order of +- 50% must be expected. This also shows up in the Cs-137/Cs-134 ratio which in the soil samples can be explained by the varying depth of the sampling since the soil around the roots was taken for analysis. The freshly deposited Cs-134 and -137 from the Chernobyl fall-out are still contained in the top soil layer (4), whereas the cesium found below 2,5 cm is the cesium deposited from the weapons' tests in the fifties. Since the Cs-134 has a half life of only 2.06 years it has disappeared, leaving only the Cs-137, and therefore the ratio Cs-137/Cs-134 increases.

The amount of deposition on the above ground parts varies according to the surface area and the texture of the plant.

The wheat grains contain a rather large amount of cesium which can only be explained by the fact that the grains are storage parts of the plant with a high percentage of dry matter.
The peas and potatoes were planted below the soil surface only 10 days before the incident and would not have been exposed to the fall-out directly and due to the low mobility of the cesium this would not have been transported to the root levels yet for transport to take place. The potatoes on the other hand were hoed sometime during growth.This operation mixes the soil and surface soil could have been brought to the vicinity of the roots. The cicorino rosso was planted outside as four week-old plants only four days before, but the leaves originally exposed to the fall out were small and would have died before harvest. A more likely form of contamination in these cases are rainsplash or wind deposition.

Table II shows the activities of plants which were planted after the Chernobyl accident on freshly tilled ground where the cesium was mixed with the soil throughout the rooting area. Included in this table are also some transfer factors soil ---> plant, calculated from our results according to the definition mentioned earlier in this paper.

Any activity in the plants in this case must result mainly from the uptake from the soil. It cannot be removed by washing the surfaces. The activities are generally low, in many cases they are below the detection limit (<d.l.). This can be expected since the availability of cesium in soil would be low. Further experiments in a follow-up period will bring more information on the migration of cesium in the ground and transfer from the soil to the food plants.

Table III shows a comparison of two different samples of peas of the same variety, one planted shortly before, and the other shortly after the fall-out occurred.

The activity of the peas planted 10 days before the Chernobyl fall-out occurred is below the detection limit. The other peas planted shortly afterwards in a freshly tilled field with the soil well mixed contain a small amount of measurable activity.

In 1987, a year after the Chernobyl accident, the total cesium activity was less than 4 Bq/kg fresh weight for all the above listed items. Since all these fruits, berries and nuts grew on well established bushes and trees the cesium cannot have been taken up from the soil through the roots in the relatively short time (6, 7, 8) and so foliar uptake and transport within the plants is the most likely pathway by which the radionuclides got into the edible parts.

Our foliar uptake experiments with vegetables showed that cesium is taken up and transported to the edible parts very efficiently, whereas the uptake of iodine and strontium was found to be very small (mostly not detectable, in one case only 5% for iodine and 2% for strontium). A concentration gradient through the plant from the leaves to the edible part could be observed.

The curve A in figure 1 shows the cesium activity in radishes after application of the radionuclide solution to the leaf at different time intervals before the harvest and under different experimental conditions. The percentage of the total activity in the radish (ordinate) has been plotted against the % weight of the radish of the whole plant (abscissa). The open circles show the percentage of the total activity in the radish which was applied to the leaves of the plants between 5 and 18 days before the harvest.

The circles marked with an S show the effect of an application of an inactive salt solution between 7 hours and 2 days prior to the foliar application of the radionuclide solution, and the circles marked with a + denote the uptake of cesium after the active mixture was diluted 1:1 with an inactive salt solution containing: KCl 2g/l; KI 0,5mg/l; $CaCl_2$ x $2H_2O$ 1g/l. The linear relationship shows that the uptake of the radionuclide is complete after less than 5 days and that inactive potassium and calcium do not seem to compete with cesium for uptake. If the activity were distributed evenly throughout the plant a straight line with a slope equal to one should result.

The curve B in figure 1 shows the effects of washing the leaves after application of the radionuclide mixture. The squares marked S show the effects of a single thorough showering for 1 minute with the salt solution 2 to 7 hours after the nuclide solution was applied to the leaves 6 to 18 days before harvest; the squares marked W show the effect of a similar showering with distilled water. Again there is a good relationship which shows that at an early stage (2 - 7 hours) after foliar contamination 30% of the radionuclides can be washed off the leaves (e.g. by heavy rain), irrespective of whether the washing water contained salt or not.

Figure 2 gives some information of the speed of uptake. The black circles show the amount of activity taken up in 6 and the open circles in 20 - 24 hours. (The dotted line is from figure 1).

Figure 3 shows the activity in the edible parts of potatoes and beans after application of the radionuclide solutions to the leaves 30 - 37 days prior to the harvest. Again the results follow the dotted line from figure 1 approximately.

Figure 4 illustrates the kinetics of the foliar uptake of the three radionuclides. A mixture of Cs-134, I-131 and Sr-85 was applied to several leaves of potato plants. At various intervals (abscissa) a treated leaf was immersed in 50 ml distilled water for 3 minutes. The black circles represent the activity remaining in the leaf after each washing and the open circles show the activity collected in the water. Each pair of points represent measured values from one single leaf. All the leaves were collected and their activities were measured simultaneously after the last leaf was washed. This procedure works very well for cesium; after a period of about 6 hours 70% of the activity is taken up irreversibly by the leaves and this does not change any more with time. In the case of iodine about 60% is taken up by the leaves, some iodine is lost from the leaves: this is presumably released to the atmosphere. With strontium only about 20% are taken up by the leaves, leaving about 70% which can be washed off. The maximum taken up seems again to be reached after about 6 hours.

4. Conclusions

After the fall-out due to the Chernobyl reactor accident the amounts of cesium-134 and -137 were measured in food plants, sowed or planted before and after this incident. In the first case the plants were partly contaminated with radionuclides by deposition, on the outside of the plant, in the latter case the contamination was due to transfer from soil to plant via the roots with the isotope deposited within the plants.

In most cases the activities were found too small to be either measured accurately, or for real differences to be found in various parts of the plants. Some of the values were within the fluctuations of the background. Normal consumption of vegetables grown in the northern parts of Switzerland would not have contributed significantly to an increase of ingested radionuclides.

Since the migration of cesium in soil is slow the experiments will be continued. Next year soil --> plant transfer should be the only mechanism by which radionuclides can contaminate the vegetation, apart from some possible minor contributions from resuspension or rainsplash.

Radionuclides, especially cesium and to a lesser extent strontium are taken up by plants from the top surface of leaves and transported through the plants to the edible portion. This process is completed after about 6 hours. Iodine is also taken up, but not to the same extent. In contrast to cesium and strontium some iodine is lost presumably to the atmosphere. Only a small amount of strontium is taken up compared with the other two radionuclides.

Showering of the plants within a few hours of deposition removes most of the radionuclides from the leaves.

5. References

1) Nagel, E., Strahlenexposition der Bevölkerung, 18. Jahrestagung des Fachverbandes fürStrahlenschutz, 6.-10. Oktober 1985, pp. 633-643.

2) Frissel, M.J. und Jakubick, A.T., Radioökologie, Berichtsband der Fachtagung des Deutschen Atomforums, 2.-3. Oktober 1979, p.131.

3) König, L.A., Langguth, K.G., Papadopoulos, D. und Radziwill, A., Vertical migration of HTO and Cs-137 in the soil, Radiation-Risk-Protection, 1, 197-200, 6th Int. Congr. IRPA,Berlin 1984.

4) Portmann, W., Crameri, R. und Görlich, W., Zeitliche und räumliche Veränderungen der Isotopenverhältnisse in radioaktivem Ausfall von Tschernobyl, Proc. Symposium Radioaktivitätsmessungen in der Schweiz nach Tschernobyl und ihre wissenschaftliche Interpretation", Bern, 20.-22. Oktober 1986.

5) Zehnder, H.J., private communication.

6) Weller, H., Bestimmung von Boden-Pflanze-Transferfaktoren für Sr-90 und Cs-137 sowie von Kenndaten der Böden und Pflanzen an Proben von den Standorten Mühlheim-Klärlich, Hamm und Neupotz, Untersuchungsberichte LUFA Speyer 1979, 1982, 1983.

7) Kampe, W., Auswirkungen radioaktiver Stoffe auf landwirtschaftliche Produktion und Nahrungsqualität, zweiter aktueller Situationsbericht Obstbau 10/86.

8) Haunold, E., Horak, O., Gerzabek, M., Umweltradioaktivität und ihre Auswirkungen auf die Landwirtschaft, 1. Das Verhalten von Radionukliden in Boden und Pflanze. Bericht Oesterreichisches Forschungszentrum Seibersdorf Ges.m.b.H., Institut für Landwirtschaft, OEFZS-4369/LA-163/86, Aug. 1986.

Table I. Radionuclide content of vegetables planted before the Chernobyl reactor accident.

		Cs-134 Bq/kg	Cs-137 Bq/kg
WINTER WHEAT	planted 7 months before, harvest 5 months after		
soil (dry weight)		59	123
roots		47	85
straw		10	20
ears		1	2
grain		4	8
BEETROOT	planted 5 weeks before, harvested 8 weeks after		
soil (dry weight)		72	135
peel		1	2
beetroot		<d.l.	<d.l.
PEAS	planted 10 days before, harvested 63 days after		
soil (dry weight)		44	85
roots		2	3
leaves and stalks		<d.l.	<d.l.
pods		<d.l.	<d.l.
peas		<d.l.	<d.l.
POTATOES	planted 10 days before, harvested 126 days after		
soil (dry weight)		32	57
roots		<d.l.	<d.l.
tubers		<d.l.	<d.l.
leaves and stalks		1	3
CICORINO ROSSO	planted 4 days before, harvested 56 days after		
soil (dry weight)		152	269
roots		4	7
leaves		1	2

Table II: Radionuclide content of vegetables planted after the Chernobyl reactor accident.

	Cs-134 Bq/kg	Cs-137 Bq/kg	Transfer factor	
			Cs-134	Cs-137
BEANS				
growing time 70 days				
soil (dry weight)	17	41		
roots	<d.l.	1	---	0.02
leaves and stalks	<d.l.	1	---	0.02
beans	<d.l.	<d.l.	---	---
FRISEE SALAD				
growing time 42 days				
soil (dry weight)	8	19		
roots	<d.l.	<d.l.	---	---
leaves	<d.l.	<d.l.	---	---
MAIZE				
growing time 140 days				
soil (dry weight)	20	51		
roots	<d.l.	2	---	0.04
leaves and stalks	<d.l.	<d.l.	---	---
cobs	<d.l.	<d.l.	---	---
kernels	<d.l.	<d.l.	---	---
PEAS				
growing time 63 days				
soil (dry weight)	52	103		
roots	3	6	0.058	0.06
leaves and stalks	1	1	0.019	0.01
pods	2	4	0.038	0.04
peas	<d.l.	1	---	0.01

Table III: Comparison of the radionuclide content of peas planted before and after the Chernobyl reactor accident.

	Cs-134 Bq/kg	Cs-137 Bq/kg	Transfer factor Cs-134	Cs-137
PEAS				
planted 10 days before, harvest 63 days after				
soil (dry weight)	44	85		
roots	2	3		
leaves and stalks	<d.l.	<d.l.		
pods	<d.l.	<d.l.		
peas	<d.l.	<d.l.		
PEAS				
planted 4 days after, growing time 63 days				
soil (dry weight)	52	103		
roots	3	6	0.058	0.058
leaves and stalks	1	1	0.019	0.010
pods	2	4	0.038	0.039
peas	<d.l.	1	---	0.010

Table IV : Radioactivity in berries, fruits, nuts and processed products in Switzerland, Summer and Autumn 1986 (5).

Product	Total Cesium activity Bq/kg fresh weight
Strawberries (open land) June	22
Strawberries (plastic tunnels) June	<2
Strawberries (open land) July	<d.l.
Strawberries (plastic tunnels) July	<d.l.
Red currants	52 - 81
Gooseberries	44 - 104
Raspberries	11 - 30
Bilberries (cultured)	<4
Elderberries	15 - 22
Cherries	74
Apples	4 - 22
Pears	4
Grapes	4 - 12
Hazelnuts	125
Walnuts	14
Apple juice	4 - 7
Apple juice concentrate	29 - 74
Apple powder	78
Pear juice concentrate	78 - 192
Honey	59

175

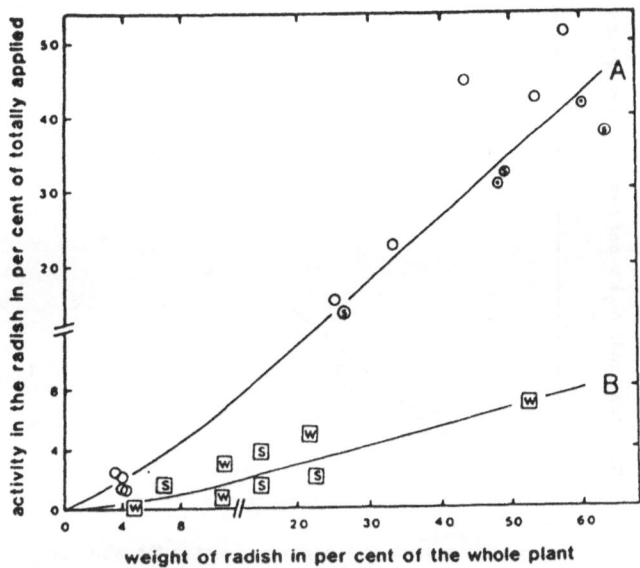

Figure 1: Cesium activity in radishes following foliar application

○ : harvest 5 - 18 days after foliar application of radionuclide solution

⊙ : leaves sprayed with an inactive salt solution 7 hours - 2 days before foliar application.

⊕ : active solution was diluted with an equal volume of the inactive salt solution

[S] : plants showered with inactive salt solution

[W] : plants showered with water

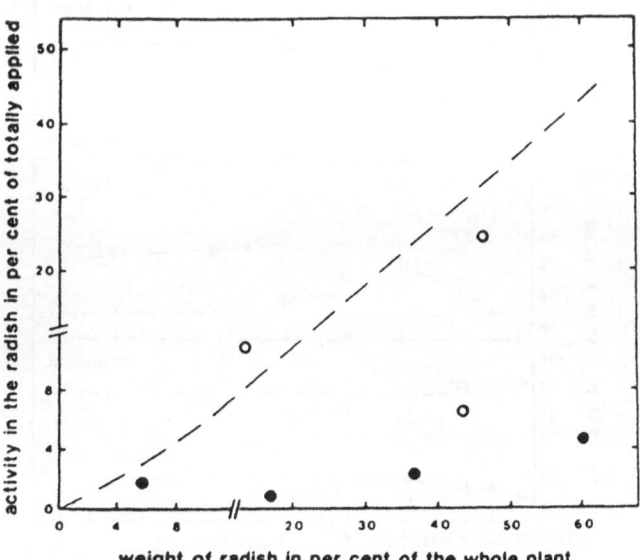

Figure 2: Time dependent uptake of radionuclides by radishes.

● : harvest 6 hours and

○ : 20 - 24 hours after application of radionuclides to the leaves.

(dashed curve from figure 2).

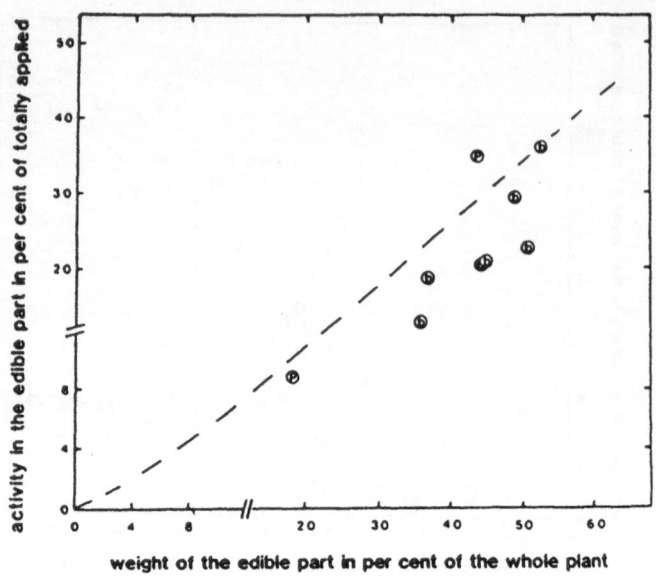

Figure 3: Cesium uptake by beans and potatoes following foliar application.
(dashed curve from figure 2 for radishes).

ⓟ : potatoes

ⓑ : beans.

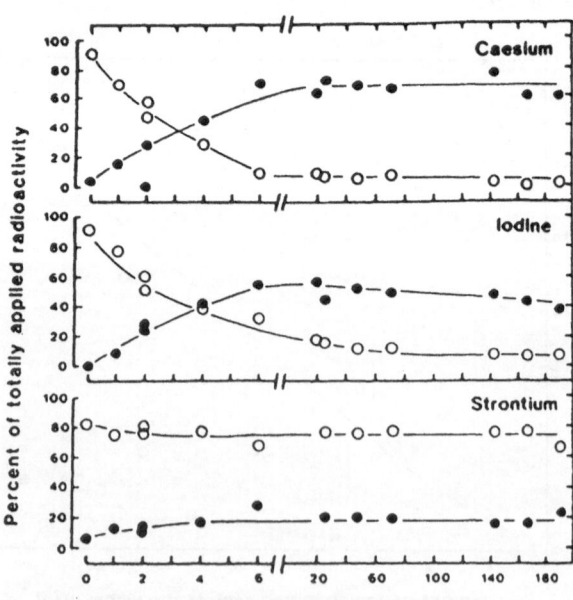

Figure 4: Time-dependent foliar uptake of radionuclides

○ : activity collected in the water

● : activity remaining in the leaf.

177

INTERCOMPARISON OF MODEL CALCULATION OF THE TURNOVER OF
Ra-226 WITHIN AN AQUATIC ECOSYSTEM

U. Bergström
STUDSVIK NUCLEAR
Safety and Systems Analysis
S-611 82 NYKÖPING
SWEDEN

ABSTRACT

Within the international BIOMOVS project a broad comparison and
testing of assessment models is undertaken for different release and
dispersion scenarios.

One of these scenarios dealt with the turnover of Ra-226 and Th-266
in a lake ecosystem. This ecosystem is continuously contaminated by inflow from a river over a period of 100 years with a concentration of
1 Bq/1 of each nuclide.

This paper discusses the results from seven models which have been
used to calculate results for Ra-226. Best estimates and associated uncertainties of the concentration in fish, sediment and drinking water are
compared as well as major contributions to the uncertainties. The reasons
for discrepancies between the results from the different models are summarized. Some recommendations on how to improve the confidence in results
and reduce the uncertainty are given.

INTRODUCTION

Mathematical models are extensively used to evaluate the potential

environmental impact of releases of radionuclides. Because models can

only approximate real world conditions, the predictions are associated

with uncertainty.

To ensure confidence in model-based decisions the uncertainty in

the predictions must be estimated and if necessary reduced. To be able

to do so a variety of tests should be carried out. These tests involve

the comparison of model predictions with independent data sets over a range of environmental conditions as well as intercomparisons of the predictions made by different models. An international project, BIOMOVS (BIOspheric MOdel Validations Study) has been initiated to address the issue of model testing /1/.

The primary objectives of BIOMOVS can briefly be described as follows:

1. to test the accuracy of the predictions
2. to explain differences
3. to recommend priorities

Two different approaches are used to fulfil these objectives. One approach uses independent data sets for formulation of test scenarios while the other uses specific test scenarios selected on the basis of priorities.

One of these scenarios treats the behaviour of radium-226 and thorium-230 released to a lake from a river. This paper is a brief summary of the results obtained for radium /2/. In addition some recommendations are given for improving data and models.

SCENARIO DESCRIPTION

The test scenario is described as follows: given an average concentration of Ra-226 (1 Bq/l) in a river over a period of 100 years, a) calculate the concentration (Bq/m^3) at times 1, 10, 50 and 100 years in the sediment of the lake, and b) calculate the concentration at times 1, 10, 50 and 100 years in the edible tissues of fresh water fish (in Bq/kg fresh wt) and in drinking water prior to human consumption.

Assume a suspended sediment load of 50 g/m^3, a sedimentation rate of 1 cm/y, an average discharge of water from the lake of 300 m^3/s with a turnover rate of the water of once per year, and that after a period of 100 years the lake has not yet been filled with sediment.

The scenario was deliberately described in general terms to ensure the participants to consider all factors influencing the results.

The best estimate prediction was requested as well as minimum and maximum values (corresponding to a 95 % confidence interval). The participants were also requested to identify the major source of the uncertainty in their results.

PRESENTATION OF THE MODELS

Seven mathematical models or computer codes have been applied to this scenario. They are:

- AMURAD; Oak Ridge National Laboratory, USA /3/
- BIOPATH; Studsvik Energiteknik AB, Sweden /4/
- BIOS; National Radiation Protection Board, UK /5/
- DETRA; Technical Research Centre of Finland, Finland /6/
- ECOS; Associated Nuclear Scenarios, UK /7/
- LASER; Risø National Laboratory, Denmark /8/
- NCRP; National Council on Radiological Protection, USA /9/

Most models are linear compartment models like AMURAD, BIOPATH, BIOS and DETRA. The LASER model uses an analytical equations for calculating time-varying concentrations. The ECOS model is based on solution of the compartment systems at equilibrium. The NCRP model is a screening model that is also based on systems at equilibrium. The transfer processes considered in the different models are shown in Table 1.

TABLE 1
Processes considered and number of compartments used in the models

Processes	AMURAD	BIOPATH	BIOS	DETRA	ECOS	LASER	NCRP
Turnover of water	X	X	X	X	X	X	X
Water - sed	X	X	X	X	X	X	
Surf sed - deeper sed	X	X	X	X			
Surf sed - water	X	X	X	X	X		
Dynamic uptake in fish and veg	X						
Filtration in drinking water			X		X	X	
Number of compartments	6	3	3	3	2	1	

Different methods have been used to obtain the uncertainties: AMURAD and BIOPATH use a statistical approach while DETRA, ECOS and LASER use parameter perturbation. Finally, judgemental estimates have been used by BIOS and NCRP. In Table 2 the values of the input parameters are given.

TABLE 2

Model parameter values

	AMURAD	BIOPATH	BIOS	DETRA	ECOS	LASER	NCRP
Depth of the lake (m)	26	10	16	24	100	20	
Depth of the surf sed (m)	0.05	0.1	0.1	0.35	1		
Mass sed rate (kg/m**2y)		2	10	10	12.5	10	
Sed dens (kg/m**3)	1930	2000	2600	2800			
Water fraction (%)	91	90	75	72			
K_d (m**3/kg) best est	4	10	5	5	5	2	
max		100		50	50	10	
min		0.5		0.5	0.5	1	
C_f (1/kg) best est	16*	25*	50	50	300*	50*	60*
max	603	100		500*	3000	100	180
min	1	10		50	30	10	0.6

* from unfiltered water.

RESULTS AND DISCUSSION

Results are presented in Figures 1-6. The transfer coefficients of importance used for the best estimate results are summarized in Table 3.

TABLE 3

Transfer coefficients (y^{-1})

	Model					
Process	AMURAD	BIOPATH	BIOS	DETRA	ECOS	LASER
Water −> surf sed	4.7E-1	1.2	2.52	1.9	7.8E-1	9.1E-1
Surf sed −> deep sed	2.0E-1	5.0E-2	1.0E-1	7.1E-1	-	
Surf sed −> Water	1.5E-1	1.0E-3	3.2E-3	7.2E-2	9.4E-10	

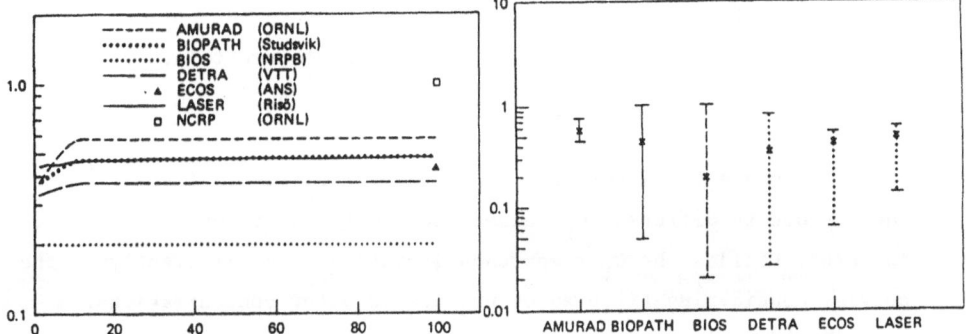

Figure 1. The concentration of radium in the water (Bq/1) as a function of time (y)

Figure 2. The uncertainties of the concentration of radium in the water (Bq/1) after 100 years of release

The best estimate results for the concentration of radium in water are within a factor of five of each other. Steady state is reached after only a few years for all models. The ranges of uncertainty vary from about a factor of three to almost two orders of magnitude.

All models except AMURAD have used the same expression to obtain the transfer rate from water to sediment. When comparing the concentration in water with the transfer coefficient from water to sediment, the agreement is quite good, with one exception, the ECOS-model. However, looking at the K_d-values, sometimes the opposite relationship is evident. For example, BIOPATH uses the highest K_d but assumes the lowest mass sedimentation rate. This assumption compensates totally for the high K_d-value. In addition, neither AMURAD, BIOPATH or DETRA consider filtration of the drinking water, which to some extent may cause their higher values compared with the others.

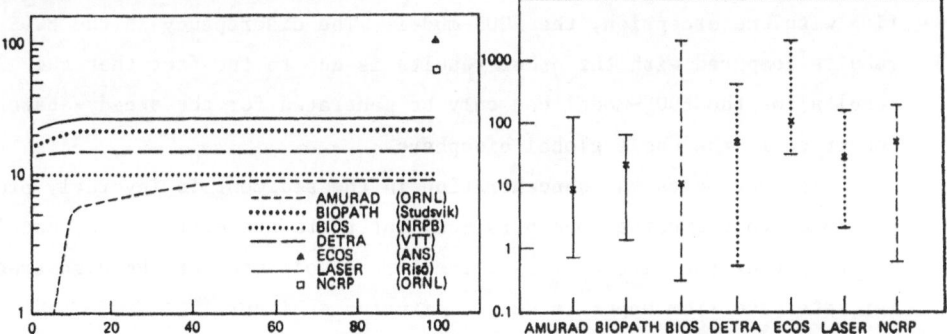

Figure 3. The concentration of radium in fish (Bq/kg) as a function of time (y)

Figure 4. The uncertainties for the concentration of radium in fish (Bq/kg) after 100 years of release

The best estimate calculations of the concentration in fish are within a factor of ten. However, the ranges of uncertainty given by the modelling groups are large, with the largest at about four orders of magnitude.

The cause to differences in the concentration in fish is consequently due to different C_f-values applied for different concentrations in water. Besides the C_f:s approach is interpreted differently by the models, considering filtered or unfiltered water concentrations.

The dynamics of the results are similar with the exception of AMURAD, which is the only model with fish as a dynamic compartment. It considers the biological halflife of radium in fish to be equal to the mean life expectancy of catchable fish.

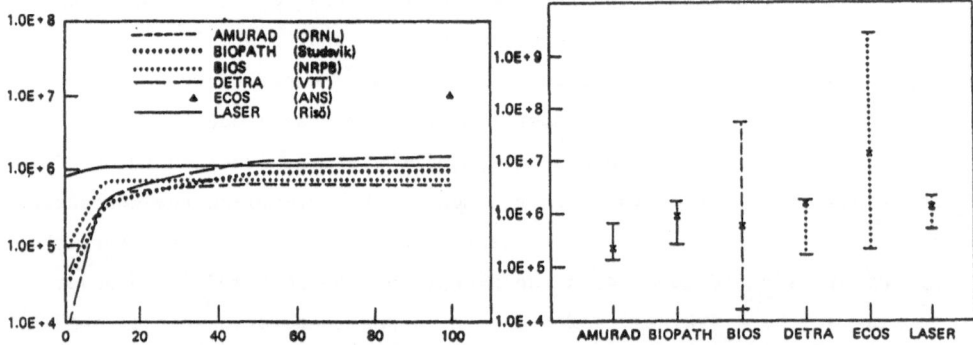

Figure 5. The concentration of radium in the sediment (Bq/m^3) as a function of time (y)

Figure 6. The uncertainties for the concentration of radium in the sediment (Bq/m^3) after 100 years of release

The results for the sediment show less variation than those for fish with one exception, the ECOS-model. The discrepancy in the ECOS results compared with the other results is due to the fact that the results for the ECOS-model can only be generated for the steady-state condition in the whole global biosphere.

In most cases the concentration in the sediment is inversely proportional to the water concentration, but there are differences, mainly due to either the size of the sediment reservoir used or the high transfer rate used from upper to deeper situated sediment.

New scenarios within the BIOMOVS study will further address the problems of modelling the processes to and within the sediment /10/.

Most participants identified the uncertainty in the K_d and the C_f

values as being of greatest importance for the uncertainty in their results.

The estimates of uncertainties for each model result were mostly very high and often exceeded several decades. On first glance these ranges of uncertainties are not especially satisfying. However, much of these large ranges are due to the use of judgement based on estimation of uncertainties in single parameters. The sensitivity of these parameters to the model result was not taken into account when the uncertainty estimates were made. Therefore, the models using statistical methods mostly produce the smallest ranges of uncertainty. These methods seem more reliable as they employ the most appropriate methods for combining estimates of parameter uncertainty.

CONCLUSIONS

The first study of model- and code-comparison performed within the BIOMOVS project confirms the absolute necessity of making comparisons especially for scenarios where there are difficulties in finding independent data for validation.

The BIOMOVS group recommends following improvements in the models and data with regard to some major points, some of which are of a general nature:

o   The data used in a modelling exercise should be carefully selected so as to be appropriate for the situation to be modelled.

o   Future assessment models ought to include statistical approaches for analysis of the effect of parameter uncertainties in the model predictions. In addition, quality assurance tests ought to be undertaken to eliminate uncertainties due to common errors in the compilation and documentation of results.

o   It is very important for the modellers to maintain close cooperation with experimentalists and other specialists within related disciplines such as hydrology, limnology and sedimentology.

o   The importance of the bioaccumulation factor to the concentration in fish has led to a proposal to improve models for the uptake in fish taking the metabolism and age of the fish population into consideration.

o   An obvious need for finding data for comparison of the water-fish pathway.

ACKNOWLEDGEMENTS

This paper is supported by the Swedish National Institute of Radiation Protection, which is gratefully acknowledged.

REFERENCES

1. BIOMOVS
   Progress Report No.1, January 1986. N.I.R.P. Stockholm, Sweden.

2. BIOMOVS Technical Report, Scenario B3 (Draft).

3. GARDNER, R H, BLAYLOCK, B G and HOFFMAN, F O. A dynamic model of Ra-226 Behaviour in an Oligotrphic Lake. In: National Uranium Tailings Program, vol 2. Prepared by SENES Consultants Limited, Ontario, Canada, 1985.

4. BERGSTRÖM, U et al. BIOPATH - A computer code for calculation of the turnover of nuclides in the biosphere and the resulting doses to man. Basic description. STUDSVIK/NW-83/261, 1982.

5. LAWSON, G and SMITH, G M. BIOS: A model to predict radionuclide transfer and doses to man following releases from geological repositories. Chilton, NRPB-R169 (1984) (London HMSO).

6. KORHONEN, R and SAVOLAINEN, I. Assessment of biospheric behaviour of releases from nuclear waste repositories. IAEA and CEC seminar on the environmental transfer to man of radionuclide released from nuclear installations. Brussels, Belgium, 17-21 Oct, 1983.

7. KANE, P and THORNE, M C. User's guide to the biosphere codes ELOS. ANS R-385. DOE report No. DOE/RW/84.121, 1984.

8. Not yet published.

9. National Council on Radiation Protection and Measurements, NCRP. Commentary No. 3. Screening Techniques for Determining Compliance with Environmental Standards. Releases of Radionuclides to the Atmosphere. March 1986.

10. BIOMOVS
    Progress Report No. 4, June 1987. N.I.R.P Stockholm, Sweden.

UNCERTAINTIES ASSOCIATED WITH ESTIMATES OF RADIUM
ACCUMULATION IN LAKE SEDIMENTS AND BIOTA

A. L. Brenkert
Science Applications International Corporation
Oak Ridge, TN, U. S. A.

R. H. Gardner, S. M. Bartell and F. O. Hoffman
Environmental Sciences Division
Oak Ridge National Laboratory
Oak Ridge, TN, U. S. A.

ABSTRACT

A dynamic model of radium transfers between water, sediments and fish-flesh was developed to compare lakes that differ in their biological and physical characteristics. Results indicate that factors associated with differences in biological productivity between lakes can result in significant differences in the predictions and uncertainties of radium in various aquatic components.

INTRODUCTION

Previous analysis [1] of a six component model of ^{226}Ra transfer between water, sediment and gamefish-flesh has shown that predictions of potential radiological exposure to humans are sensitive to changes in the transfer rates that directly link these components. Therefore, the model and methods presented here use detailed information and estimates of uncertainties to: (1) compute these transfer rates based on the biological and physical properties of individual lakes; (2) simulate the dynamics of radium through time; and (3) determine the parameters and processes associated with model uncertainties.

MODEL DESCRIPTION

Estimates of the rates of transfer of radium between water, fish and sediments (Fig. 1) have been developed as a function of the physical and biological properties of individual lakes and empirical measurements of the concentration ratios of radium in aquatic systems. The transfer rate from water to sediments is given by:

$$Wa{>}Sed = (\sigma_s K_{ds} + Kl) / h \, . \tag{1}$$

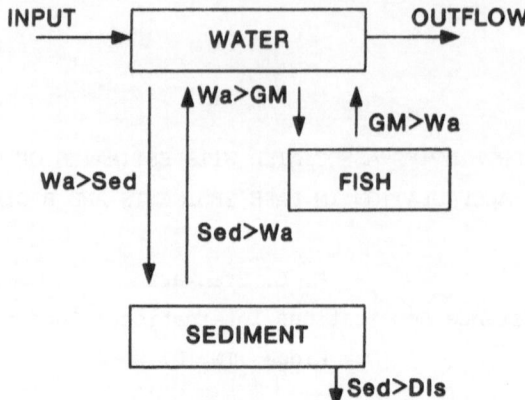

Figure 1. Diagrammatic representation of the ordinary differential equations used to simulate the transfer of radium between water, gamefish-flesh, and sediments (symbols are defined in the text).

σ_s is the flux of sediment removal from the water column (g m^{-2} yr^{-1}), h is the effective mixing height of the water column (m), K_{ds} is the average distribution coefficient (m^3 g^{-1}) of radium in sedimenting material, and Kl is the mass transfer coefficient (m yr^{-1}) for movement of radium from sediments into the water column. K_{ds} is a weighted average of k_o and k_i, the distribution coefficients for organic and inorganic matter, respectively. The weights given to k_o and k_i are based on the fraction, F_{wo}, of sedimenting material which is organic. The value of Kl is based on diffusion related processes and the renewal time of the sediment surface layer. Two key parameters in this process are u, the current velocity (m y^{-1}) and d_f, the diffusion coefficient (m^2 y^{-1}) (see [2] for further details). The transfer rate from sediment to water is:

$$\text{Sed>Wa} = (c \ Kl) \ / \ \{ \ \lambda \ [K_{dd} \ \rho_s \ (1.0 - e) + e] \ \} \ . \qquad (2)$$

c is based on an equilibrium relationship between C_λ, the concentration of radium in the interstitial water (Bq m^{-3}), and C_w, the concentration of radium in the water column (Bq m^{-3}). K_{dd} is the average distribution coefficient of radium for stabilized sediment (m^3 g^{-1}), calculated as a weighted average of k_o and k_i. Thus $K_{dd} = F_{so} \ k_o + (1.0 - F_{so}) \ k_i$, where F_{so} is the fraction of organic matter in the sediments. λ is the assumed thickness of the sediment exchange zone (0.05 m). The volume fraction of liquids in the sediments, e (m^3 / m^3), is estimated from the moisture content M_c of the sediments and ρ_s, the density of the stabilized sediments (g m^{-3}). The rate constant for loss of radium by sediment burial is:

$$\text{Sed>Dls} = (c \ \sigma_d \ K_{dd}) \ / \ \lambda \ [K_{dd} \ \rho_s \ (1.0 - e) + e] \quad , \qquad (3)$$

where σ_d is the flux of sediment burial (or removal from the active exchange layer, g m^{-2} y^{-1}) and is calculated as $\sigma_d = \sigma_i / (1.0 - F_{so})$. Other terms in eqn (3) are as defined in eqs (1) and (2).

The rate constants for uptake (Wa>GM) and loss (GM>Wa) of radium from gamefish-flesh are:

$$Wa{>}GM = CF \ FiBio \ / \ (MLE \ WaVol) \qquad (4)$$

and

$$GM{>}Wa = 1.0 \ / \ MLE \ . \qquad (5)$$

CF is the ratio of concentration in fish-flesh and water at equilibrium. FiBio is the total quantity of gamefish-flesh (kg) in the lake. MLE is the mean life expectancy (d) of the dominant fish species, and WaVol (l) is the volume of water in the lake.

SENSITIVITY AND UNCERTAINTY ANALYSIS

The analysis of the radium model was performed with PRISM [3], a program for Monte Carlo estimation of sensitivity indices and analysis of model uncertainties. PRISM, using latin hypercube sampling, generated random samples from prespecified frequency distributions (Table 1), and performed the model simulations by assuming that the water flowing into each lake (the Input parameter in Table 1) contained 1 Bq of radium per liter. After 500 Monte Carlo samples were collected, results were analyzed to determine the uncertainties of the predictions. The classical sensitivity index [4] analytically examines the model predictions, y, and model parameters, p, to obtain dy/dp, the derivative of y with respect to p. Precise numerical estimates of the sensitivities can also be made using Monte Carlo methods by making the parameter perturbations very small (i.e., 1%) and then [5] regressing y on each of the p's. The uncertainties in the model predictions were calculated in a similar manner, except that parameters were perturbed over the full range of values (Table 1). Relationships between h and WaVol and between σ_{do} and FiBio were expressed in the Monte Carlo simulations as correlations which were lake dependent (correlation coefficients ranged from 0.6 to 0.9).

The Monte Carlo process results in a 24% to 40% coefficient of variation (C.V. = [standard deviation / mean] * 100.0) for Wa>Sed, 48% to 56% for Sed>Wa and 40% - 58% for Sed>Dls. The C.V. for Wa>GM was 160-180% and for GM>Wa was 45 - 52%. In general the rate constants for the Canadian Shield Lake were the least variable, while those for Trobbofjärden were the most variable. The number of shared parameters (see eqns 1-3) results in correlations among these calculated rate constants ranging from

TABLE 1

Distribution Type, Mean Value and Relative Variability
of Lake Specific Model Parameters[1]

Parameter Name	Dist. Type	Canadian Shield[2]		Lake Uri[3]		Trobbofjärden[4]	
		Mean	C.V.	Mean	C.V.	Mean	C.V.
Input	c	4.3E7	...	3.4E9	...	2.3E7	...
Outflow	t	2.7E-3	10%	9.4E-4	7%	2.5E-3	19%
h	t	10.0	10%	30.0	14%	3.1	21%
σ_{do}	t	100.0	16%	70.0	16%	970.0	23%
F_{wo}	t	0.55	24%	0.08	28%	0.399	20%
F_{so}	t	0.23	9%	0.02	46%	0.099	54%
CF	l	16.0	145%	16.0	147%	4.0	122%
FiBio	n	500.0	10%	568.8	10%	933	10%
MLE	t	2920.0	43%	2920.0	43%	1460	47%
WaVol	n	1.6E10	10%	3.6E12	10%	9.3E9	20%

[1] Distribution types are the probability distributions from which random values were selected. The types used are c = constant, n = normal, t = triangular, l = lognormal. C.V. is the coefficient of variation (standard deviation / mean * 100) expressed as a percent. Additional parameters in eqn 1-3 were also varied but their values were not lake specific. Additional information on the computational methods can be obtained from the authors.

[2] An oligotrophic lake with a surface area of 160 ha, volume of 1.6E10 liters, and turnover time of approximately 0.7 y.

[3] A lake in Switzerland with a surface area of 2600 ha, volume of 3.6E12 liters, and turnover time of approximately 2 y.

[4] A eutrophic lake in Sweden with a surface area of 300 ha, volume of 9.3E9 liters, and a turnover time of approximately 0.75 y.

−0.32 (Wa>Sed vs Sed>Wa) to 0.34 (Sed>Wa vs Sed>Dls). The correlations are dependent upon properties of the lake being modelled and the variances of the parameters used in eqn 1-3 to calculate these transfer rates. For instance, when all variances are set to 1% of the nominal value (the sensitivity case), the correlation between Sed>Wa and Sed>Dls can be as large as 0.96.

TIME DEPENDENT MODEL SENSITIVITIES AND UNCERTAINTIES

The variance of the parameters of the lakes are of similar magnitudes

(Table 1), therefore differences in geometric mean values and 95% confidence intervals (Table 2) can be related to the physical and biological differences in the lakes. The high concentration of radium in the water of the Canadian Shield Lake is due to a low rate of sedimentation and a high turnover rate of water in the lake. As a result, the Canadian Shield Lake also has the highest concentration of radium in fish–flesh. The high rate of sediment renewal in Trobbofjärden (1.4 cm y^{-1} compared to 0.70 for Lake Uri and 0.09 for the Canadian Shield Lake) combined with a high transfer rate of radium from the water column to the buried sediments (eqn 2–3) cause this lake to have the lowest concentrations of radium in sediments. The high residence time, high inorganic content and low biological productivity of Lake Uri results in high sediment concentrations and lower uncertainties for this system when compared to the Canadian Shield Lake.

TABLE 2

Predicted Radium Concentrations in Water, Fish–flesh and Sediments[1]

| | Geometric Means (95% Confidence Intervals) | | |
	YEAR .1	YEAR 1.0	YEAR 50.0
CANADIAN SHIELD			
water	0.94 (0.7– 1.2)	0.96 (0.7– 1.3)	0.94 (0.7– 1.2)
fish	15.1 (12.3–18.6)	16.1 (11.1–23.4)	19.8 (3.3–120.2)
sediment	5.8 (4.2– 8.3)	5.8 (4.2– 8.0)	5.3 (2.3– 12.0)
LAKE URI			
water	0.8 (0.6– 0.9)	0.8 (0.6– 1.0)	0.8 (0.6– 1.1)
fish	12.3 (10.2–14.8)	13.1 (9.0–19.1)	16.2 (2.6–100.8)
sediment	13.6 (9.7–19.2)	13.3 (9.3–19.1)	13.2 (6.3– 27.7)
LAKE TROBBOFJÄRDEN			
water	0.2 (0.1– 0.4)	0.2 (0.1– 0.5)	0.2 (0.1– 0.6)
fish	0.9 (0.7– 1.2)	1.2 (0.6– 2.5)	2.6 (0.4– 15.4)
sediment	1.2 (0.9– 1.7)	1.2 (0.9– 1.6)	1.3 (0.6– 2.7)

[1] Statistics for the concentration of radium in water (Bq/l), fish–flesh (Bq/Kg), and sediments (Bq/g dry wt) are based on the geometric means and geometric standard deviations calculated from 500 Monte Carlo iterations. All simulations began with the state values set at equilibrium.

The time–dependent changes in the percent of the variance of model sensitivities and uncertainties that can be explained by individual parameters are illustrated in Figure 2. Initially the biomass of the fish

(FiBio) is the only parameter which affects the predicted concentration of radium in fish-flesh. By the end of the first year of simulation the models are no longer sensitive to this value. Other parameters, related to the physical dimensions of the lake (WaVol, Input, Outflow, h) are more

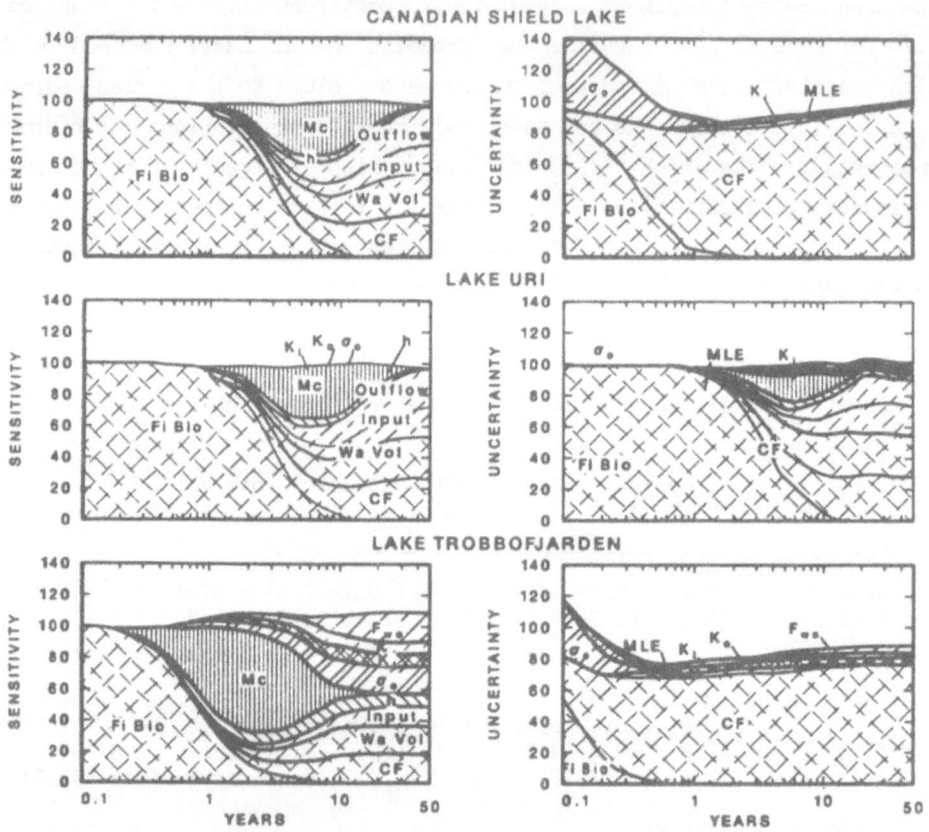

Figure 2. Percent variance that individual parameters explain of model sensitivities and uncertainties of the concentration of radium in fish-flesh through time.

sensitive after the first year of simulation because it takes the system longer to adjust to small changes in these parameters. The high quantities of organic matter in the water column and the rapid burial of radium in the sediments in Trobbofjärden cause parameters associated with these processes to play a larger role in this lake than in Lake Uri or the Canadian Shield Lake.

The time-dependent changes in parameters which explain the largest percent of model uncertainties are different from those which explain the model sensitivities (Fig. 2). In each of the three lakes the radium bioconcentration factor (CF) is the major parameter explaining the predicted uncertainty of radium in fish-flesh: 85% in Trobbofjärden after half a year and 95% in Lake Uri and the Canadian Shield Lake after two years. Because the estimates of CF are very uncertain (Table 1), perturbations of this parameter by the Monte Carlo simulations result in large changes in the predicted concentrations of radium in fish-flesh. Differences in the behavior of the lakes are also evident in the list of parameters which explain the uncertainty of predicted concentration of radium in the water and sediments at the end of the 50 year simulation. Parameters which describe the physical characteristics of the lake (e.g., WaVol, Outflow and h) account for 95%, 50% and 8% of the uncertainty in radium concentration in water of the Canadian Lake, Lake Uri and Lake Trobbofjärden respectively, while parameters associated with the sedimentation process (σ_{do}, f_{wo}, k_i, k_o) account for 8%, 73% and 79%, respectively.

CONCLUSIONS

The comparison of three different lake systems by Monte Carlo methods allows the identification of the physical and biological factors which most affect model sensitivities and uncertainties. One important effect that has been noted is that the physical and biological properties of a lake cause strong dependencies among calculated transfer rates. Simulation of these dependencies by correlations among transfer rates may not be satisfactory because the strength of the relationship can depend on the statistical properties of the underlying parameters (eqn 1-3) used to calculate the transfer rates. Because transfer rates can be calculated as functions of more fundamental processes, it is possible to avoid this dilemma by methods similar to those presented here.

A second observation is that uncertainties associated with the prediction of radium in water, fish-flesh, and sediments are highly dependent on within-lake processes. Thus, the design of parsimonious measurements to improve the predictability of these systems (i.e. reduce uncertainties) must be both time and lake specific. For instance, sedimentation processes in Lake Uri are dominated by inorganic chemistry while those of Trobbofjärden are related to the high organic content of the sediments.   If it is important to improve the predictions of radium concentration in fish-flesh, then species specific information (i.e. CF and MLE) will be needed, especially for short term predictions.

The methods presented here are intended to be illustrative rather than definitive. Although all the variances were not directly estimated from data, the general trends of model predictions indicate the importance of the physical and biological differences between lakes.   The disaggregation of the transfer rates into more detailed processes allows this simple model to be easily applied to a diversity of lake systems. Results indicate that similar releases of radioisotopes to different aquatic systems can result in large differences in potential radiological exposure to humans.

ACKNOWLEDGEMENTS

This research was sponsored by the Office of Health and Environmental Research, U. S. Department of Energy, under Contract No. DE-AC05-84OR21400 with Martin Marietta Energy Systems, Inc., supported by NSF grant no. BSR 8614981.

REFERENCES

[1] Gardner, R. H., B. G. Blaylock and F. O. Hoffman. 1985. A dynamic model of radium-226 behavior in an oligotrophic lake. IN: "Probabilistic Model Development for the Assessment of the Long-term Effects of Uranium Mill Tailings in Canada - Phase III."   Research report prepared by SENES Consultants Limited for the National Uranium Tailings Program, CANMET, EMR, Ontario.

[2] Scharer, J. M. and B. E. Halbert. 1985. Modelling Radium Exchange Between the Water Column and Lake Sediment. IN: "Probabilistic Model Development for the Assessment of the Long-term Effects of Uranium Mill Tailings in Canada - Phase III." Research report prepared by SENES Consultants Limited for the National Uranium Tailings Program, CANMET, EMR, Ontario.

[3] Gardner, R. H., B. Röjder, U. Bergström. 1983. PRISM: A systematic method for determining the effect of parameter uncertainties on model predictions. Studsvik Energiteknik AB report/NW-83/555, Nyköping Sweden.

[4] Tomovic, R. 1963. Sensitivity Analysis of Dynamic Systems. McGraw-Hill, New York. 142 pp.

[5] Gardner, R. H., R. V. O'Neill, J. B. Mankin and J. H. Carney.   1981. A comparison of sensitivity analysis and error analysis based on a stream ecosystem model. Ecological Modelling, 12: 177-194.

FINITE ELEMENT MODELLING OF TRANSPORT OF RADIUM-226 AND URANIUM FROM PORT GRANBY RADIOACTIVE WASTE MANAGEMENT SITE TO LAKE ONTARIO

S.R. Joshi and A.G. Bobba
National Water Research Institute
Canada Centre for Inland Waters
Environment Canada
P.O. Box 5050
Burlington, Ontario   L7R 4A6
Canada

ABSTRACT

The development and verification of a finite element model for delineating radioactive contaminant plumes via groundwater are described. The model is then used to predict the migration of ^{226}Ra and U from the near-shore Port Granby waste management site to Lake Ontario. The model predicts that the continuous migration, via groundwater, of both contaminants toward Lake Ontario is likely to persist even after the waste is removed from the site.

INTRODUCTION

During 1955-77 residual radioactive wastes from the uranium refinery at Port Hope, Ontario were disposed of by Eldorado Resources Limited at its Port Granby waste management site, located about 13 km west of Port Hope. Subsequent surface lake water [1] and site groundwater [2] measurements indicated that the leachate plume moves parallel to Lake Ontario shoreline in the direction of the prevailing wind. The present communication introduces a study of groundwater measurements begun in 1981 and compares the field data with that predicted by a finite element model.

The site (Fig. 1) covers a 10-hectare area sloping down from the north toward Lake Ontario. The land surface along the shore rises abruptly as bluffs above the lake level. The dominant topographic features on the site are two deep valleys which are formed due to erosion from surface runoff and groundwater seepage along the bluffs. These valleys are shown as East Pond and West Pond in Fig. 2 which also depicts elevations (in metres) and borehole locations. The geology of the site is assumed to be very similar to the Bowmanville-Newcastle area [3]. Along most of the shore line, beach deposits of gravel and sand occur between the base of bluffs and the lake.

The groundwater flow at the site is generally influenced by the geology and topography of the site. Two permeable aquifers exist in the overburden deposits, separated by relatively semipermeable deposits that contain sand and gravel layers with microfractures. The hydraulic gradients are predominantly downward and springs are formed at the contact of the bluffs. The hydrogeological parameters used in the simulation of numerical model to delineate the groundwater flow have been given earlier [2].

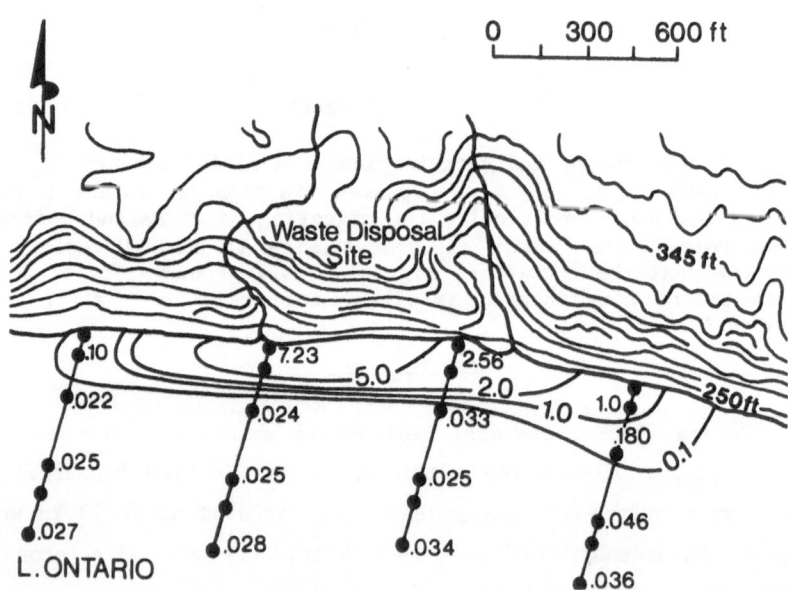

Figure 1. General location of the Port Granby radioactive waste management site. The numbers refer to the observed [1] migration of ^{226}Ra (pCi/L) in Lake Ontario surface waters.

MATERIALS AND METHODS

Field work and radioanalysis

A number of piezometers, or standpipes, were installed in a row along the shore beginning in 1981. In addition, several piezometers, installed

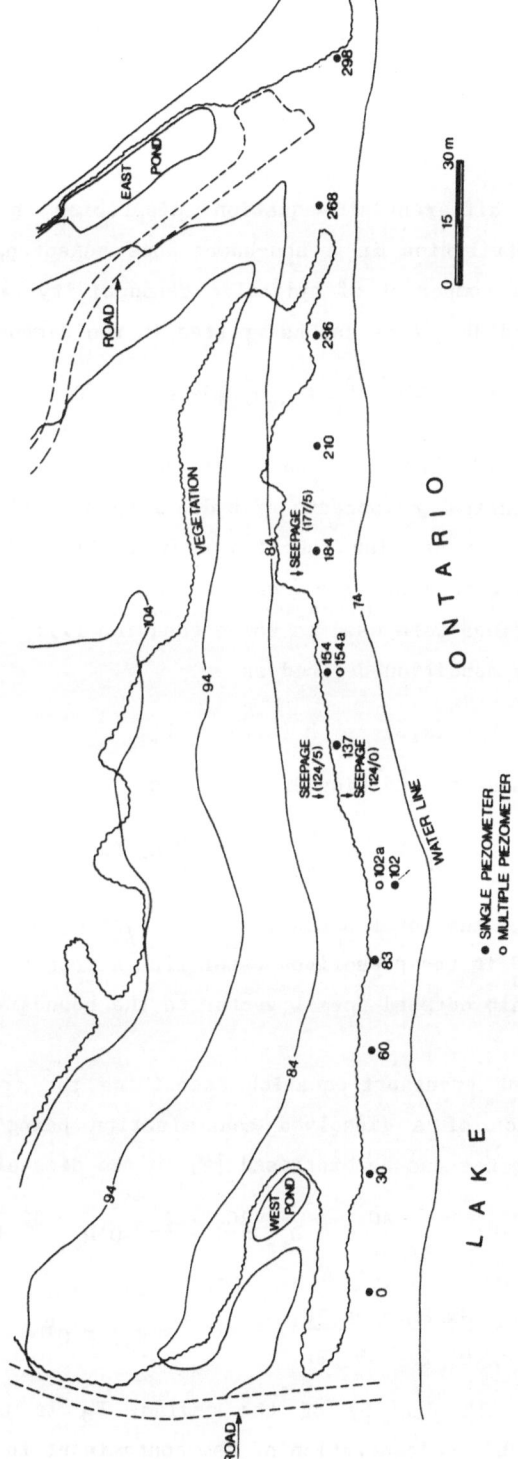

Figure 2. Detailed borehole locations and corresponding elevations at the study site.

earlier by Eldorado Resources Limited, were also available for sampling. Water samples were assayed for total uranium and [226]Ra as described earlier [1,2].

Mathematical model

The general differential equation describing the steady-state hydraulic head distribution in a non-homogenous anisotropic flow regime where the principal component of hydraulic conductivity tensor is coincident with the co-ordinate axes can be written in two dimensions [4] as

$$\frac{\partial}{\partial x} [K_{xx} \frac{\partial h}{\partial x}] + \frac{\partial}{\partial z} [K_{zz} \frac{\partial h}{\partial z}] \pm Q^h = 0 \tag{1}$$

where h is the hydraulic head, x and z are the cartesian directions, K_{xx} and K_{zz} are the principal components of hydraulic conductivity tensor, and Q^h is a liquid source of sink function. It is assumed that K_{xx}, K_{zz} and Q^h can vary spatially but are constant within each element. Following boundary conditions were used to solve equation (1): the Dirichlet or fixed head boundary condition defined as

$$h(x,z) = \tilde{h}(x,z) \text{ or } \Gamma_1, \tag{2a}$$

and the Neumann boundary condition expressed as

$$[K_{xx} \frac{\partial h}{\partial x} + K_{zz} \frac{\partial h}{\partial x}] \bar{n} = q_n^h \text{ or } \Gamma_2 \tag{2b}$$

where $\Gamma_1 + \Gamma_2 = \Gamma$, the total boundary of the system, $\tilde{h}$ is the prescribed hydraulic head, q_n^h is the prescribed water flux across the Neumann boundary, and $\bar{n}$ = the unit outward normal vector to the boundary.

The contaminant transport-equation describing the transient concentration distribution of a dissolved contamination being transported in groundwater flow system can be expressed [4] in two dimensions as:

$$(\theta + \rho_b K_d) \frac{\partial C}{\partial t} = \frac{\partial}{\partial x} (\theta D_{xx} \frac{\partial C}{\partial x} + \theta D_{xz} \frac{\partial C}{\partial z}) + \frac{\partial}{\partial z} (\theta D_{zx} \frac{\partial C}{\partial x} + \theta D_{zz} \frac{\partial C}{\partial x}) -$$

$$\frac{q_x}{\theta} \frac{\partial C}{\partial x} - \frac{q_z}{\theta} \frac{\partial C}{\partial z} - \lambda (\theta + \rho_b K_d) \mp Q^c \tag{3}$$

where ρ_b is the bulk density of the medium, K_d is the distribution coefficient, C is the concentration of the contaminant in the water, D_{xx}, D_{xz}, D_{zx} and D_{zz} are the components of the hydrodynamic dispersion tensore

λ is the first order reaction constant (or radioactive decay constant) and Q^c the mass rate per unit volume for addition or removal at a source or sink. It is assumed that λ is constant and ρ_b, K_d and Q^c can vary spatially. The following conditions are used to solve the transport equation (3). The initial condition necessary for the solution of equation (3) is

$$C(x,z,o) = \tilde{C}_0(x,z), \qquad (4)$$

where C_0 is the initial prescribed concentration. The Dirichlet boundary condition expressed as,

$$C(x,z,t) = \tilde{C}(x,z,t) \text{ or } \Gamma_3 \qquad (5a)$$

and the Neumann boundary condition expressed as,

$$\left[(\Theta D_{xx} + \Theta D_{zx})\frac{\partial C}{\partial x} + (\Theta D_{xz} + \Theta D_{zz})\frac{\partial C}{\partial z}\right] n = q_n^c \text{ or } \Gamma_4, \qquad (5b)$$

where C is the prescribed concentration, $\Gamma_3 + \Gamma_4 = \Gamma$, and q_n^c, the prescribed dispersive solute flux across the Neumann boundary, are the other conditions.

Numerical solution

To solve the model equations, the Galerkin finite element-method is used to determine approximate solutions to equations (1) and (3) under the prescribed initial and boundary conditions. The cross-sectional representation of the flow system is divided into an equivalent system of subregions, which in this case are triangular elements. The solutions are obtained by using the Gaussian elimination methods and are compared with analytical solutions [5].

RESULTS AND DISCUSSION

The model runs were made simulating either one or two stratigraphic layers and simulating varying degrees of anisotropy in each layer. The difference between simulated hydraulic heads and those observed was least when the model included two layers, the upper layer being less permeable than the lower layer and both layers being anisotropic with a 100:1 ratio of horizontal to vertical hydraulic conductivity. Figure 3 gives the simulated hydraulic head distribution of the north-south geological cross section. Both the simulated and the field data showed that hydraulic gradients were predominantly downward. Two more geological cross-sections were similarly used to simulate the groundwater flow system at different stations. The Darcy velocities were computed from the simulated hydraulic heads and then used in the contaminant transport model to predict the ^{226}Ra and uranium concentrations at these locations before reaching Lake Ontario. These simulations show that after 15 years, the contaminants will reach the Lower Sandy aquifer and move towards Lake Ontario under the influence of regional groundwater flow system. The model-predicted ^{226}Ra and uranium concentrations are found to be in good agreement (Table 1) with those actually observed.

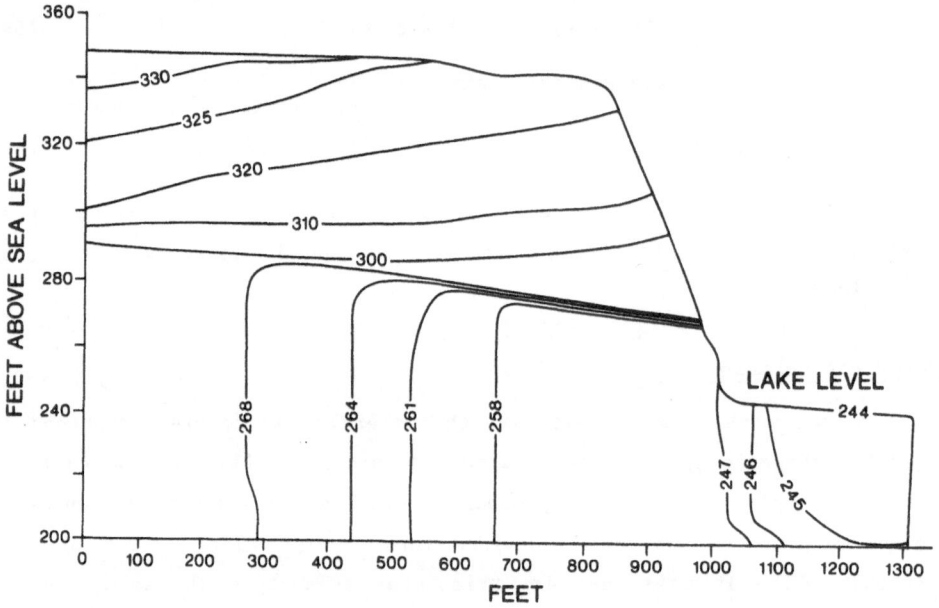

Figure 3. Computed hydraulic head distribution of the north-south geological cross-section.

Assuming that the applicable site parameters remain unchanged, the concentrations of the two contaminants were computed for the next 50 years or so. The observed patterns (Fig. 4) suggest consistent migration of ^{226}Ra and uranium to Lake Ontario over the projected time period. The previous migration of the predicted plume on Lake Ontario waters is discernible from the 1977 surface water ^{226}Ra measurements [1] where the plume is observed to traverse along the coastal zone (Fig. 1). Thus it appears reasonable to infer that, barring unforeseeable circumstances, both these long-lived radionuclides will continue their slow migration toward Lake Ontario in small but measureable amounts.

TABLE 1
Computed and observed levels of ^{226}Ra and uranium
in the Port Granby groundwater

	Observed	Computed	Borehole
^{226}Ra (pCi/L)	228.00	203.09	206
	3.00	2.81	206
	13.48	12.06	154
	1.15	1.78	154 (Multilevel piezometers)
	0.48	0.83	154 (Multilevel piezometers)
	9.63	9.78	Top Seepage
	7.95	7.82	184
Uranium (µg/L)	25.05	19.20	Top Seepage
	13.00	15.50	Shore piezometers
	6.19	5.20	154 (Multilevel piezometers)
	2.15	4.50	154 (Multilevel piezometers)

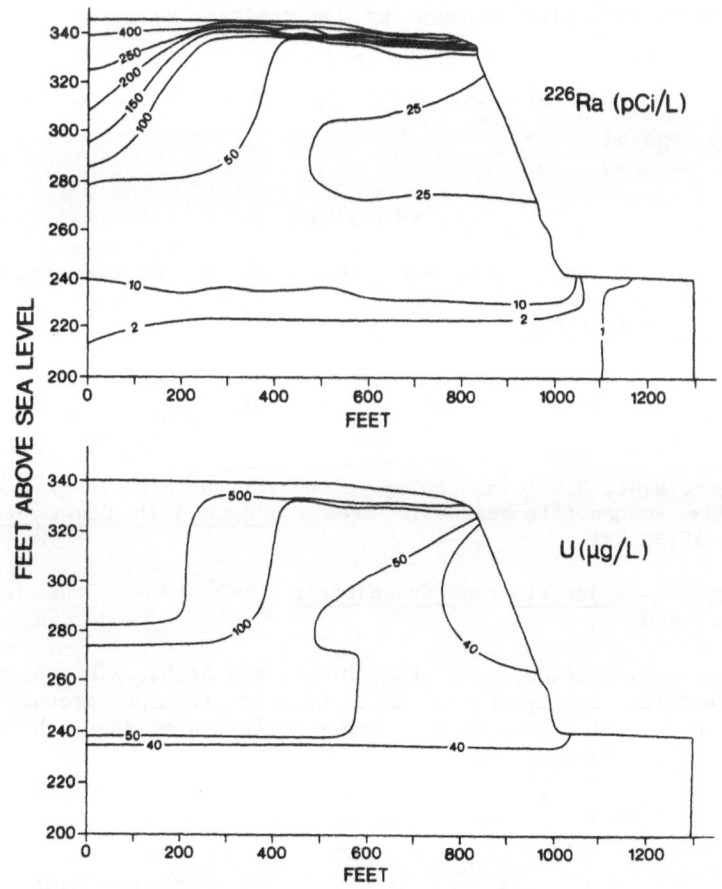

Figure 4. Model-predicted concentrations of ^{226}Ra and
uranium at 50 years from present.

Relocation of the waste from the Port Granby site is one of the remedial actions being considered by Eldorado Resources Limited. To assess the effectiveness of this remedial measure, the model was used to predict the recovery of ^{226}Ra concentrations at the site and in beach deposits. The results show that contaminant buildup will continue for 165 years before receding at the beach locations.

CONCLUSION

In conclusion, the results from the present investigation clearly demonstrate the migration of ^{226}Ra and uranium from the Port Granby waste management site via the groundwater. The computations further show that this migration is a slow process and the radioactive contaminants already present in groundwater will slowly keep on moving toward Lake Ontario for many years even if all the waste currently stored at the site is completely removed.

REFERENCES

1.   Durham, R.W. and Joshi, S.R., Investigation of Lake Ontario water quality near Port Granby radioactive waste management site. Water, Air, and Soil Pollut. 1980, 13, 17-26.

2.   Platford, R.F., Bobba, A.G. and Joshi, S.R., The Port Granby radioactive waste management site. Water Poll. Res. J. Canada 1908, 19, 90-96.

3.   Singer, S.N., Surficial geology along the north shore of Lake Ontario in the Bowmanville-Newcastle area. Proc. 16th Conf. Great Lakes Res. 1973, 441.

4.   Bear, J., Hydraulics of Groundwater, McGraw-Hill, New York, 1979, pp. 60-300.

5.   Bobba, A.G., Durham, R.W., Lam, D.C.L. and Joshi, S.R., Simulation of contaminant transport to Lake Ontario through groundwater flow systems: Port Granby study. Paper presented at the 26th Conf. Great Lakes Res., Oswego, 1983.

BIOSPHERE MODEL VALIDATION BY INTERCOMPARISON TO OBSERVED BEHAVIOUR OF FALLOUT RADIONUCLIDES IN THE AQUATIC ENVIRONMENT

Riitta Korhonen and Seppo Vuori
Technical Research Centre of Finland
Nuclear Engineering Laboratory
P.O.Box 169, SF-00181 Helsinki
Finland

ABSTRACT

To improve the reliability of environmental transport models an extensive validation effort is necessary. The deposition brought about by the Chernobyl accident gave new insights and possibilities for model validation. A dynamic compartment model DETRA is employed in this study for the consideration of radionuclide transport in a large watercourse taking into account both direct deposition and runoff from the terrestric environment. The results are shortly compared with the experimental observations, but a comprehensive quantitative analysis can be performed only after the careful evaluation of extensive raw data has been completed.

INTRODUCTION

For preplanning of countermeasures it is important to have reliable computational capabilities to predict the long-term radiation exposures to the population. An important part of forecasting process is the analysis of radionuclide transport in the biosphere. A typical feature in the Finnish environment is the abundance (about 10 % of the surface area) of relatively shallow lakes. Consequently the transport of radionuclides in the inland watercourses can be an important pathway as regards the radiation exposures of the population. Especially in situations like the Chernobyl fallout on vast areas before the actual growing period, the aquatic exposure pathway gets increased importance.

In this study a conceptual model for the Kymijoki river basin has been developed and the transfer of radionuclides deposited after the Chernobyl accident has been analyzed with a dynamic compartment model DETRA. This geographical area was chosen due to several reasons: average deposition was among the highest in Finland, extensive observations were made in this area, and the largest lake in the watercourse is utilized as a supply for drinking water to the capital area.

As regards the usefulness in view of model validation, the compiled surveillance data after the Chernobyl accident has certain benefits as the source-term had only a restricted temporal duration and consequently the dynamic behaviour of radionuclides in the environment can more easily be followed.

DESCRIPTION OF MODELS APPLIED IN THE DEMONSTRATIVE CASE STUDY

The computer model DETRA [1, 2] applied in this study comprises two modules: the dynamic analysis of environmental transport of pollutants and the analysis of radiation exposures with the employment of the concentration factor method.

The conceptual model for the analyzed region (Kymijoki river basin) has been developed in two parts. The first part describes the runoff of radionuclides deposited on typical rural areas. The pertinent compartment model depicted in Table 1 includes the most important soil layers and their interactions. Deposited activity is assumed to be transferred both along with the water flow (surface/soil groundwater) and with the solid material erosive flow. Table 1 describes the key parameters of the runoff model.

TABLF 1
Parameters of the runoff scenario

Parameter	Dimension		Scenario
x_r (rainfall)	m/a	*	0.75
x_e (evaporation)	m/a	*	0.40
x_{12}	"	*	0.50
x_{21}	"	*	0.40
x_{23}	"	*	0.50
x_{32}	"	*	0.40
x_{34}	"	*	0.50
x_{43}	"	*	0.40
x_{45}	"	*	0.10
x_{16}	"	*	0.25
x_{16}^s (solids)	kg/m^2/a		0.08+
K_d Sr	ℓ/kg		20
Cs	"		3000

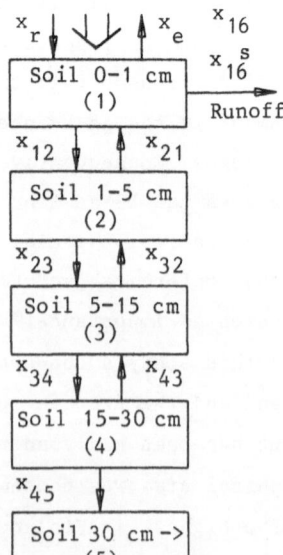

* Water volume equivalent to rainfall (m^3/m^2/a)
+ Based on material balance with the sedimentation rate

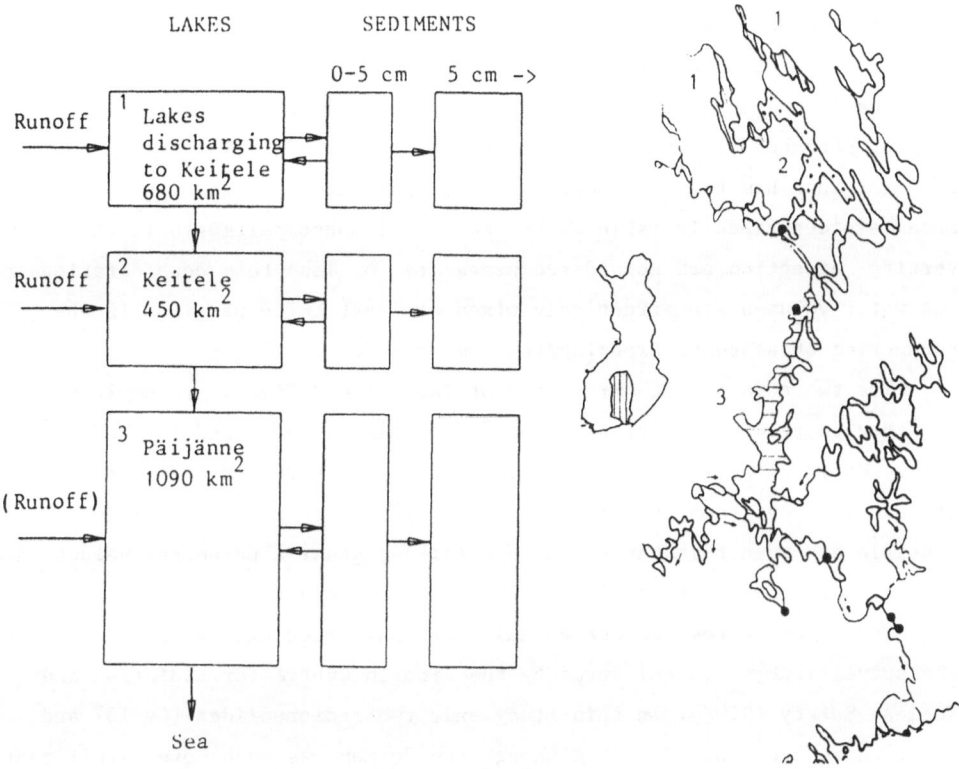

Figure 1. The compartment model for the Keitele drainage scenario on Kymijoki watercourse

TABLE 2
Parameters of the Keitele drainage scenario

	Lakes discharging to Keitele	Lake Keitele	Lake Päijänne
Drainage area (km^2)	3800	1350	1000
Lake area (km^2)	680	450	1090
Water volume (m^3)	$4,1\cdot 10^9$	$2,7\cdot 10^9$	$1,8\cdot 10^{10}$
Water flow from lake (m$^3\cdot$a^{-1})	$1,1\cdot 10^9$	$1,5\cdot 10^9$	$7,1\cdot 10^9$
Amount of solids in water (kg$\cdot$m^{-3})	$5,0\cdot 10^{-3}$	$5,0\cdot 10^{-3}$	$2,0\cdot 10^{-3}$
Sedimentation rate (kg m^2 a^{-1})	0,5	0,25	0,1
Deposition of Cs-137 (TBq)	130	36	(84)

The second part of the conceptual model shown in Fig. 1 describes in a simplified way the further transport of the runoff and initial deposition on lakes in the upper course of the Kymijoki river basin. As illustrated in the map attached to Fig. 1 only one of the main routes discharging to the Lake Päijänne has been included. The main parameters of this part of the model are described in Table 2. No additional subcompartments in the vertical direction are considered necessary for long-term considerations as the water volumes are effectively mixed at least twice per year in the connection of seasonal freezing/thawing process.

The two compartment structures of Table 1 and Fig. 1 are employed together to predict the transfer of radionuclides contained in the drainage in the Kymijoki watercourse. The transport characteristics assumed in the calculation have been chosen from the literature as region-specific as possible and when this has not been possible, generic parameter values have had to be chosen corresponding to Finnish mean values.

The source terms for the calculations have been chosen on the basis of the surveillance data collected by the Finnish Centre for Radiation and Nuclear Safety (STUK). In this study only two radionuclides (Cs-137 and Sr-90) have been considered. Although the former was much more significant in the case of Chernobyl, it was decided to perform parallel calculations also for Sr-90, because its transport characteristics differ markedly from those of Cs-137 and thereby a broader spectrum of transport characteristics of radionuclides can be covered. In the Kymijoki region included in our analysis the total amount of Cs-137 deposition on land and lake surfaces was 380 TBq and 120 TBq respectively [3]. The corresponding deposition on the Keitele drainage was 140 TBq and 26 TBq respectively. Based on less comprehensive surveillance data the Sr-90 deposition was assumed in our calculations to be one percent of the corresponding value for Cs-137.

RESULTS

In the present calculations the spatial discretation has been quite coarse both as regards the fallout pattern and the description of lake reservoirs. The primary aim of this study has been to derive general insights of the possibilities and reliability of predicting radionuclide behaviour during aquatic transfer in large watercourses. In the following some comparisons of the predicted and observed behaviour of Cs-137 and Sr-90 will be made.

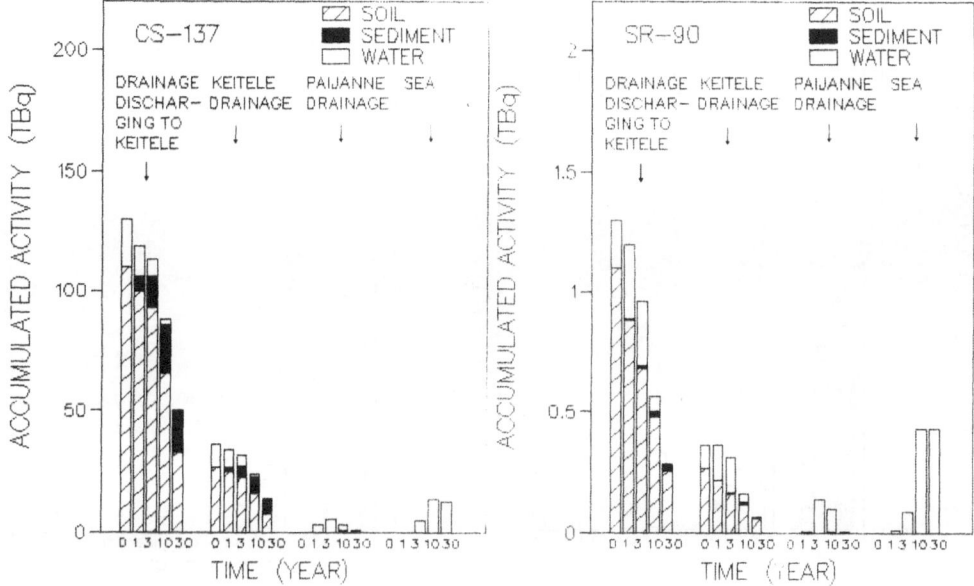

Figure 2. The calculated transfer of the Chernobyl fallout deposited on the drainage discharging to Keitele and Keitele drainage (deposition on soil and on water). Time points 0, 1, 3, 10 and 30 years after the deposition.

In this study we have attempted to show the results in such a way that the temporal and regional distribution of radionuclides as well as their partition between soil, water and sediment can be clearly illustrated. In Fig. 2 it has been assumed that deposition has occured only on Lake Keitele and its upstream drainage area (both on land and water surfaces). Excluding radioactive decay, only slightly more than 20 per cent of Cs-137 has been removed during 30 years from the whole Keitele drainage area by transport processes further into the watercourse. Also about 20 per cent has been transported to the sediments of this drainage after 30 years. In the case of Sr-90 a much larger fraction (55 per cent) of activity is transplaced from the initial deposition area. Only a very small fraction of the removed activity has been distributed on the Päijänne, almost all has been further transported to the sea. In the case of caesium, soil and sediment of the whole Keitele drainage area are important sinks for radionuclides, whereas for strontium the sediments are not important. Together with the easier mobility by runoff, the transfer of strontium to sea is consequently much faster than for caesium.

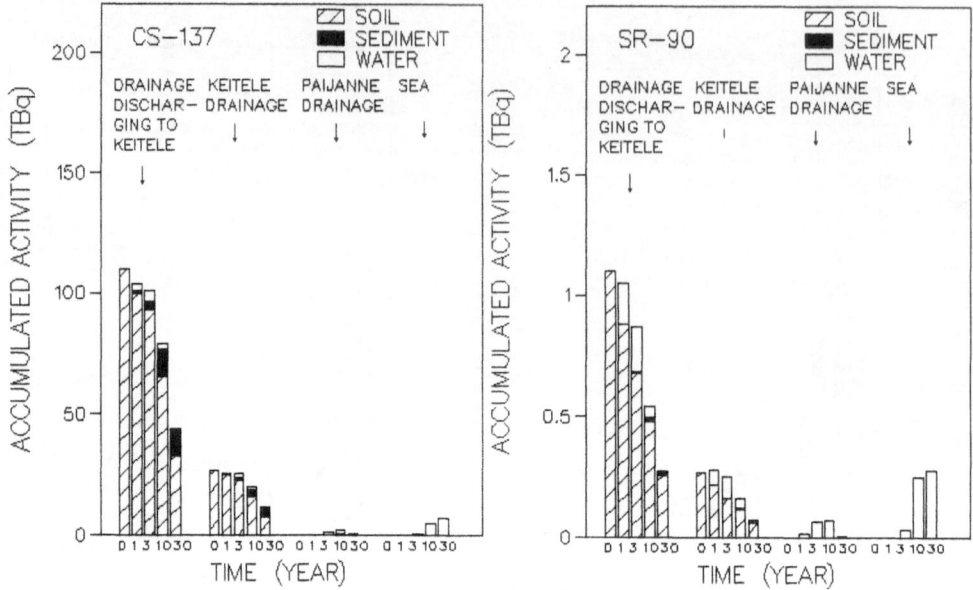

Figure 3. The calculated transfer of the Chernobyl fallout deposited on the
drainage discharging to Keitele and Keitele drainage (deposition
on soil). Time points 0, 1, 3, 10 and 30 years after the
deposition.

In Fig. 3 the transfer of deposition occurring only on land surfaces
is considered separately. The transport by runoff is according to
calculations so effective that (by comparison of Figs 2 and 3) one can find
out that even after 10 years, when the initial deposition on Lake Keitele
has already been transferred to sea through the watercourse, the runoff
from soil causes an increase to the accumulation of the activity inventory
in the sea compartment.

The modelling of the whole drainage discharging to Päijänne has been
less detailed than the modelling of the Keitele drainage. The other
watercourses discharging to Päijänne have been assumed to be similar to
Keitele watercourse.

Limited comparisons of predicted and observed concentrations of the
considered radionuclides have been made. Initially the calculated and
measured concentrations do not differ much from each other. In the Lake
Päijänne the calculated and measured concentrations at the end of May 1986
have been 3300 Bq/m^3 and 2700 Bq/m^3 for Cs-137 and 33 Bq/m^3 and 50 Bq/m^3
for Sr-90. When considering the concentrations in the Lake Päijänne the

model predictions show that the concentrations of both Cs and Sr remain rather constant during the first years, but the observations indicate that Cs-concentration is rapidly decreasing [4]. However, before drawing any conclusions one should study the experimental observations as a whole taking the different accumulation possibilities into account. It seems that the behaviour of Cs-137 in the Chernobyl fallout has according to measurements been very different in the aquatic environment from that of the old fallout. This conclusion should be confirmed with careful evaluation on the basis of all available information.

The predicted long-term behaviour of caesium and strontium can be compared to the observed behaviour of the same radionuclides of the weapon testing fallout. The surveillance of the transfer to sea has been rather extensive, but only isolated measurements about the removal of Sr-90 and Cs-137 from soil have been available [5]. According to the measurements [5] only 5 % of caesium has been removed from the soil up to 1981. The fraction of Cs-137 transferred to sea of the total deposition on the Kymijoki drainage up to 1981 has been 8 %. The corresponding figures for strontium are 37 % and 52 % [5]. According to calculations the transfer of Cs-137 has been more significant and in the case of Sr-90 slightly less significant than the observed transfer. However, the model predictions and observations describe consistantly the different behaviour of these radionuclides. Observations from another Finnish main watercourse (Kokemäenjoki) show quite similar behaviour and one can conclude that for both watercourses - besides differences in run-off rates - the sedimentation processes have an important role in the transplacement of deposited Cs-137 downstream the watercourses.

CONCLUSIONS AND OBJECTIVES OF FURTHER STUDIES

In this paper our preliminary validation efforts as regards radionuclide transfer in a large watercourse have been described. As there rarely exists a really good independent data set for strict validation purposes, we have chosen a gradual iterative approach, where different sets of surveillance data are used. The presently employed parameter values in our computer model are mostly based on generic information from literature and consequently direct measurement results corresponding Finnish site-specific conditions are needed in the model validation. So far only a restricted fraction of the Chernobyl surveillance data has been evaluated and therefore a more comprehensive and quantitative validation process should be iteratively continued when more evaluated data become available.

Independent information from the weapon testing fallout surveillance programme and also measurements of natural radionuclide concentrations in various parts of the biosphere can be employed as a complementary reference data - especially as regards long-term or steady state conditions. To be able to more completely employ the measurements of the weapon testing fallout one has first to calculate the behaviour of this fallout and then to analyze it to find suitable data about the relative behaviour of some typical unit deposition on different drainage areas.

In future studies we aim to present separately the removal of activity from the parts of the biosphere accessible to man taking into account the terrestrial and aquatic plants and animals as well.

Modelling provides an efficient tool to present the experiences from previous fallout episodes in a quantitative form that enables the prediction of the long-term behaviour of radionuclides in the environment and the resulting radiation exposures to the population. Consequently a continuous validation process, preferably in broad international cooperation, is necessary to improve the reliability of environmental transport models. Furthermore, a close interaction with the corresponding modules in probabilistic consequence assessment models have to be maintained.

REFERENCES

1. Korhonen, R., Savolainen, I., Assessment of biospheric behaviour of releases from nuclear waste repositories. IAEA and CEC Seminar on the Environmental Transfer to Man of Radionuclides Released from Nuclear Installations, Brussels, Belgium, 17-21 Oct. 1983.

2. Korhonen, R., Savolainen, I., Assessment of radiation doses due to releases to the biosphere from nuclear waste repositories. 6th International Congress of the International Radiation Protection Association, Berlin (West), May 7-12, 1984.

3. The radiological impact of the Chernobyl accident in the OECD countries. NEA, Paris, September 1987.

4. Saxen, R., Aaltonen, H., Radioactivity of surface matter in Finland after the Chernobyl accident in 1986. Supplement 5 to Annual Report STUK-A55, Finnish Centre for Radiation and Nuclear Safety, Helsinki, June 1987.

5. Salo, A., Saxen, R., Puhakainen, M., Transport of airborne Sr-90 and Cs-137 deposited in the basins of the five largest rivers in Finland. Aqua Fennica, 1984, 1, 21-31.

DOSE ASSESSMENT AND UNCERTAINTY WITH RESPECT
TO LIQUID EFFLUENT DISCHARGES.

Th. Zeevaert, C.M. Vandecasteele
G. Volckaert, R. Kirchmann
S.C.K./C.E.N.
MOL - BELGIUM

ABSTRACT

This paper is dealing with the uncertainty of dose assessments, due to the variability of transfer parameter values. Committed effective dose equivalents from the discharge of the liquid effluents of a nuclear waste treatment facility into a river are being considered. The radionuclides in those effluents, contributing most to the doses, include Cs-137, Co-60, Sr-90 and I-131. The value ranges of the transfer parameters, to which the dose assessment is very sensitive, have been derived from the literature.

The median values and the uncertainty ranges (95 % confidence interval) of the doses are calculated, following a Latin Hypercube sampling procedure of the parameter values. The parameters, which contribute most to the dose uncertainty are identified and the influence of the introduction of site-specific values of these parameters on the uncertainty ranges is shown.

INTRODUCTION

Although much effort has been spent on the experimental determination of transfer parameter values for radionuclides released in the environment (i.e. post-Chernobyl measurements), much uncertainty about these values is persisting. The inaccuracy or irreproducibility of sampling and measuring procedures may be contributing to this uncertainty, however the variability of the environmental conditions constitutes the major source of uncertainty.

The aim of this paper is to demonstrate the influence of the variability of these transfer parameters on the uncertainty of dose assessments for the case of the routine discharges of liquid effluents from the S.C.K./C.E.N. into a small river. The relative contributions of the most important radionuclides to the total βγ-radioactivity released amount to 68 % for Co-60, 20 % for Cs-137, 5 % for Sr-90 and 7 % for I-131. Committed effective dose equivalents to a critical individual over the 30th year of routine releases are calculated, assuming a resulting concentration of 1 MBq.l^{-1} in the raw riverwater for the most important βγ-radionuclides at the composition indicated above.

As a first approximation, only the most conventional and therefore the best documented exposure pathways are taken into account. The resultant important sources for the internal exposure of man are :
- Drinking water (after filtration of the raw riverwater);
- Fresh water fish;
- Food crops (through spray-type irrigation);
- Milk and meat (through spray-type irrigation of pasture and watering of the cattle);

The resultant important sources for the external exposure of man are :
- Sediment on the river banks;
- Surfaces of irrigated fields.

METHODS OF CALCULATION

The dose assessment methodology consists of a set of algebraic equations, in which three categories of parameters or factors may be distinguished :
- Transfer parameters, characterizing the transfer of the radionuclides in the biosphere;
- Parameters concerning the dietary and residential habits of man, affecting his exposure;
- Dose factors.

The uncertainty analysis, carried out is only a preliminary one, since only the values of the transfer parameters, listed in table 1, are being varied. Measured, site-specific values of these parameters are indicated in table 2, when available.

The variability of the transfer parameter values is taken from the literature and indicated in table 3. The main sources were [1], [2], [3], [4], [5] and [6].

For the external exposure from irrigated surfaces, assessed according to [7], no variability ranges for the transfer factors and coefficients could be given. This was also the case for the translation velocity of the mobile bed sediment layer, a parameter in the Schaeffer model [8], according to which the external exposure from the sediment on the river banks has been calculated. This velocity was determined experimentally and amounts to approximately 1000 m.y^{-1}.

The ingestion rates and durations of exposure of man are taken from [9] and [10]. The dose factors applied are derived from [11] (ingestion), [12] (external exposure from soil) and [13] (external exposure from sediment).

The uncertainty ranges of the dose values are calculated, applying a modified Monte Carlo sampling technique, called the Latin Hypercube method. According to this technique, values of the parameters concerned are sampled in a random way, from different intervals with equal probability of occurence. In this way 500 runs of the dose assessment programme were performed with 500 different values of the transfer parameters concerned. From the resulting distribution of the dose values, the 95 % confidence interval and the median are derived and indicated in table 4.

The calculations have been performed without and with the introduction of the site-specific values available.

Also the results of a deterministic dose calculation are shown in table 4. The calculations were performed, introducing best-estimate values for the transfer parameters for which site-specific values were not available.

Sensitivity analyses have been carried out, applying the Smirnov test on the dose value distributions obtained from the uncertainty analysis. For each transfer parameter concerned, two sets of dose values were compared ; one set was associated with the lower parameter values, the other set with the higher parameter values. According to the Smirnov test the difference between the distributions of the two sets of dose values is compared with a critical value in order to determine whether the distributions are different and consequently whether the dose assessment is sensitive to the parameter considered.

TABLE 1

Variable transfer parameters

K_d	: Distribution Coefficient			$(1.kg^{-1})$
K_s	: Sedimentation Coefficient			(m^{-1})
CF_{fish}	: Concentration Factor-Freshwater fish			$(1.kg^{-1})$
B	: Root Uptake Factor :	B_{gr}	: Grains	$(kgDW.kg^{-1}FW)$
		B_{gv}	: Gr. Veget.	(")
		B_{rc}	: Root Crops	(")
		B_p	: Pasture	$(kg.DW.kg^{-1}.DW)$
λ_s	: Root zone Removal Rate :	$\lambda_{s,v}$	: Food Crops	(y^{-1})
		$\lambda_{s,p}$	: Pasture	(y^{-1})
R/Y	: Mass Interception Factor :	R/Y_{gr}	: Grains	$(m^2.kg^{-1}FW)$
		R/Y_{gv}	: Gr. Veget.	$(m^2.kg^{-1}FW)$
		R/Y_{rc}	: Root Cr.	$(m^2.kg^{-1}FW)$
		R/Y_p	: Pasture	$(m^2.kg^{-1}DW)$ λ_w
	: Wheathering Removal Rate	$\lambda_{w,v}$	: Food Crops	(y^{-1})
		$\lambda_{w,p}$	: Pasture	(y^{-1})
t_e	: Crop Exposure Period	$t_{e,v}$	: Food Crops	(y)
		$t_{e,p}$	: Pasture	(y)
F_m	: Milk Transfer Factor			$(d.1^{-1})$
F_f	: Meat Transfer Factor			$(d.kg^{-1})$
P	: Soil Density	P_v	: Food Crops	$(kgDW.m^{-2})$
		P_p	: Pasture	$(kgDW.m^{-2})$
Q_p	: Daily Grass Intake-Cow			$(kgDW.d^{-1})$
Q_w	: Daily Water Intake-Cow			$(1.d^{-1})$

TABLE 2

Site-specific Values

	Co-60	Cs-137	Sr-90	I-131
K_d (l.kg^{-1})	$2 \ 10^4$	$4 \ 10^4$	$9 \ 10^2$	
K_s (m^{-1})	$2 \ 10^{-5}$	$5 \ 10^{-5}$		
CF_{fish} (l.kg^{-1})	290	580	58	15
B_{gr} (kgDW.kg^{-1}FW)	$1.5 \ 10^{-2}$	$8.5 \ 10^{-2}$		
B_{gv} (")	$7 \ 10^{-3}$	$2.1 \ 10^{-2}$		
B_{rc} (")	$1.8 \ 10^{-2}$	$1.2 \ 10^{-1}$		
B_p (kgDW.kg^{-1}DW)		1.0		
R/Y_p (m².kg^{-1}DW)	1.0	1.0	1.0	1.0
F_m (d.l^{-1})	$7.5 \ 10^{-5}$	$7.1 \ 10^{-3}$	$9 \ 10^{-4}$	$4 \ 10^{-3}$
F_f (d.kg^{-1}DW)	$7.0 \ 10^{-4}$	$2.1 \ 10^{-2}$		

DISCUSSION OF THE RESULTS

The results in table 4 show the external exposure from the sediment to be the most important exposure pathway for the critical individual (an angler on the river banks) and Co-60 to be the critical radionuclide. This is true for the relative βγ-composition of the liquid effluents assumed. For equal quantities of Co-60, Cs-137, Sr-90 and I-131 discharged, Cs-137 would be the critical radionuclide. The next important exposure pathway is constituted by the ingestion of fish. Those two most important exposure pathways also show the largest relative ranges of uncertainty (max./min. ratios) as far as no site-specific values are introduced. After the introduction of the site-specific values, indicated in table 2, the ranges of uncertainty are considerably reduced as well for the most important exposure pathways as for the most important radionuclides.

The sensitivity analysis indicated the parameters K_s (Co-60 and Cs-137) and CF_{fish} (Cs-137) to be the most important sources of uncertainty, not only for the doses from the external exposure to sediment and from ingestion of fish, but also for the sum total of the doses from all pathways.

TABLE 3
Transfer Parameters : Uncertainty Ranges

	Co-60			Cs-137			Sr-90			I-131		
K_d	2.0	4.7	(L.U)	2.7	5.3	(L.U)	0.0	3.0	(L.U)	-1.0	2.0	(L.U)
K_s	-6.7	-4.3	(L.U)	-6.0	-3.7	(L.U)						
Cf_{fish}	1.0	2.85	(L.U)	2.3	3.85	(L.U)	-0.15	2.3	(L.U)	1.0	2.15	(L.U)
B	-1.82	0.45	(L.N)	-2.30	0.65	(L.U)	-1.07	0.59	(L.N)	-2.30	0.65	(L.N)
B^{gr}	-1.64	0.72	(L.N)	-2.26	0.65	(L.N)	-0.48	0.52	(L.N)	-1.58	0.71	(L.N)
B^{gv}	-1.22	0.49	(L.N)	-1.83	0.53	(L.N)	-1.21	0.43	(L.N)	-1.58	0.71	(L.N)
B^{rc}_{p}	-1.40	0.40	(L.N)	-1.36	0.58	(L.N)	0.146	0.53	(L.N)	-0.82	0.36	(L.N)
$\lambda^{s,v}$	-2.4	0.90	(L.N)	-3.21	0.83	(L.N)	-1.61	0.87	(L.N)		n.a.	
$\lambda_{s,p}$		$= 2\lambda_{s,v}$			$= 2\lambda_{s,v}$			$= 2\lambda_{s,v}$			n.a.	
R/Y	-1.22	0.37	(L.N)		see Co-60			see Co-60			see Co-60	
R/Y^{gr}	-1.00	0.26	(L.N)		"			"			"	
R/Y^{gu}	-1.22	0.37	(L.N)		"			"			"	
R/Y^{rc}_{p}	0.255	0.19	(L.N)		"			"			"	
$\lambda^{w,v}$	1.09	0.25	(L.N)		"			"			"	
$\lambda_{w,p}$	1.32	0.16	(L.N)		"			"			"	
$t^{e,v}$	0.10 0.20 0.50		(T)		"			"			"	
$t_{e,p}$	0.04 0.08 0.50		(T)		"			"			"	
F	-2.92	0.50	(L.N)	-2.17	0.25	(L.N)	-2.92	0.21	(L.N)	-1.97	0.25	(L.N)
F^{m}_{f}	-2.01	0.41	(L.N)	-1.68	0.30	(L.N)	-3.24	0.52	(L.N)	-2.35	0.33	(L.N)
P	200	400	600(T)		see Co-60			see Co-60			see Co-60	
P^{v}_{p}		$= P_v/2$			"			"			"	
Q	16	2.6	(N)		"			"			"	
Q^{p}_{w}	60	10	(N)		"			"			"	

Legend : L.U = Log-Uniform : Values indicated are logs of min. and max. values.
L.N = Log-Normal : Values indicated are means (μ) and standard deviations (σ) of logtransformed data.
T = Triangular : Values indicated are min., modus and max. values.
N = Normal : Values indicated are means and standard deviations. K_d and K_s are positively correlated.

TABLE 4
Uncertainty Ranges

30th year of routine discharges.
(Sv.y^{-1}/MBq.l^{-1})

Exposure Pathways	Uncertainty without site-specific values			Uncertainty with site-specific values			Best-estimate values
	95 % Confid. Interv.		Median	95 % Confid. Interv.		Median	
Drink. water	2.5	4.1	3.5	2.8	2.8	2.8	2.8
Fish	2.7	80	15	8.6	8.6	8.6	8.6
Grain	0.21	5.6	1.1	0.34	5.4	1.2	2.3
Gr. Veget.	0.24	3.3	0.96	0.22	3.3	0.90	1.7
Root Crops	0.25	3.5	0.82	0.42	3.6	0.91	1.0
Milk	0.58	5.8	1.7	1.1	4.7	2.5	1.6
Meat	0.41	7.7	1.7	0.88	4.0	2.2	1.4
Soil	2.3	2.3	2.3	2.3	2.3	2.3	2.3
Sediment	2.6	81	24	69	69	69	69
Radionuclide							
Co-60	8.1	84	24	64	67	65	66
Cs-137	9.4	87	28	26	33	29	27
Sr-90	1.2	5.6	2.2	1.5	5.8	2.2	2.3
I-131	0.65	1.6	1.0	0.56	0.92	0.63	0.68

The large variability of the K_d values have only minor effects on the uncertainty ranges of the dose values for drinkwater- and fish ingestion (the two exposure pathways in which they play a role), since in these dose assessments they are multiplied by a very low value, namely the concentration of suspended solid matter in the raw riverwater which amounts to 30 mg.1^{-1}.

The uncertainty ranges of the doses from the ingestion of food crops are not significantly influenced by the introduction of the site-specific transfer parameter values. The sensitivity analyses on these pathways proved that this dose assessment was very sensitive to the parameters related to the foliar deposition (especially the mass interception factor R/Y) and not to the root uptake factors, to which much experimental effort has been spent.

The dose assessments from ingestion of milk and meat were also significantly influenced by the transfer parameters related to foliar deposition, but they were the most sensitive to the milk and meat transfer factors. Consequently the introduction of the site-specific values of these transfer factors reduced the uncertainty ranges of these doses significantly. After this the soil density (P_p) and the grass-uptake by the cattle (Q_p) remained the most important sources of uncertainty for these dose assessments.

The exposure pathways considered here are not the only possible ones, which may be important from a radiological point of view. The contamination of food crops, milk and meat due to the application on the fields or pastures of sediment or aquatic plants dredged from the river and deposited on the banks are also important exposure pathways which, moreover have been observed. However due to the lack of quantitative information about the possible transfers, they were not considered here.

CONCLUSION

The usefulness of uncertainty and sensitivity analyses in the case of dose assessments from liquid effluent discharges, has been demonstrated. Although the results are only preliminary, since not all possible exposure pathways have been examined and not all parameter values were varied, they do give a certain indication about the uncertainty ranges of the doses and about the transfer parameters to which the dose assessment is very sensitive.

REFERENCES

[1] Ng, Y.C., A Review of Transfer Factors for Assessing the Dose from Radionuclides in Agricultural Products. Nuclear Safety, 1982, 23, 57 - 71.

[2] Hoffman, F.O., Baes III, C.F., A Statistical Analysis of Selected Parameters for Predicting Food Chain Transport and Internal Dose of Radionuclides. Report NUREG-CR-1004 (1979).

[3] Blaylock, B.G., Radionuclide Data Bases Available for Bioaccumulation Factors for Freshwater Biota. Nuclear Safety 1982, 23, 427-438.

[4] Hoffman, F.O. Gardner, R.H., Eckerman, K.F., Variability in Dose Estimates Associated with the Food-Chain Transport and Ingestion of Selected Radionuclides. Report NUREG/CR-2612 (1982).

[5] Coughtrey, P.J., Thorne, M.C., Radionuclide Distribution and Transport in Terrestrial and Aquatic Ecosystems. A Critical Review of Data. A.A. Balkema, Rotterdam, 1983.

[6] Steffens, W., Mittelstädt, W., Führ, F., The Transfer of Sr-90, Cs-137, Co-60 and Mn-54 from Soils to Plants-Results from Lysimeter Experiments. In Fifth International Congress of the IRPA, Vol. III Jeruzalem, March 9-14, 1980. Report INIS-mf-5876.

[7] Anon., Calculation of Reactor Accident Consequences. App. E. In Reactor Safety Study. Report WASH-1400 (1975).

[8] Schaeffer, R., Conséquences du Déplacement des Sédiments sur la Dispersion des Radionucléides. In Impacts of Nuclear Releases into the Aquatic Environment. Proceedings of a Symposium, Otaniemi, 30/6-4/7/1975. Report STI/PUB/406, pp. 263-276.

[9] Anon., Limits for Intakes of Radionuclides by Workers. ICRP Publication 30. Ann ICRP, 1979, 2, n° 3/4.

[10] Kocher, D.C., Dose-rate conversion factors for external exposure to photons and electrons. Health Physics, 1983, 45, 665-686.

[11] Hunt, G.J., Simple Models for Prediction of External Radiation Exposure from Aquatic Pathways. Rad. Prot. Dosim., 1984, 8, 215-224.

[12] Cresta, M., Ledermann, S., Garnier, A., Lombardo, E., Lacouply, G., Etude des Consommations alimentaires des Populations de onze Régions de la Communauté Européenne en vue de la Détermination des Niveaux de Contamination radioactive. Rapport EUR. 4218f (1969).

[13] Anon., The radiological exposure of the population in the Meuse basin. Report EUR 10670 EN (1986).

A MODEL PERFORMANCE TEST FOR THE AQUATIC DISPERSION
OF RADIONUCLIDES IN THE NORTH-EAST ATLANTIC WATERS

Sven P. Nielsen and Asker Aarkrog
Health Physics Department
Risø National Laboratory
DK-4000 Roskilde
Denmark

ABSTRACT

Several aquatic dispersion models have been developed for the North-East Atlantic seas in order to assess radiation doses to humans received via marine pathways due to radioactive wastes from nuclear installations. This paper describes a performance test of such a model. The model is an extended and adjusted version of the compartment model developed by the National Radiological Protection Board in the United Kingdom for the coastal seas around North-Western Europe. The model provides an improved simulation of the water transport from the North Sea to the Baltic Sea. This improvement was obtained by calibrating the model to obtain the best possible agreement between predictions and observations of seawater concentrations of ^{137}Cs. For the purpose of testing the performance of the model two other radionuclides have been studied: ^{134}Cs and ^{99}Tc. Discharges from Sellafield and Cap de La Hague covering the time period 1966-85 have been used as source terms for the model predictions. These predictions are compared with environmental measurements of seawater concentrations in different water regions.

INTRODUCTION

Models simulating the dispersion and transport of radioactive effluents in the environment are important tools in connection with radiological risk assessment of nuclear facilities. Several aquatic-dispersion models have been developed for the purpose of assessing the radiation doses to humans received via marine pathways in respect of radioactive wastes released from European nuclear installations. The structure and physical processes considered by the models vary considerably. Three dimensional models may provide quite realistic simulations of water transport and dispersion, but they tend to require large amounts of input data and involve relatively large computational expenditure. Compartment models are much simpler and easier to use, but may sometimes not be able to predict the water movements in sufficient detail.

MATERIAL AND METHODS

The present model, the LEUS model [1,2], is a compartment model which covers the water regions shown in Figure 1. Figure 2 shows the schematic box structure and transfer mechanisms of the model. The model is a combination of three other compartment models: the NRPB model [3], the Grimwood model [4] and the Evans model [5]. The NRPB model was developed in 1980 covering the North-West European water regions adjacent to the United Kingdom. Grimwood extended this model in 1982 to include the Barents Sea, the Spitsbergen Waters and the Norwegian Waters. Finally Evans in 1985 presented a box model of the Baltic Sea, which was merged into the LEUS model. A number of the transfer coefficients was altered from their original values in order to improve the model. This was accomplished by using the reported releases of ^{137}Cs in 1960-1984 from Sellafield, Cap de La Hague and Dounreay as a calibration tracer. The model considers removal by sedimentation as insignificant, which is a reasonable assumption for Cs outside the Baltic.

For the model performance test the releases of ^{134}Cs and ^{99}Tc from Sellafield and Cap de La Hague have been used covering the time period of 1966-1985. Reported release data have been used where available [6,7], but for ^{99}Tc it has been necessary to use estimated release data from both Sellafield and Cap de La Hague. For Sellafield releases of ^{99}Tc before 1978 have been estimated in ref. [8], and for Cap de La Hague the ^{99}Tc releases have been estimated from ref. [9] in addition to qualified judgements of data from seaweed samples collected near La Hague. The release data used are given in Table 1. From these input data we have calculated predicted time-dependant seawater concentrations for the different compartments.

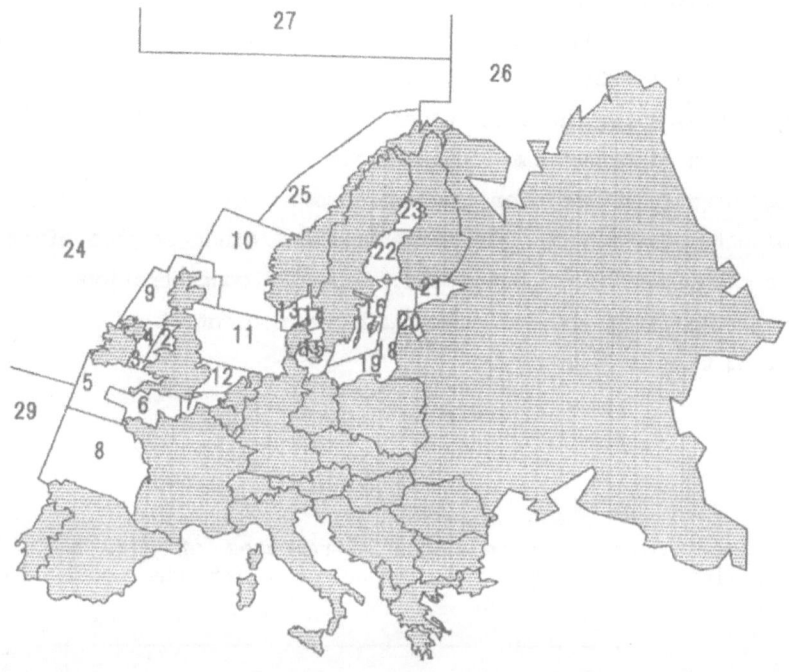

Figure 1.   The water regions and corresponding compartment numbers used by the LEUS model.

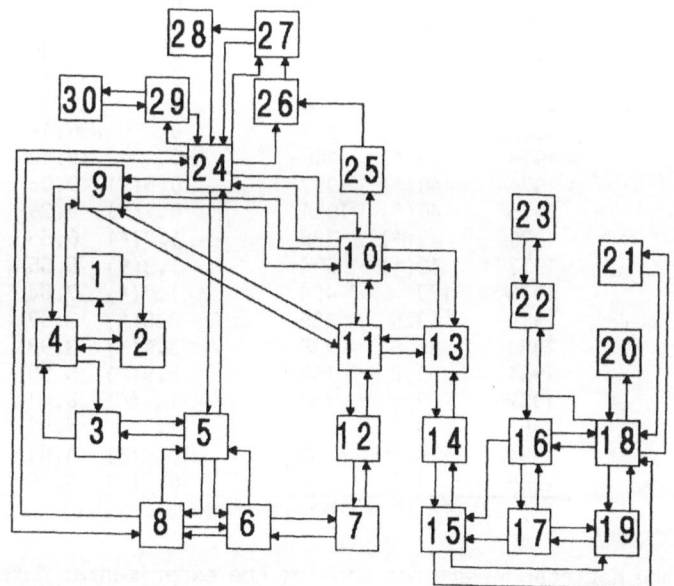

Figure 2.   The schematic box structure and transfer mechanisms of the LEUS model.

The experimental seawater concentrations used for the performance test are from samples taken and measured by our laboratory [10-21] and cover the time period 1978-1985. Most of the data are from the North Sea and the Danish waters, while the other water regions are covered more sporadically, except for an intense sampling campaign in 1983. The same samples of seawater have been used for the measurements of the two Cesium isotopes. For the determination of ^{99}Tc concentrations other samples of seawater as well as of seaweed have been used. Seawater concentrations of ^{99}Tc have been estimated from the latter by using a concentration factor of 10^5 Bq m^{-3} per Bq t^{-1} dry weight [8].

TABLE 1

Reported and estimated aquatic releases of ^{99}Tc and ^{134}Cs from Sellafield and Cap de La Hague used for the model calculations

Year	Sellafield		Cap de La Hague	
	^{99}Tc (TBq/y)	^{134}Cs (TBq/y)	^{99}Tc (TBq/y)	^{134}Cs (TBq/y)
1966	0	114	0	0.28
1967	0	96.6	0	1.63
1968	0	245	0	3.03
1969	0	300	0	1.30
1970	40(*)	251	0	13.8
1971	40(*)	236	0	48.0
1972	40(*)	215	0	6.14
1973	40(*)	166	0	8.40
1974	40(*)	997	0.5(*)	9.03
1975	40(*)	1081	0.7(*)	4.26
1976	40(*)	738	1.2(*)	6.55
1977	40(*)	594	1.3(*)	9.55
1978	179	404	1.7(*)	7.84
1979	43.5	235	3.0(*)	3.57
1980	56.8	239	3.1(*)	3.93
1981	5.8	168	5.9(*)	5.96
1982	3.6	138	11 (*)	8.41
1983	4.4	89	11.7	4.92
1984	4.3	35	10 (*)	4.81
1985	1.9	30	10 (*)	8.18

(*) estimated releases

For the Kattegat we are not showing the experimental data for ^{99}Tc for the years 1980 and 1981. This is because these data were used to estimate the unreported ^{99}Tc discharges from Sellafield [8].

RESULTS

The results of the model calculations are given together with the experimental observations in Figure 3 and 4 for the boxes for which we have the most experimental data. Figure 3 shows the results for ^{134}Cs for the compartments 9, 10, 11, 13, 14, 15, 25 and 27, while Figure 4 shows the results for ^{99}Tc for the compartments 9, 10, 11, 12, 14, 15, 25 and 27. In the figures the model calculations are shown as full lines and the observed values are shown as square marks.

DISCUSSION

As shown in Table 1 a significant part of the release data for ^{99}Tc is based on estimated values. For Sellafield the data prior to 1978 derive from observations of ^{99}Tc in seaweeds in the Danish Straits and in the waters around Greenland [8]. For Cap de La Hague the only reported release is from 1983 and the remaining data have been estimated from levels of ^{99}Tc in seaweed collected near the point of release [9]. The lack of reported release data is a significant draw-back of the data material. However, we believe that the quality of the estimated data is acceptable for the present investigation.

The concentration factor of 10^5 for the steady-state transfer of ^{99}Tc from seawater to seaweed is used for estimates of seawater concentrations. In the Danish Straits the observed variation of the concentration factor amounts to a relative error of 20% [8].

In Figure 3 showing the results for ^{134}Cs the agreement between model predictions and experimental values appears to be very good. Considering the use of ^{137}Cs as a calibration tracer, the results for ^{134}Cs were expected to be very good, since the only significant differences between the two isotopes in this context are the source terms and the radioactive half-lifes. Only for box 9, the Scottish Waters, there is a significant difference between the model predictions and the observations. This is due to the inhomogeneous conditions in this water region in connection with the

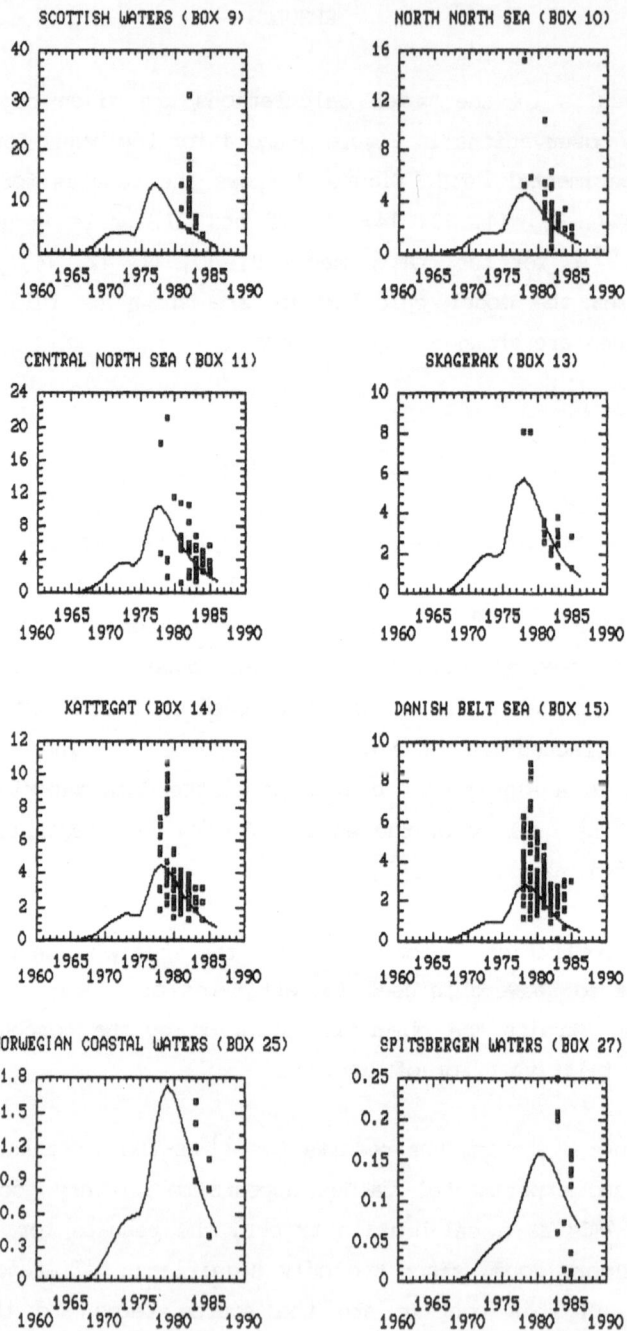

Figure 3. Calculated and measured values of ^{134}Cs in seawater (Bq/m^3) in 8 water regions. The calculated values are given as full lines and the measured values are marked individually as squares.

223

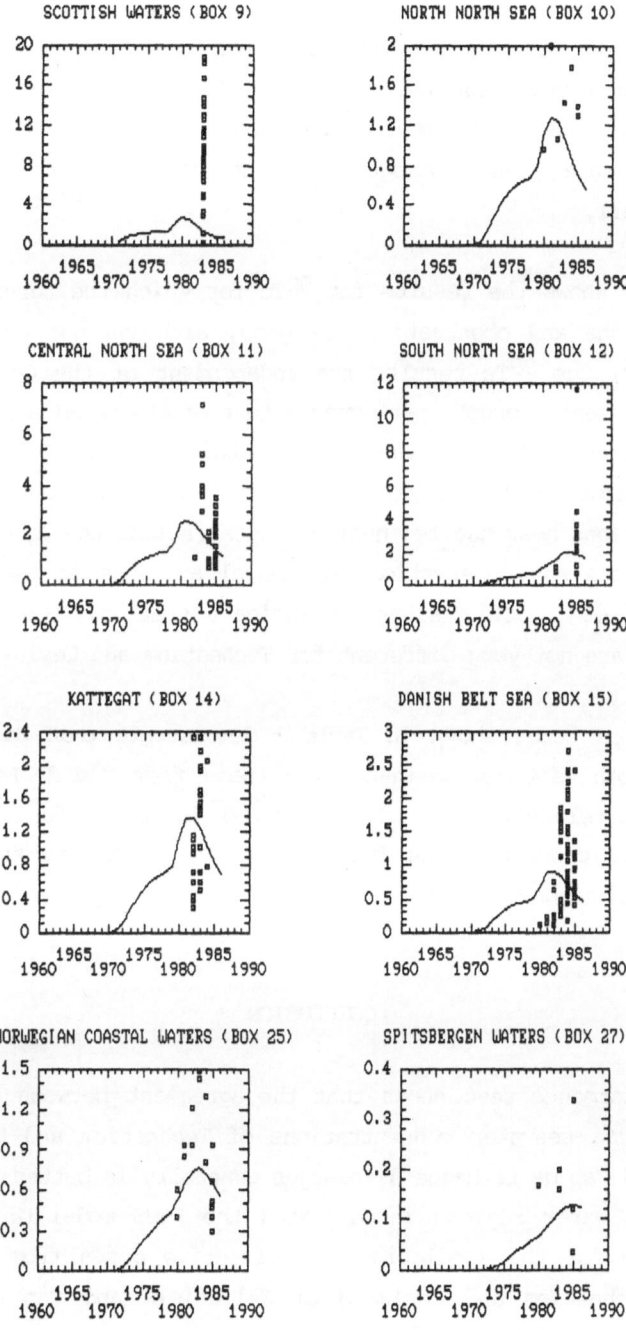

Figure 4. Calculated and measured values of ^{99}Tc in seawater (Bq/m^3) in 8 water regions. The calculated values are given as full lines and the measured values are marked individually as squares.

relative proximity to Sellafield in addition to the biased sampling along the coast in that box. The compartment model assumes instantaneous mixing throughout each compartment for every time step, which is a very unrealistic assumption close to the source. But further away from the source where the conditions are less inhomogeneous, the simulation of the transport works quite well.

Figure 4 shows the results for ^{99}Tc for which the agreement between model predictions and observations is good, although not as good as for ^{134}Cs. However, the ^{99}Tc results are independent of the Cesium data and therefore represent a proper performance test of the model and as mentioned previously none of the experimental ^{99}Tc data have been used for the estimation of unreported discharges. Apart from water regions with inhomogeneous conditions near the two nuclear installations the agreement between model predictions and observations of annual averages is generally better than a factor of 2. This seems to indicate that conditions related to sedimentation are not very different for Technetium and Cesium.

The source terms shown in Table 1 demonstrate that Sellafield discharges for both isotopes dominate over those from Cap de La Hague. This means that the reliability of the LEUS model is better for the prediction of radionuclide concentrations from Sellafield discharges than that from other site discharges.

CONCLUSION

The performance test shows that the agreement between the predicted and the observed seawater concentrations of Technetium and Cesium due to Sellafield and Cap de La Hague discharges generally is better than a factor of 2. For the water regions investigated the LEUS model thus provides a reliable basis for the prediction of collective doses from aquatic discharges of Technetium and Cesium from Sellafield and Cap de La Hague.

ACKNOWLEDGEMENT

This work was supported by the Commission of the European Communities.

REFERENCES

1. Hallstadius, L., Garcia-Montano, E., Nilsson, U. and Boelskifte, S., An Improved and Validated Dispersion Model for the North Sea and Adjacent Waters. J. Environ. Radioactivity, 1987, 5, pp. 261-274.

2. Nielsen, S.P. and Boelskifte, S., Dispersion of Radioactive Substances in Marine Waters. EEC draft final report of contract 85E1007, 1987.

3. Clark, M.E., Grimwood, P.D. and Camplin, W.C., A Model to Calculate Exposure from Radioactive Discharges into Coastal Waters of Northern Europe. NRPB-R109, National Radiological Protection Board, Harwell, England, 1980.

4. Grimwood, P.D., The Estimation of Collective Doses per Unit Discharge to Sea from Sellafield. Proc. 3rd Int. Symp. Radiological Protection, Inverness, Scotland, June 6-11, 1982.

5. Evans, S., A Box Model for Calculation of Collective Dose Commitment from Waterborne Releases to the Baltic Sea. J. Environ. Radioactivity, 1985, 2, pp. 41-57.

6. BNFL 1978-85, Annual Report on Radioactive Discharges and Monitoring of the Environment. British Nuclear Fuels Ltd, Risley, Warrington, Chesire, England.

7. Calmet, D. and Guegueniat, P., Les Rejets d'Effluents Liquides Radioactifs du Centre de Traitement des Combustibles Irradies de La Hague (France) et l'Evolution Radiologique du Domaine Marin. In Behaviour of Radionuclides Released into Coastal Waters, IAEA-TECDOC-329, International Atomic Energy Agency, Vienna, 1985.

8. Aarkrog, A., Boelskifte, S., Dahlgaard, H., Duniec, S., Hallstadius, L., Holm, E. and Smith J.N., Technetium-99 and Cesium-134 as Long Distance Tracers in Arctic Waters. Estuarine, Coastal and Shelf Science, 1987, 24, pp. 637-647.

9. Patti, F., Masson, M., Vergnaud, G. and Jeanmaire, L., Activités du Technétium 99 Mesurées dans les Eaux Résiduaires, l'Eau de Mer et Deux Bioindicateurs (Littoral de la Manche, 1983). In Technetium in the Environment, ed. G. Desmet and C. Myttenaere, Elsevier Applied Science Publishers, London, 1986, pp. 37-51.

10. Aarkrog, A., Bøtter-Jensen, L., Dahlgaard, H., Hansen, H., Lippert, J., Nielsen, S.P. and Nilsson, K., Environmental Radioactivity in Denmark in 1978. Risø-R-403, Risø National Laboratory, Denmark, 1979.

11. Aarkrog, A., Bøtter-Jensen, L., Dahlgaard, H., Hansen, H., Lippert, J., Nielsen, S.P. and Nilsson, K., Environmental Radioactivity in Denmark in 1979. Risø-R-421, Risø National Laboratory, Denmark, 1980.

12. Aarkrog, A., Bøtter-Jensen, L., Dahlgaard, H., Hansen, H. Lippert, J., Nielsen, S.P. and Nilsson, K., Environmental Radioactivity in Denmark in 1980. Risø-R-447, Risø National Laboratory, Denmark, 1981.

13. Aarkrog, A., Bøtter-Jensen, L., Dahlgaard, H., Hansen, H. Lippert, J., Nielsen, S.P. and Nilsson, K., Environmental Radioactivity in Denmark in 1981. Risø-R-469, Risø National Laboratory, Denmark, 1982.

14. Aarkrog, A., Bøtter-Jensen, L., Dahlgaard, H., Hansen, H., Lippert, J., Nielsen, S.P. and Nilsson, K., Environmental Radioactivity in Denmark in 1982. Risø-R-487, Risø National Laboratory, Denmark, 1983.

15. Aarkrog, A., Boelskifte, S., Bøtter-Jensen, L., Dahlgaard, H., Hansen, H. and Nielsen, S.P., Environmental Radioactivity in Denmark in 1983. Risø-R-509, Risø National Laboratory, Denmark, 1984.

16. Aarkrog, A., Boelskifte, S., Bøtter-Jensen, L., Dahlgaard, H., Hansen, H. and Nielsen, S.P., Environmental Radioactivity in Denmark in 1984. Risø-R-527, Risø National Laboratory, Denmark, 1985.

17. Aarkrog, A., Boelskifte, S., Bøtter-Jensen, L., Dahlgaard, H., Hansen, H. and Nielsen, S.P., Environmental Radioactivity in Denmark in 1985. Risø-R-540, Risø National Laboratory, Denmark, 1987.

18. Aarkrog, A., Dahlgaard, H., Hallstadius, L. and Holm, E., Environmental Radioactivity in Greenland in 1982. Risø-R-489, Risø National Laboratory, Denmark, 1983.

19. Aarkrog, A., Dahlgaard, H., Hallstadius, L., Holm, E., Hansen, H. and Lippert, J., Environmental Radioactivity in the Faroes in 1982. Risø-R-488, Risø National Laboratory, Denmark, 1983.

20. Aarkrog, A., Boelskifte, S., Buch, E., Christensen, G.C., Dahlgaard, H., Hallstadius, L., Hansen, H., Holm, E., Mattsson, S. and Meide, A., Environmental Radioactivity in the North Atlantic Region. The Faroe Islands and Greenland included. 1983. Risø-R-510, Risø National Laboratory, Denmark, 1984.

21. Aarkrog, A., Boelskifte, S., Buch, E., Christensen, G.C., Dahlgaard, H., Hallstadius, L., Hansen, H. and Holm, E., Environmental Radioactivity in the North Atlantic Region. The Faroe Islands and Greenland included. 1984. Risø-R-528, Risø National Laboratory, Denmark, 1985.

RADIOACTIVITY OF FRENCH COAST OF THE CHANNEL DUE TO THE RELEASE OF TECHNECIUM 99 AND IODINE 129 : MODELISATION AND MEASUREMENTS

D. ROBEAU - F. PATTI - S. CHARMASSON

Institut de Protection et de Sûreté Nucléaire
Département de Protection Sanitaire
B.P. n° 6
92 265 - FONTENAY AUX ROSES CEDEX

- ABSTRACT -

Radioactive releases of Iodine 129 are controlled with measurements of the radioactivity of the liquid effluents before to be released in sea from the outlet of the reprocessing plant of La Hague. The effects on marine environment are contolled with a radioactive survey of Technecium 99 and Iodine 129 in Fucus (common seaweed).
This radioactivity is measured along north Coast of France from Roscoff in the west of Britany to Wimereux close to Belgium frontier. On other hand, the theoritical study of dispersion of radionuclides in the Channel has permitted to make a model of simulation of the transfer of pollutants and particularly Technecium 99 and Iodine 129.

Introduction

In the framework of its actions regarding environmental protection and assessment of the radiological nuisance that industrial development of nuclear energy could cause, the Institute of Protection and Nuclear Safety of the French Atomic Commission has developed a numerical model which enables to assess the hourly tide flow and simulate the transfer of radionuclides in the Channel. That was done in the beginning of 1980 's.

The objective of these studies was to estimate the doses received by those submitted to the hypothetical accidental radioactive leaks in the Channel; These leaks could come from the wreck of a freighter conveying nuclear fuel, or from the normal releases of the French and British reprocessing nuclear plants.

This kind of simulation model is interesting if it can be validated. With the aim of validating this model, the department of health protection has set up a network of measurements, with the purpose of doing a radiological monitoring of the French coast of the Channel.On one hand, and of obtaining data necessary to validate the model, on the other hand. This network was based on the radiological survey of a common seaweed of the french coast, the fucus serratus. Five sampling points (called survey stations) make up this radiological network survey.

From West to East, the first survey station is located close to Roscoff on the North Coast of Britany ; the second survey station is located close to Granville on the West station is located close to Herquemoulin and the outlet of the Reprocessing Nuclear Plant of La Hague on the norhwest Coast of the Cotentin peninsula ; the fourth survey station is located close to Ouistreham and the Seine estuary ; the fifth survey station is located close to Wimereux in the North Coast of France. These five stations are indicated on the referenced map.

The fucus is sampled monthly. Seaweeds are collected and conditioned before being measured by gamma spectrometry ; this spectrometry allows to determine its specific radioactivity due to ^{99}Tc (Technecium 99) and ^{129}I (Iodine 129) and other radionuclides.

The aim of this particular study is the simulation of the transfer of radionuclides released by the Nuclear reprocessing plant of La Hague in the Channel. The transfer of ^{99}Tc was simulated in the Channel using the effective radioactive releases of the ^{99}Tc between october 1984 and April 1986. The transfer of ^{99}Tc was simulated during ten months : July 1986 to April 1986.

The transfer of ^{129}I was simulated in the Channel from the effective radioactive releases of ^{129}I between June 1983 and Januray 1985. The transfer of ^{129}I was simulated during ten months : March 1984 to December 1984.

"Mean monthly theoritical radioactive concentrations " of ^{99}Tc and ^{129}I in the sea water were computed, particularly close to the survey stations. Afterwards, mean monthly theoritical radioactive concentrations in fucus serratus were calculated using concentrations in sea water and the most realistic concentration factor. These radioactive concentrations in fucus serratus were compared with those measured at the same place and same time.

Modelisation

The model describing water flow in the Channel was schematized by a single mean cycle of tide (twelve hourly flow fields) established for an intensity coefficient tide equal to 80. These flow fields were computed by solving the fundamental bi-dimensional equations of hydrodynamic discretized on a regular grid covering all of the Channel from West to East and from North to South with a step equal to ten kilometers.

The resolution of the linear system arising from the discretization of the equations in accordance with the principle of the centered finite differences method permits to define the components of the flow on every node of the grid at every hour of the tide cycle.

The boundary conditions on the the opened frontiers (water) are calculated from the measurements in-situ of water heights ; the boundary conditions on the closed frontiers (coast) are conditions of null current.

The method used to solve the diffusion advection equation is a "random walk" method which allows to describe the process of the cahotic movement of a particle in suspension into a fluid. Particles move under the advective action of a fluid, and under the cahotic action of this same fluid. The pathway of these particles can be described by a continuous serie of straight segments. The directions of each straight segment follow a distribution of probability depending on the medium and on the stochastic process describing the movement of the particles. Numerous random walks are simulated, and distributions of particles in time and space are calculated. If the random walks of particles are sufficiently numerous, the distributions of presence and transition of particles are well described and get close to exact solutions of diffusion-advection equations.

Simulations of the transfer of ^{99}Tc

The transfer of ^{99}Tc released from the reprocessing nuclear plant of La Hague, was simulated during ten months from July 1985 to April 1986. The Channel is discretized on a grid identical to that used for the flow computations. The ^{99}Tc is assumed totaly soluble.

- The boundary conditions on the opened frontier are conditions of null entering flow ; those taken on the closed frontier are conditions of total reflexion.

- The initial condition is an instantaneous and punctual release, two hours before the tide.

The equation is solved on a regular grid twice as fine as that used to define the medium. The grid is represented by 48 x 96 meshes.

The dynamics of the contamined spot can be described in three phases. During the first phase (the two first months after the release) the spot grows and extends quickly Westward to Britany and the Channel Islands, and Eastward to the Bay of Seine. This quick extension can be easily explained by strong flows in the North of the Cotentin peninsula due to the narrowing of the French and English coasts of the Channel.

During the second phase (the next ten months) a slow growing of the spot westward , and in Bay of Seine eastward, can be observed. This visible stagnation is due to the widening of the Channel implying an important decrease of the flow intensity. The concentration of ^{99}Tc in this spot decreases little by little.
During the third phase, the spot is slowly suched up in the direction of the North sea : the narrowing of Channel in the strait of Dover leads to a flow increase .

The following results are "mean montly radioactive concentrations" calculated in the three meshes containing the three most interesting stations of the network of radiological survey : Herquemoulin, Granville and Ouistreham. The Herquemoulin, Granville and Ouistreham survey stations are included, respectively, in the (41,28) mesh, (47,8) mesh and (67,21) mesh of the referenced grid.

Technecium 99 taken into account in this exercice of simulation was released from the outlet of reprocessing plant of La Hague between October 1984 and April 1986. Released radioactivity is indicated in the following table (expressed in TBq) :

1984 October	1,163	May	1,3	December	4,2
November	1,320	June	0,3	1986 January	1,4
December	3,63	July	1,3	February	0,7
1985 January	3,8	August	0,4	March	2,2
February	3,3	September	1	April	1,3
March	3,6	October	2,5	May	1
April	1,6	November	2,1		

The mean montly radioactive concentrations of ^{99}Tc in the sea water from July 1985 to April 1986 are caculated using the methodology described above. The radioactive concentration of ^{99}Tc in the fucus (units of Bq per kilogram of dry fucus) is calculated using a concentration factor equal to 3.10^4. This factor was determined from measurements in situ by Patti and Al. All these results are indicated in the following table :

Date	Herquemoulin		Granville		Ouistreham	
	Sea Water	Fucus	Sea Water	Fucus	Sea Water	Fucus
07.85	20,64	3096	2,32	348	5,94	8,91
08.85	24,61	3691	2,48	372	7,11	1066
09.85	18,43	2764	1,54	231	7,48	1110
10.85	19,86	2979	1,24	186	7,16	1065
11.85	26,7	4005	2,1	315	5,2	780
12.85	25,80	3870	1,66	249	5,08	762
01.86	34,2	5130	3,06	459	14,9	2235
02.86	20,7	3105	1,62	243	17,1	2565
03.86	17,2	2580	1,34	201	9,7	1455
04.86	26,7	4005	3,08	462	7,4	1110

The results of the "mean monthly radioactive concentrations" of fucus serratus sampled in the three survey stations between July 1985 and April 1986 and measured by Patti and Al are indicated in the following table (units of Bq per kilogram of dry Fucus) :

Date	Herquemoulin (Fucus)	Granville (Fucus)	Ouistreham (Fucus)
07.85	3654	331	761
08.85	3667	544	940
09.85	NM	525	1029
10.85	3567	570	1367
11.85	3644	NM	1375
12.85	3140	377	NM
01.86	4564	516	1319
02.86	4444	370	NM
03.86	3626	454	NM
04.86	4227	519	1063

The measurements of ^{99}Tc and the estimated theoretical values of releases are consistent if the incertainties of the numerous parameters of the problem are taken into account.

Simulation of the tranfer of ^{129}I

Using the same methodology as for ^{99}Tc, the transfer of ^{129}I was simulated from the effective releases of the reprocessing nuclear plant of La Hague. This transfer was simulated during three months from March 1984 to December 1984. The mean monthly radioactivity of ^{129}I released from June 1983 to January 1985 is indicated in the following table (radioactivity in units of GBq).

1983 June	20,4	1984 January	14,4	August	9,1
July	8,9	February	12,1	September	8,7
August	4,9	March	19,3	October	10,6
September	2,5	April	-	November	23,5
October	8,2	May	-	December	33,9
November	6,8	June	12,7	1985 January	40,9
December	7,6	July	7,7		

The "mean montly radioactive concentrations" of ^{129}I were calculated in mesh (41,28) of the reference grid, where the Herquemoulin survey station is located. The factor of concentration in the fucus serratus was taken equal to 3.10^4 to caculate the radioactive concentrations of ^{129}I in the fucus resulting from the radioactive concentrations of ^{129}I in the sea water (expressed in units of Bq/1). All these calculated results are included in the following table. The results of measurements of radioactive concentrations of ^{129}I in fucus sampled in Herquemoulin during the period of simulation are also indicated in the following table.

These results are expressed in units of Bq per kilogram of dry fucus and have been obtained from Patti and al.

Date	Sea Water radioactive Concentration	Fucus radioactive Concentration	Fucus radioactive measurement
03.84	$3.85.10^{-2}$	5,77	3,4
04.84	$4,03.10^{-2}$	6,04	4,8
05.84	$3,9.10^{-2}$	5,85	7
06.84	$4,01.10^{-2}$	6,01	7,1
07.84	$3,95.10^{-2}$	5,92	5,4
08.84	$3,57.10^{-2}$	5,35	2,6
09.84	$3,61.10^{-2}$	5,41	4,0
10.84	$3,94.10^{-2}$	5,91	3,4
11.84	$4,06.10^{-2}$	6,09	4,5
12.84	$5,7.10^{-2}$	8,55	5,9

Conclusions

This kind of model not rudimentary but simple (since it simplifies water movements down to twelve tide current fields) shows that it is possible to make assessments of radiological concentrations in sea water, biological indicators and seafood with fair approximation. This type of modelisation is interesting for the study of main parameters as concentration factors, transfer duration from point of releases to survey station, and estimation of the variation of radioactive measurements in fonction of future releases.

COMPARISON OF MARINE DISPERSION MODEL PREDICTIONS WITH ENVIRONMENTAL RADIONUCLIDE CONCENTRATIONS

C.E. JOHNSON
Department of Mechanical Engineering
Imperial College of Science and Technology
Exhibition Road, London SW7 2BX
UK

and

W.A. McKAY
Environmental and Medical Sciences Division
Harwell Laboratory
United Kingdom Atomic Energy Authority
Oxfordshire OX11 0RA
UK

ABSTRACT

The comparison of marine dispersion model results with measurements is an essential part of model development and testing. The results from two residual flow models are compared with seawater concentrations, and in one case with concentrations measured in marine molluscs. For areas with short turnover times, seawater concentrations respond rapidly to variations in discharge rate and marine currents. These variations are difficult to model, and comparison with concentrations in marine animals provides an alternative and complementary technique for model validation with the advantages that the measurements reflect the mean conditions and frequently form a useful time series.

INTRODUCTION

The Dounreay Nuclear Power Development Establishment, operated by the United Kingdom Atomic Energy Authority is situated on the North Coast of Scotland (Fig. 1). It consists primarily of a prototype fast reactor, a reprocessing plant and a fuel production plant. Discharges to the sea are made through a pipeline extending 600 m offshore, and include a minor contribution from an adjacent naval nuclear propulsion test establishment. A summary of previous discharges has been published [1].

A combined programme of dispersion modelling and seawater measurements has been set up in order to describe the marine dispersion of radionuclides from Dounreay and to contribute to dose assessments. Further measurements of radionuclides in marine foodstuffs are made regularly [2] as part of the routine monitoring around the site.

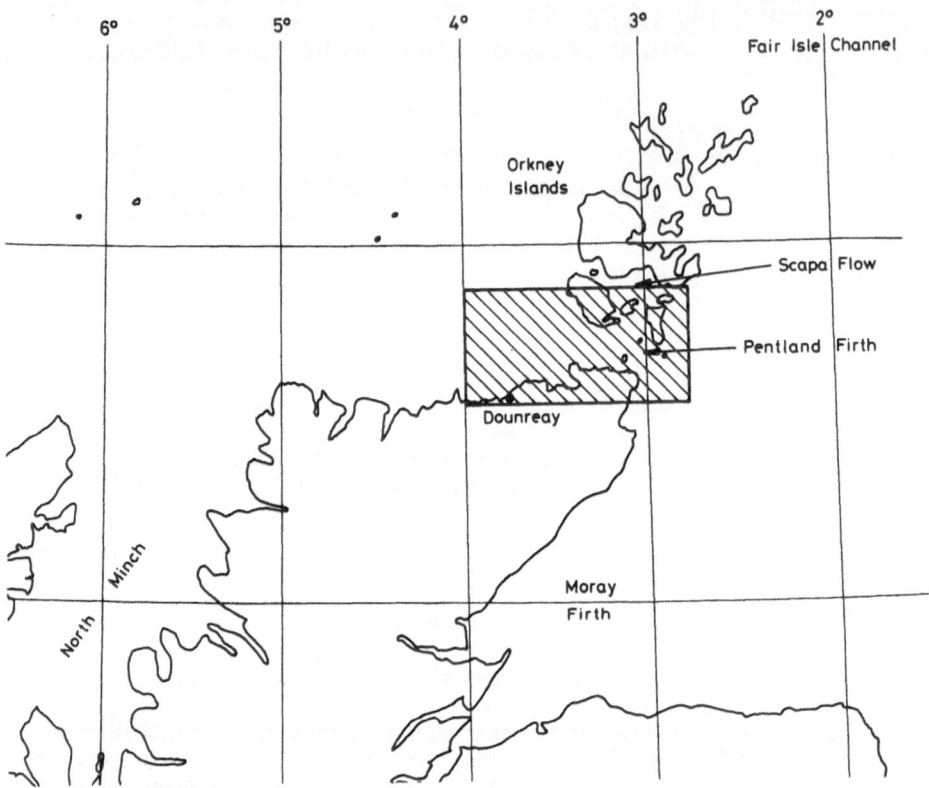

Figure 1. Map showing location of the model area.

Careful interpretation of these radionuclide measurements needs to be made as some of the activity may be due to advection of material from the Irish Sea by the Scottish coastal circulation [3,4]. A seperate model has therefore been developed in order to predict concentrations at Dounreay from discharges into the Irish Sea.

Model results can be compared with:
1) Measurements repeated at one place to form a time series.
2) Measurements taken at a single time over an extensive area.
3) Measurements of concentration in sediments and marine organisms, often taken as part of routine monitoring programmes.

The comparison of different types of data gives alternative and complementary information about model behaviour. For example, time series comparison may provide information about exchange rates, while measurements over the whole model area should give information about the direction of dispersion. Practical considerations of collection and analysis limit the amount of seawater data available, and the information available from routine foodstuff monitoring programmes is often more extensive.

Measurements taken of concentrations in marine sediments and organisms are frequent modelling targets, but are considerably more difficult to model than seawater concentrations. In particular, care should be taken with marine organisms because of different biological half times for each organism and radionuclide and because of differing routes of uptake.

DISPERSION MODELS

A residual flow model of the Pentland Firth has been developed in order to describe the dispersion of radionuclides from Dounreay. The model calculates the dispersion from the application of the advection-diffusion equation to a two dimensional depth-averaged grid with 200 elements of 4 km size.

The model boundaries are shown on Fig. 3. The equation is numerically integrated with first order discretisation of the diffusion terms, and a donor cell technique for the advection. Residual current directions were taken from a hydrodynamical model [5]. The diffusivities in the x and y directions were taken as constant and set at 3×10^6 m^2s^{-1}. The model was not very sensitive to the value of the diffusivity in the x direction which was aligned to the principle direction of the residual currents. Both the diffusivity and the magnitude of the advection were set to give good comparison with concentrations of Co-60 measured in seawater along a transect across the Pentland Firth. The model formulation does not allow for adsorption onto bottom sediments, however, little sediment finer than sand is present in the area modelled, and thus adsorption is expected to be small.

A similar model was used to describe the advection of soluble isotopes from the Irish Sea in the Scottish coastal circulation. The model grid and residual currents were taken from the hydrodynamical model of dispersion of Prandle [6]. The formulation of the diffusivities was different in the present model, as they were related to the magnitude of the residual rather than tidal currents.

MODEL COMPARISONS WITH MEASUREMENTS

Two data sets are available to compare with the Scottish coastal circulation model, as Cs-137 and Sr-90 concentrations in the Pentland Firth have been measured routinely for a number of years [4,7]. The model was calibrated against the Cs-137 data by adjusting the overall diffusivity levels to give agreement between the model and the maximum concentrations recorded. The Sr-90 data provided an independent set of values to compare with the model.

Fig. 2 shows these comparisons, and it appears that the time trend matching in both radionuclides is good. However, the Cs-137/Sr-90 ratio predicted by the model is equal to the Sellafield discharge ratio, around 6 in recent years, while the measured values in the Pentland Firth in 1984 and 1985 are about 4. This shift in favour of Sr-90 has been noted previously [7], and is attributed to weapons fallout. The background concentrations of Sr-90 present in Atlantic water (2-3 mBq/ℓ) do not seem sufficient to produce such a strong change in this ratio, and the continued input of weapon fallout Strontium from land runoff must also be proposed to account for this change.

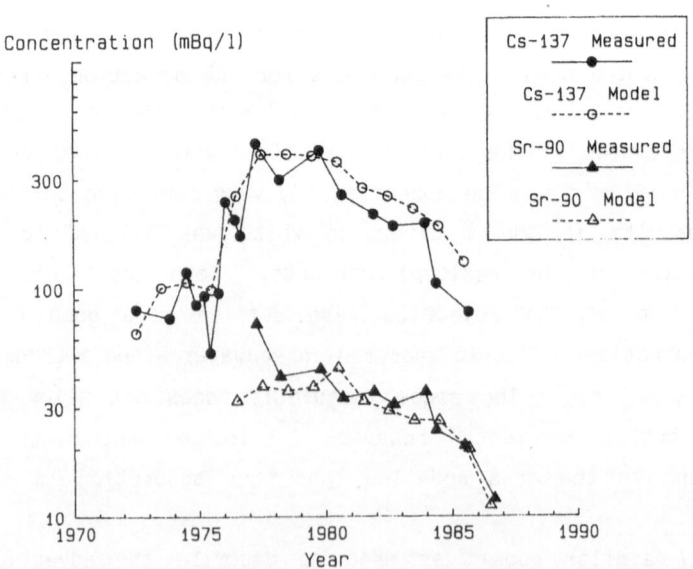

Figure 2. Model and measured Cs-134 and Sr-90 concentrations in the Pentland Firth

Many of the measurements made in the area cannot be used to compare with models of dispersion from Dounreay due to the presence of radionuclides advected from the Irish Sea. The radionuclides present in the Scottish coastal circulation include 238-Pu, 239+240-Pu and 241-Am [8], as well as the more soluble isotopes 90-Sr and 137-Cs. This limits the extent to which model validation may be carried out. However, one radionuclide, 60-Co, has been measured in seawater which has a negligible contribution from other sources.

Fig. 3 shows a comparison between model contours and spot measurements of Co-60 in the Pentland Firth. The turnover time of the model area is in the order of weeks and measurements may thus be affected by the spring-neap tidal cycle and currents induced by wind forcing. For accurate model to measurement comparisons, time scale matching between model and turnover time of the sea volumes must be achieved. Close to Dounreay, the comparison may also be affected by the day to day variations in the discharge pattern, and by tidal currents.

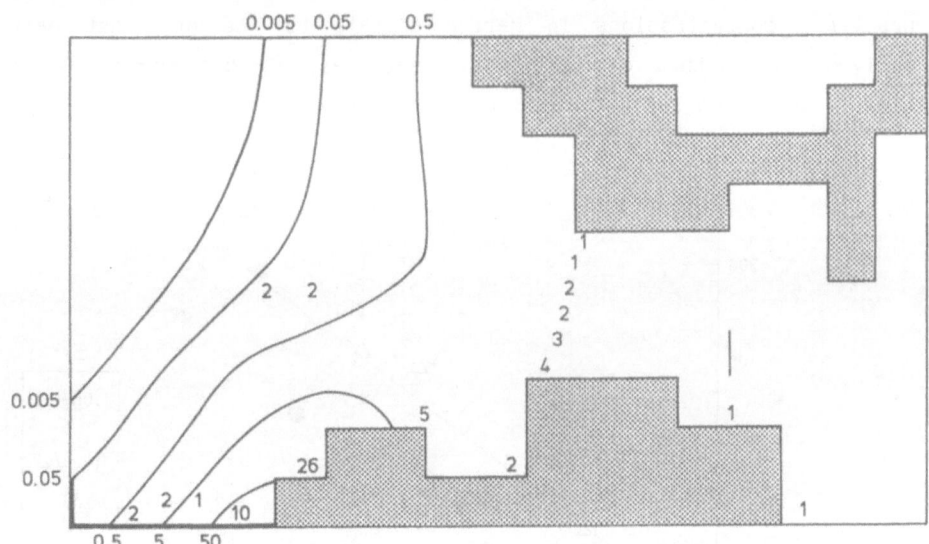

Figure 3. Measured (data) and model (contours) Co-60 concentrations, (mBq/ℓ) August 1985

Routine monitoring of radionuclide concentrations present in marine sediments and living organisms is nearly always carried out around nuclear sites. Although these measurements are not undertaken for the purpose of model validation, they often present the longest time series for model comparison. Concentrations in marine molluscs and crustacea are frequently considerably higher than in seawater, thus allowing a larger range of radionuclides to be measured. A further advantage of model comparison with biological concentrations is that certain organisms such as Winkles (<u>Littorina</u> sp.) tend to reflect the mean seawater concentration with reduced scatter from short term fluctuations in source rate and dispersion parameters.

Models of radionuclide concentrations in foodchains are frequently made via the concentration factor concept. The concentration factors used [9] are ones derived from a variety of sources, and apply to soluble seawater concentrations. In the model, these have been calculated from the total seawater concentrations using a Kd [9] and a mean sediment load figure. This estimate of the dissolved fraction will be sensitive to the sediment load figure for elements with a high Kd. In practise, concentrations in marine animals may be at least partly derived from uptake of particulate material, so concentrations will clearly depend on sediment load.

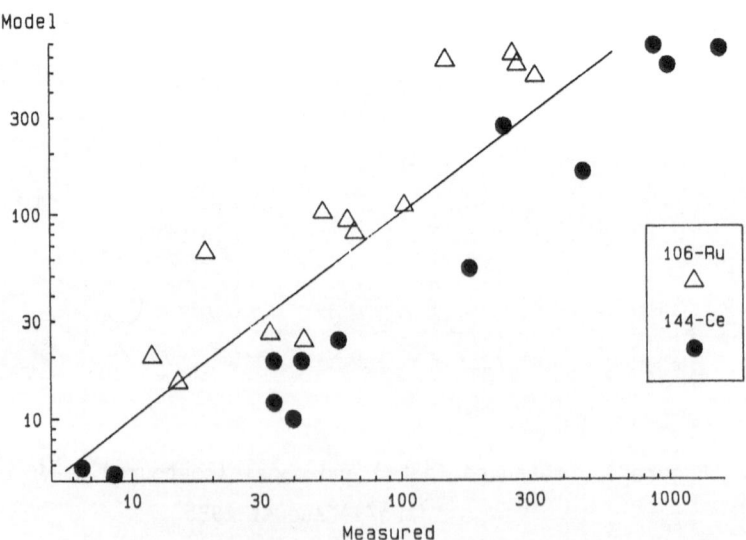

Figure 4. Radionuclide concentrations in Winkles (Bq/kg)

Fig. 4 shows a comparison of model and measured [2] mean annual concentrations in Winkles for 106-Ru and 144-Ce at Sandside Bay, which is about 3 km from the Dounreay discharge point, and in the source box of the model. The relation between the measured and modelled concentrations appears to be linear, with correlation coefficients of 0.90 for 106-Ru and 0.93 for 144-Ce. However, modelled concentrations are to high for 106-Ru and too low for 144-Ce, with mean model: measured ratios of 1.8 and 0.5 respectively. This difference indicates the need for reliable site-specific information on concentration factors to be available for this type of comparison.

CONCLUSIONS

Comparison of model results with time-trends of seawater concentrations is a powerful technique for model validation, as model response to changes in discharge rates may be verified, and the effects of variations in dispersion rates with time may occur. Such measurements made over a sufficiently long time period are often lacking, and the concentrations measured routinely in marine foodstuffs provide an alternative for model comparison. In this case however, the time delay and concentration factors between the marine organisms and the seawater must be recognized, and the use of site-specific information may become necessary.

ACKNOWLEDGEMENTS

This work was performed at the request of the Dounreay Nuclear Power Development Establishment, and was supported by the Department of Energy via the United Kingdom Atomic Energy Authority Fast Reactor Programme. We would like to thank J Garland and G Tyler of the UKAEA for help during this study; and D Prandle of the Institute of Oceanographic Sciences for supplying hydrodynamical model results.

240

REFERENCES

1.   Hill, M.D. and Cooper, J.R., Radiation Doses to Members of the Population of Thurso. NRPB R-195, Chilton, 1986.

2.   Anon, Environmental monitoring for radioactivity in Scotland: 1981 to 1985. Scottish Development Department, Statistical Bulletin No. 1(E), 1987.

3.   Jefferies, D.F., Preston, A. and Steele, A.K., Distribution of Cs-137 in British coastal waters. Marine Pollution Bull., 4, 1973, pp 118-122.

4.   Kautsky, H., Jefferies, D.F. and Steele, A.K., Results of the radiological North Sea programme RANOSP 1974 to 1976. Dt. hydrogr. Z., 33, 1980, pp 152-157.

5.   Pingree, R.D. and Griffiths, D.K., Currents driven by a steady uniform wind stress on the shelf seas around the British Isles. Oceanologica Acta, 3, 1978, pp 227-236.

6.   Prandle, D. A modelling study of the mixing of Cs-137 in the seas of the European continental shelf. Phil. Trans. R. Soc. Lond. A 310, 1984, pp 407-436.

7.   Kautsky, H., Distribution and content of different artificial radionuclides in the water of the North Sea during the years 1977 to 1981. Dt. hydrogr.Z. 38, 1985, pp 193-224.

8.   Hallstadius, L., Aarkrog, A., Dahlgaard, H., Holm, E., Beilskifte, S., Duniec, S. and Persson, B. Plutonium and americium in Artic waters, the North Sea and Scottish and Irish coastal zones. J. Environ. Radioactivity 4, 1986, pp 11-30.

9.   Anon, Sediment Kds and Concentration Factors for Radionuclides in the Marine Environment. IAEA, Vienna, 1985.

MODELLING THE BEHAVIOUR OF LONG-LIVED RADIONUCLIDES
IN THE IRISH SEA - COMPARISON OF MODEL PREDICTIONS
WITH FIELD OBSERVATIONS

P. J. Kershaw, R. J. Pentreath, P. A. Gurbutt, D. S. Woodhead,
J. A. Durance and W. C. Camplin
Ministry of Agriculture, Fisheries and Food
Directorate of Fisheries Research
Fisheries Laboratory
Lowestoft, Suffolk NR33 0HT, United Kingdom

ABSTRACT

A multi-compartmental box model of the Irish Sea has been developed
to predict the distribution and radiological consequences of radionuclides
discharged from the Sellafield reprocessing plant. The box structure was
based on observations of radionuclide distributions in the sea bed and the
water circulation was generated from extensive time-series data on ^{137}Cs
concentrations in seawater. Measurements of naturally-occurring nuclides
provided both data on the extent and rate of these processes and a means
to validate the model assumptions.

The model structure is briefly outlined, comparisons are made between
model predictions and field observation, and some of the difficulties in
making such comparisons are discussed.

INTRODUCTION

Low-level radioactive waste has been discharged, under authorization,
into the eastern Irish Sea since 1952 from the BNFL reprocessing plant at
Sellafield. The transuranium elements plutonium and americium have been
important constituents of this waste. Both have a high affinity for
particle surfaces, to the extent that a large proportion of the total
quantity discharged is presently residing in the sea bed close to the
Cumbrian coast [1]. An integral part of the assessment of the radiological consequences of these releases has been the development of a multi-
compartmental box model to predict the behaviour of long-lived radionuclides within the Irish Sea, and the resultant future dose to man on an
annual basis. This paper describes the development and validation of the
model.

MODEL DEVELOPMENT

The regional box structure (Figure 1) was initially based largely on the observed distribution of plutonium in the sea bed, as of 1977/78 [1], and the observed distribution of sediment types – with finer resolution near the radionuclide source along the Cumbrian coast. The sea bed was represented by up to four layers and an interface box was included to accommodate sea-bed–water column exchanges. The initial box thicknesses were based on estimated sediment mixing depths and rates, and were revised as additional field data became available.

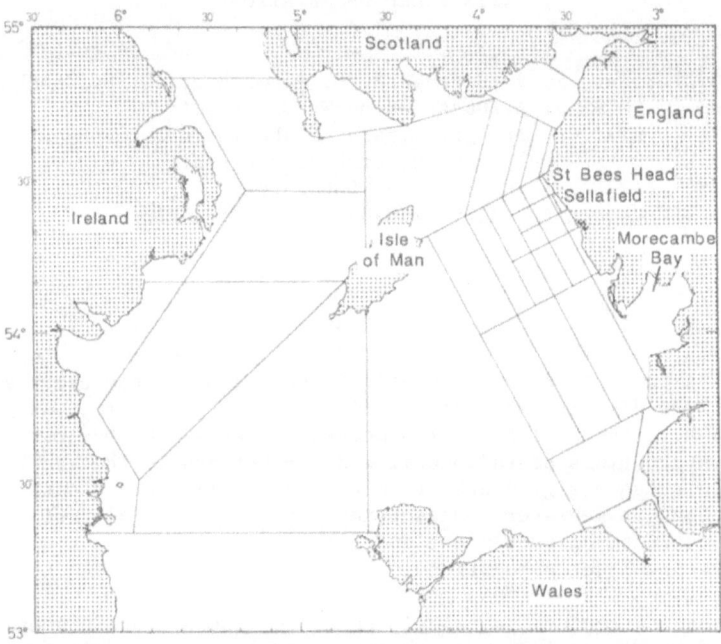

Figure 1. Location map and outline of box regions.

The water circulation was generated from ^{137}Cs discharge and time-series observation data [2,3] by trial and error to obtain a 'good fit' of the predictions with the observations [4], incorporating the gross features of the annual mean circulation and introducing an eddy diffusivity of 10^5 cm^2 s^{-1} to calculate mixing between boxes. The ^{137}Cs concentrations calculated from this data set fitted the observed values reasonably well but in the eastern Irish Sea were overestimated and in the southwest were underestimated by a factor of about five. Discrepancies occurred near the discharge point, illustrating the difficulty of comparing calculated concentrations, averaged annually over a relatively large area, with 'snapshot' observations which varied significantly in line with the reported monthly variation in discharge rate and with distance from the pipeline.

The model was designed to incorporate a number of sedimentary processes to allow the modelling of the non-conservative behaviour of plutonium. These included particle scavenging in the water column, particle mixing within the sea bed by bioturbation, pore-water diffusion and advection, and particle resuspension and sedimentation. Suspended loads in the water boxes were allowed to vary regionally, based on observed loads over several years, and the interface box had a constant, but very high (e.g. 100 g l^{-1}), suspended load to simulate repeated particle resuspension in a thin layer (0.1 m) above the sea bed. The partitioning of radionuclides between sediment and seawater was described by a distribution coefficient (K_D), assuming reversible adsorption/desorption. Particles in the water boxes were allowed to move laterally with the water, but communication in the interface box was restricted to vertical movement. A number of other variables, such as the chemical form and associations of plutonium, were considered but were excluded on the grounds that more fundamental factors, such as the flow field and simple sediment-radionuclide interactions, needed to be thoroughly understood before additional sophistication was warranted. A fuller description of the model and an explanation of the methods used is given by Gurbutt et al. [4].

Running the model with these initial assumptions and parameter values produced a reasonable match of model prediction with the 1977/78 observed plutonium inventory distribution (Figure 2) [4]. But there were a number of discrepancies, with the model predicting relatively high sea-bed inventories southwards along the English coast but failing to reproduce the characteristic northwards tail of higher inventories around St Bees Head.

There followed a series of model runs in which various parameter values were altered to see their effect on the plutonium distribution. It was then decided to test some of the model assumptions using field measurements of naturally-occurring radionuclides.

MODEL SIMULATIONS USING NATURALLY-OCCURRING RADIONUCLIDES

The time-dependent, point-source nature of the Sellafield discharge has an overwhelming influence on the pattern of distribution of plutonium in the Irish Sea. To counter the possibility that model inaccuracies were being masked by this factor, a number of model assumptions and parameter values were tested by modelling the behaviour of the naturally-occurring radionuclide ^{234}Th ($t_{1/2}$ = 24 d). This radionuclide is particle-reactive (coastal water $K_D \sim 2$ x 10^6 l kg^{-1}) and is produced by α-decay of ^{238}U ($t_{1/2}$ = 4.5 x 10^9 yr). The latter is relatively soluble and behaves conservatively in seawater. For the model runs the ^{238}U concentration in the water boxes, and hence the rate of supply of dissolved ^{234}Th, was assumed to be constant.

Measurements of ^{234}Th/^{238}U disequilibria in filtered seawater, suspended particulate and sea-bed sediment [5,6] provided information on dissolved ^{234}Th scavenging rates and mean particle residence times, indicating that the removal of ^{234}Th was not limited by the rate of adsorption and there was sufficient time to establish an equilibrium K_D. This supported the use of the K_D approximation in the model for other particle-reactive nuclides.

Further measurements of the vertical distribution of unsupported ^{234}Th and unsupported ^{210}Pb ($t_{1/2}$ = 22 yr) within the sea bed were used to redefine the thicknesses of the sediment boxes. The uppermost layer was reduced from 20 cm to 5 cm thickness - the depth range over which most of the unsupported ^{234}Th was found - and the layer beneath was increased from 20 cm to 66 cm thickness. The latter interval was selected particularly

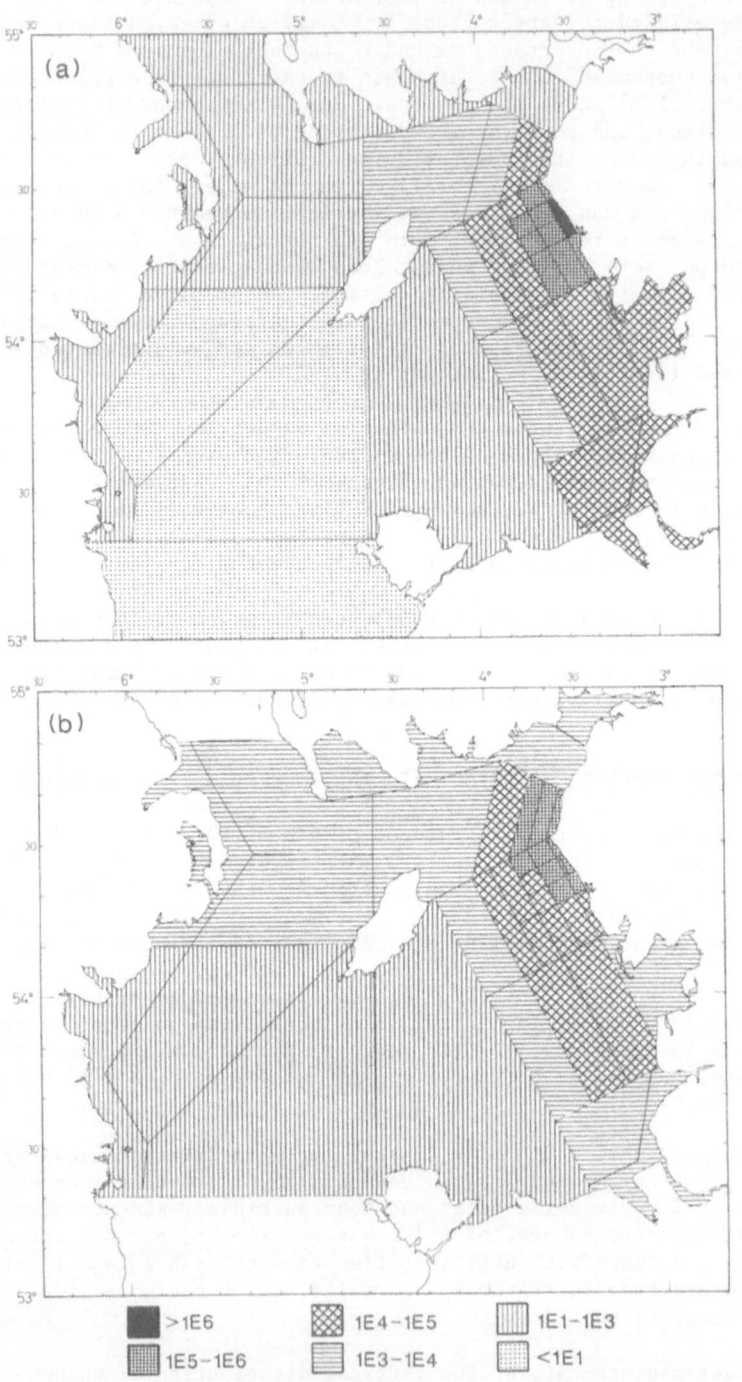

Figure 2. Comparison of plutonium sea-bed inventory
(Bq m^{-2}) distribution for 1979: (a) model
prediction; (b) field observation.

to fit an extensive data set of plutonium concentrations, from samples collected in 1983, which provided coarse resolution depth profiles at a large number of sites [7]. The third layer, with an interval of 66 to 200 cm, was included in regions of muddy sediments where bioturbation was more extensive. The fourth layer acted as a source/sink of particles to conserve particle mass.

In the initial estimates of bioturbation rates, obtained from 1-D modelling of the ^{234}Th and ^{210}Pb profiles, it was assumed that bioturbation could be modelled as a solid-particle biodiffusion process. Detailed field observations of the life-styles of the principal bioturbating organisms [8,9] brought this assumption into question, prompting attempts to reproduce the 'conveyor-belt' type of mixing which had been observed [4, 6,10]. When bioturbation rates obtained using the diffusive mixing assumption were applied to the model, plutonium was transferred downwards at too high a rate. In the latest series of model simulations, the vertical exchanges between the sediment boxes were altered in a rather arbitrary manner to obtain a best fit between the model predictions and field observations. This demonstrated the need to adopt a broad approach to the obtainment and interpretation of field data, and the need to be wary of unquestionably accepting the products of even simple mathematical treatments.

Radiocarbon dating of sediment cores was undertaken to provide information on sediment accumulation rates - probably the most critical factor in determining whether the sea bed will act as a permanent sink or a continuing source of Sellafield-derived radionuclides. The sediment ^{14}C age profile was modified by bioturbation down to at least 150 cm [11] . Present best estimates, from ^{14}C age dating of Turrittella communis shells, give slow accumulation rates of about 0.05 cm yr^{-1} over the past 3000 years. In view of these slow rates, and the uncertainty as to their general applicability, net sedimentation was assumed to be zero in the model simulations.

The use of differential suspended loads in the water boxes produced 'deficits' and 'surpluses' of suspended particulates, as the particles tended to move down particle concentration gradients. This resulted in spurious 'erosion' and 'deposition', to conserve particle mass, which tended to dominate the early simulations of the plutonium distribution [4]. For simulations of ^{234}Th behaviour, and later simulations of plutonium, the water box suspended loads were kept constant (1 mg l^{-1}). The effects of enhanced resuspension could still be included by increasing the suspended load in the interface boxes, which had the effect of increasing the transfer of ^{234}Th to the sea bed.

Measured sea-bed inventories of unsupported ^{234}Th indicated that enhanced scavenging occurred in the shallower (< 25-30 m) water of the eastern Irish Sea [5,6]. Therefore, higher suspended loads were given to those interface boxes in regions where the mean depth of the water box was less than 20 m. This increased the scavenging capacity of these box regions and raised the sea-bed ^{234}Th inventories, but it was not possible to reproduce the observed high sea-bed inventories whilst maintaining the observed ^{234}Th water concentrations. This discrepancy may have represented a deficiency in the data - for example that the study area was too restricted - but equally may demonstrate that the time and space scales used in the model were inadequate to reproduce events such as storm-induced resuspension and intra-basinal erosion and deposition which may dominate the redistribution of ^{234}Th and be responsible for the apparent imbalance between the sources and sinks of ^{234}Th.

MODEL SIMULATIONS OF THE PLUTONIUM DISTRIBUTION

Some of the problems encountered in the earlier runs at matching model predictions with the observed plutonium distribution have already been mentioned. The adjustments made to the model's vertical structure, suspended load distribution and vertical mixing rates during the ^{234}Th simulations were applied to a later series of plutonium simulations, with mixed results.

The excessive southerly spread of plutonium was removed but the tail of relatively high sea-bed inventories to the north of the pipeline around St Bees Head, a prominent feature of the observed distribution, was still poorly developed (Figure 3). For reasons discussed elsewhere [1,5], a closer match may be unlikely until consideration is given to the asymmetry of the sea-tank discharge, which is released around high water when the tide is flowing to the northwest, and to introducing a two-layer structure in the water column, following recent current meter measurements which indicated a de-coupling of flow around St Bees Head.

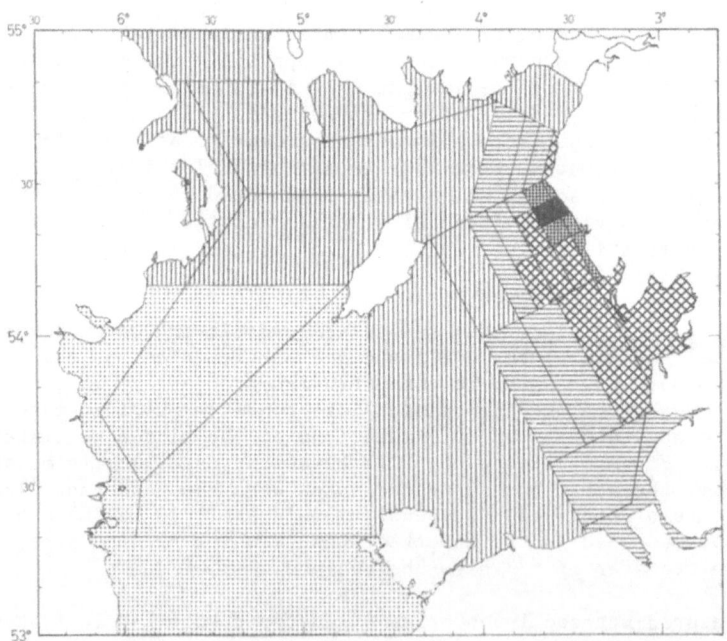

Figure 3. Modified model prediction of plutonium sea-bed inventory (Bq m^{-2}) distribution for 1979 (for key, see Figure 2).

Predicted sediment concentrations in the immediate vicinity of the discharge point were higher than measured values. Predicted seawater concentrations in these regions were up to an order of magnitude too high. The vertical, sea-bed distribution was better matched by keeping the interface box suspended loads constant (Figure 4). It should be noted

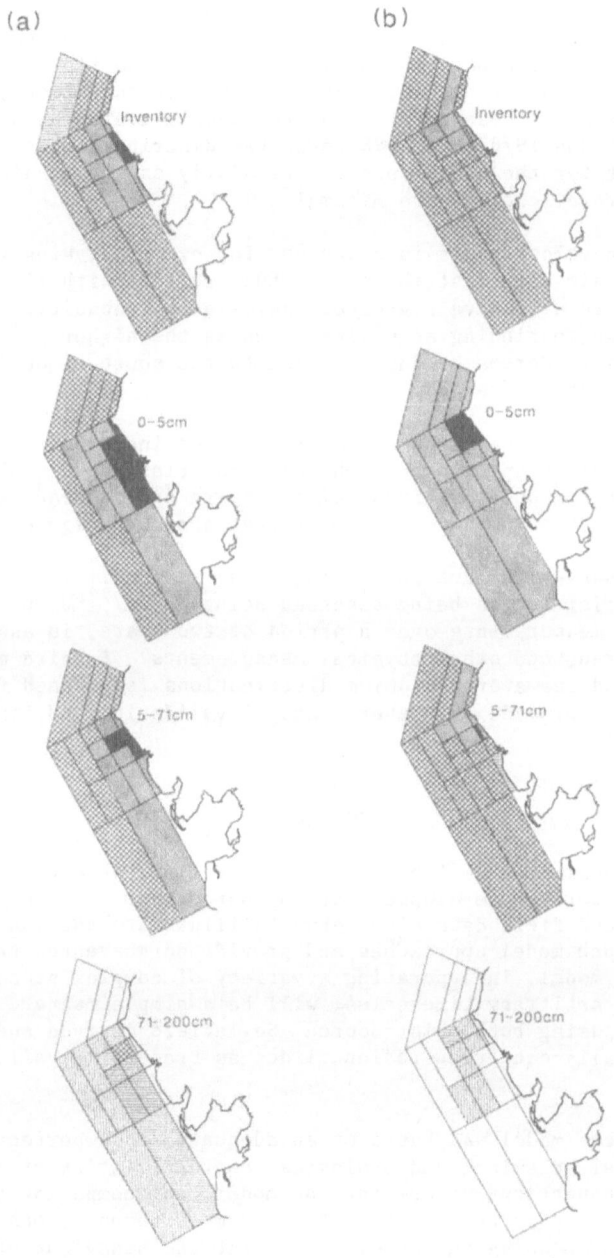

Figure 4. Comparison of plutonium sea-bed inventory (Bq m^{-2}) concentrations (Bq m^{-3})
in sediment layers for 1983: (a) model
prediction; (b) field observation
(for key, see Figure 2).

that the estimated total seawater plus sea-bed inventory of plutonium, based on extensive field measurements, can only account for about 70% of the total quantity reported as being discharged [1]. For this reason alone, a perfect match between model predictions and field observations is highly unlikely. Nevertheless, with some further adjustment, the present model may adequately describe the future, yearly-averaged, distribution of plutonium with the 1978/79 or 1983 observed distribution used as the starting point for the model runs, particularly as the predicted quantities for future discharges are minimal.

Further development is in progress, for example taking into account the varying grain size distribution. Thus, regions with fine-grained bottom sediments will have a greater assimilative capacity. This may help to resolve some continuing anomalies, such as the higher predicted plutonium inventory in Morecambe Bay, an area to the south of Sellafield composed predominantly of sands.

Re-analysis of the ^{137}Cs seawater data has indicated a need to consider a post-1976 step change in the eastern Irish Sea circulation. Initial runs with the updated circulation pattern provided an improvement in the predicted plutonium inventory over that shown in Figures 2-4.

Field observations are continuing. The seasonality of scavenging in the eastern Irish Sea is being assessed using ^{234}Th/^{238}U and ^{210}Pb/^{238}U disequilibria measurements over a period of two years, in association with long-term current and other physical measurements. A third major survey of sediment and seawater plutonium distributions is planned for 1988. These data will provide a further means of validating and improving the existing model.

CONCLUSION

The possession of reliable data sets of field observations provides justification for the development of an appropriate environmental transfer model. But good field data also serve to illustrate the inadequacies inherent in such model approaches and provide no guarantee that the establishment of a model, incorporating a variety of complex processes, perhaps averaged over arbitrary timescales, will be a simple matter. There are advantages in using both point-source, Sellafield-derived and multi-source, naturally-occurring radionuclides as tracers to validate the model.

The present model was based on an adequate, if imperfect, knowledge of the physical, chemical and biological characteristics of the Irish Sea. Testing the assumptions underlying the model, and comparing the model predictions with a large data set of field measurements, has given increased confidence in using the model to predict the behaviour of long-lived radionuclides in coastal waters, but there is still scope for further refinement.

REFERENCES

1. Pentreath, R. J., Lovett, M. B., Jefferies, D. F., Woodhead, D. S., Talbot, J. W. and Mitchell, N. T., The impact on public radiation exposure of transuranium nuclides discharged in liquid wastes from fuel element reprocessing at Sellafield, UK. Rep. Int. Atom. Energy Agency, Vienna, 1984, CN-43, 315-329.

2. Jefferies, D. F., Steele, A. K. and Preston, A., Further studies on the distribution of ^{137}Cs in British coastal waters - I. Irish Sea. Deep-Sea Res., 1982, 29 (6A), 713-738.

3. Hunt, G. J., Radioactivity in surface and coastal waters of the British Isles, 1985. Aquat. Environ. Monit. Rep., MAFF Direct. Fish. Res., Lowestoft, 1986, 14, 48pp.

4. Gurbutt, P. A., Kershaw, P. J. and Durance, J. A., Modelling the distribution of soluble and particle-adsorbed radionuclides in the Irish Sea. In Radioactivity and Oceanography : Radionuclides - a Tool for Oceanography, Elsevier Applied Science Publishers, London, (in press).

5. Kershaw, P. J. and Young, A., Scavenging of ^{234}Th in the Eastern Irish Sea. J. Environ. Radioactivity, (in press).

6. Kershaw, P. J. Gurbutt, P. A., Young, A. K. and Allington, D. J., Scavenging and bioturbation in the Irish Sea from measurements of ^{234}Th/^{238}U and ^{210}Pb/^{226}Ra disequilibria. In Radioactivity and Oceanography : Radionuclides - a Tool for Oceanography. Elsevier Applied Science Publishers, London, (in press).

7. Woodhead, D. S., Mixing processes in nearshore marine sediments as inferred from the distributions of radionuclides discharged into the northeast Irish Sea from BNFL, Sellafield. In Radioactivity and Oceanography : Radionuclides - a Tool for Oceanography, Elsevier Applied Science Publishers, London, (in press).

8. Kershaw, P. J., Swift, D. J., Pentreath, R. J. and Lovett, M. B., The incorporation of plutonium, americium and curium into the Irish Sea sea bed by biological activity. Sci. Total Environ., 1984, 40, 61-81.

9. Swift, D. J. and Kershaw, P. J., Bioturbation of contaminated sediments in the north-east Irish Sea. ICES C.M. 1986/E:18, 12 pp. (mimeo).

10. Gurbutt, P. A. and Kershaw, P. J., Biological mixing of shelf seas sediments with implications for modelling. ICES C.M. 1987/C:22, 13 pp. (mimeo).

11. Kershaw, P. J., Radiocarbon dating of Irish Sea sediments. Estuar. Cstl. Shelf Sci., 1986, 23, 295-303.

COMPARISON OF MODELS FOR ASSESSING THE RADIOLOGICAL IMPACTS
OF DEEP-SEA DISPOSAL OF RADIOACTIVE WASTES

M. Chartier, X. Durrieu de Madron
COMMISSARIAT A L'ENERGIE ATOMIQUE
IPSN/DPS/SEAPS/LES
B.P. n° 6, 92265 FONTENAY-aux-ROSES Cedex
FRANCE

and

M. Poulin
Ecole des Mines de PARIS
32, rue St Honoré, 77305 FONTAINEBLEAU
FRANCE

ABSTRACT

The radionuclides dispersion from the low-level waste drums dumped in the North Atlantic NEA site has been assessed by three independent box models developed in UK, in USA and in France within the international "Co-ordinated Research and Environmental Surveillance Programme".

These models take into account the time variations of the release rates of the nuclides out of the drums, a great deal of physical and geochemical marine processes and the transfer to man via various pathways. In order to assess the reliability of these complex transfer models, an intercomparison of UK and USA models had been performed on a benchmark problem.

The french model has been run on this problem and the results were compared with UK and USA models results. Concentration differences of up to 2 or 3 orders of magnitude have been observed in boxes potentially important for the dose calculations. These differences originate in the different parameterizations of the scavenging by the particles and of the transfer at the water/sediment interface, the different vertical diffusion and advection by the ocean fluid, and the different geometry of the boxes.

A simplified sensitivity analysis has been performed with the CEA model to crudely assess the reliability of the dose calculations. This allowed to identify the features of the CEA model which would require careful reconsideration if more precise and realistic dose assessment were requested : the sedimentation and resuspension of the particles, the nuclides transfer between the coastal and ocean boxes, the nuclides transfer within the sediment and the parameterization of specific coastal processes.

INTRODUCTION

The dumping of radioactive waste at sea is a method of disposal which has been carried out in the deep Atlantic since 1949. In 1977, a Decision of the Organisation for Economic Co-operation and Development (OECD) Council established a Multilateral Consultation and Surveillance Mechanism for Sea Dumping of Radioactive Waste. Under the terms of the Mechanism, NEA is requested to periodically assess the suitability of dumping sites proposed or used by participating countries. It has been recommended that for the assessments of the suitability of the North-East Atlantic Site, a concerted effort should be made to increase the scientific data base related to the oceanographic and biological characteristics of the dumpsite area and to develop a site specific model of the transfers of radionuclides from the dump area to human populations. To fulfil these objectives, NEA established in 1981 an international "Co-ordinated Research and Environmental Surveillance Programme" (CRESP).

CRESP experts defined a modelling framework for the radiological assessment : models of release from waste forms, benthic boundary layer, sedimentation, dispersions in oceans and coastal waters, biological transports and pathways to man are integrated in a composite model which assesses the risk associated with the sea dumping practice.

In France, a numerical integrated model has been recently developed to provide a radiological assessment of the risk, according to the modelling framework recommended by CRESP. This article focuses on the geochemical and geophysical transfer submodels of this composite model.

DESCRIPTION OF THE FRENCH MODEL

A classical numerical modelling method has been followed : the world ocean and the upper layer of bottom sediment have been divided in 55 compartments. The horizontal geometry of the model (figure 1) has been designed to increase the resolution in the North-Atlantic and to fit the topography of this ocean (the Mid-Atlantic Ridge is a natural barrier). Several coastal boxes have been added since a number of potentially important radiological pathways originate in the coastal waters. The vertical geometry of the ocean model is based on the vertical structure of the water masses and on the vertical shear of the circulation. A sediment box corresponding to the bioturbated layer has been added under all the bottom ocean and coastal boxes. A Bottom Boundary Layer (BBL) box has been inserted between the sediment and ocean boxes.

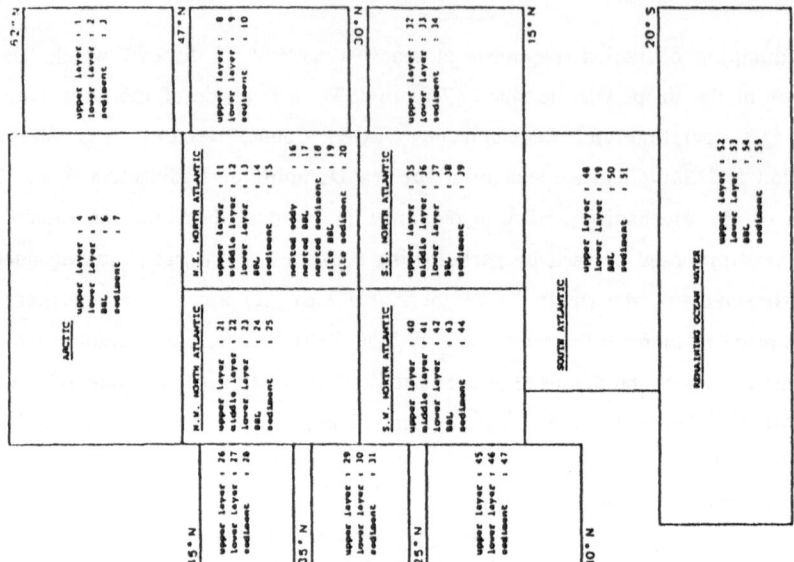

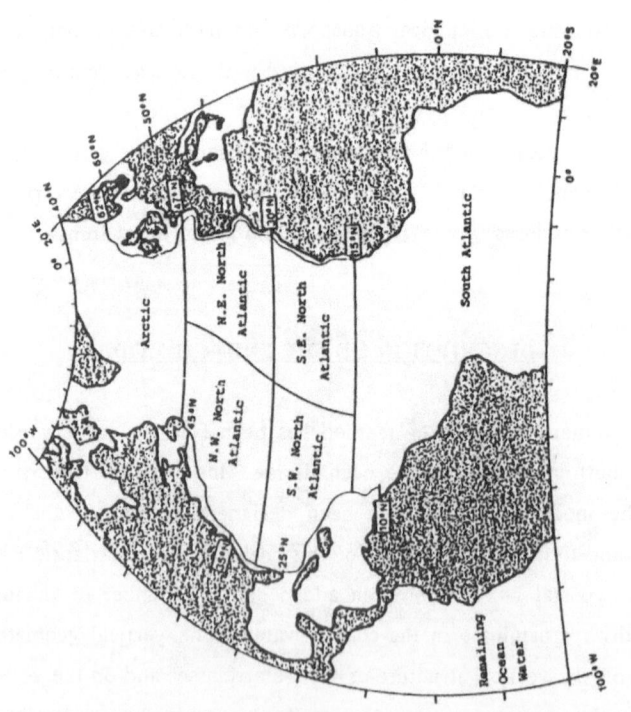

Figure 1

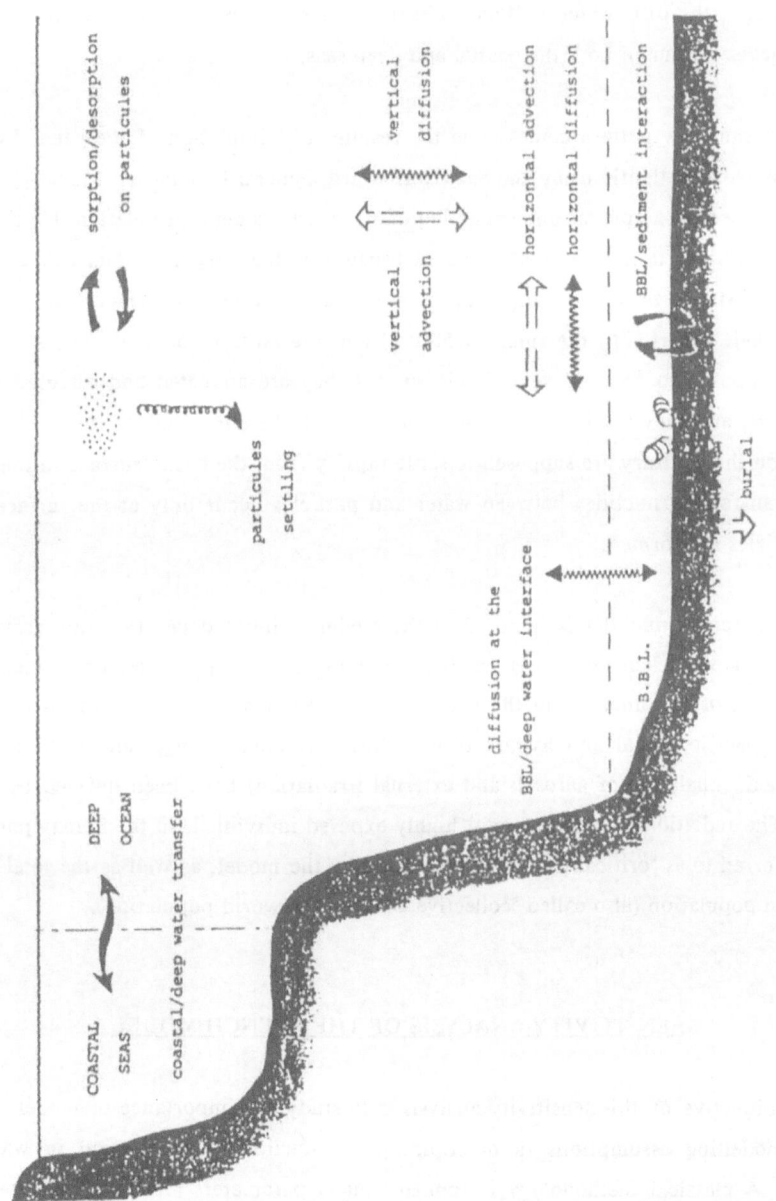

Figure 2

The model has been designed to take into account the main transport processes of the nuclides. Figure 2 is a schematic representation of the explicitely modelled processes which transfer the nuclides from one box to another. These processes are the horizontal and vertical diffusion by the water turbulence, the sorption-desorption on particles, the settling of these particles, the BBL water-bottom sediment interaction and the burial in the sediment. These processes occur in both the coastal and deep seas.

The ocean flow pattern comes from the results of a simulation of the general circulation in the North Atlantic using the Sandia-Harvard General Circulation Model [1], mixed with data based on a general understanding of the world general circulation. No advection is allowed between ocean and coastal boxes because of the very few data available. Two classes of particles have been defined, as it is recommended by the GESAMP working group in their report [2] : the small (< 50 um) and the large (> 50 um). The small particles are supposed to descend very slowly so that they are advected and diffused like the ocean water, and they are in constant equilibrium with the surrounding water. The large particles on the contrary are supposed to settle rapidly from the ocean surface to the bottom so that transfers of nuclides between water and particles occur only at the surface where large particles are formed.

The source of nuclides is imposed in the model as input data : the time variation of the release rate of each nuclide is specified. For radiological purposes, doses resulting of the spreading of the nuclides in the sea water and sediment are calculated. An extended series of specific actual and hypothetical pathways (including ingestion of contaminated marine food, inhalation of aerosols and external irradiation) have been defined by CRESP experts. The radiation dose to the most highly exposed individuals in the human population (often referred to as "critical group") is calculated by the model, as well as the total dose to the human population (also called "collective dose" to the world population).

SENSITIVITY ANALYSIS OF THE FRENCH MODEL

The objective of this sensitivity analysis is to study the importance of model parameters or modelling assumptions in determining the nuclides concentration in water and sediment. A classical methodology is applied : many parameters are varied independently and the model is run each time on a long period of simulation : the influence of each parameter is thus studied for both the short and long term. The value of the parameters considered is the extreme low or high value of the possible range of variation of the parameter, as defined by CRESP experts. The results are compared to the "best estimate" case, calculated with the best estimate value recommended by CRESP experts for all the parameters. The radionuclides considered in the sensitivity analysis are ^{14}C, ^{137}Cs, ^{129}I, ^{230}Th and ^{239}Pu, which span a large range of radioactive half-lives and partition coefficients K_d. The parameters and assumptions investigated are the following : porosity, pore water diffusivity, sediment thickness of the bioturbated layer, bioturbation coefficient, settling velocities of the small and large particles, suspended sediment loads in small and large particles, vertical eddy diffusivity and advection by the ocean circulation.

The results show that among the investigated nuclides ^{230}Th (medium half life and very high K_d) and ^{137}Cs (short half life and rather low K_d) are the most sensitive. ^{230}Th is sensitive to sediment loads and particles settling velocities (difference of one order of magnitude for the peak concentration in water boxes) and ^{137}Cs is sensitive to suspended sediment loads, ocean mixing and advective fluxes (variation of one order of magnitude of the maximum concentration in far field ocean boxes).

The concentration in the sediment boxes is influenced by the bioturbated layer thickness and the porosity (variation of concentration of one up to four orders of magnitude for ^{14}C, ^{129}I and ^{239}Pu), and is sensitive to the pore water diffusion coefficient (variations of concentration of one order of magnitude for all the nuclides). In the water boxes, the concentration is affected by variations of the advective fluxes (variations of concentration of more than one order of magnitude in the far field boxes for ^{230}Th and ^{137}Cs). In the near field boxes, the concentration varies quasi-linearly with respect to the vertical eddy diffusivity. This coefficient also influenced the concentration in the far field boxes for highly sorbed nuclides.

When the advective fluxes are set to zero, the peak activity in the water boxes bordering the site is reached 100 years later than for the base case, and for the sediment boxes, the difference is about 200/300 years. The time of maximum activity also depends strongly on the vertical eddy diffusivity coefficient.

The conclusions of this sensitivity analysis are that the most sensitive features of the transfer processes in the marine environment concerns the parameterization of the interactions between BBL and sediment, and the value of the associated "sediment parameters". It is worthy to note that variations of one or two orders of magnitude have been currently observed. Moreover, none of the possible parameter values recommended by the CRESP experts has been found to give a maximum of concentration for all the nuclides and for all the boxes concerned by the various radiological pathways.

COMPARISON OF THE F/CEA MODEL WITH THE UK/MAFF-NRPB AND US/TASC-SDP MODELS ON A BENCHMARK PROBLEM

Composite assessment models had been developed in the United Kingdom by the Ministry of Agriculture, Food and Fisheries (MAFF) and the National Radiological Protection Board (NRPB), and in the United States by the Analytic Sciences Corporation (TASC) for the Subseabed Disposal Programme (SDP). Following CRESP Executive Group recommendations, the UK/MAFF-NRPB and US/TASC-SDP box models of radionuclides transfer in the world marine environment had been compared on a purposeful benchmark problem [3]. The F/CEA model has also been run on the same benchmark problem. This section displays the results of the comparison of the french models with the other two models.

The benchmark problem consists in the release of five radionuclides (^{14}C, ^{137}Cs, ^{129}I, ^{239}Pu, ^{230}Th) at a constant rate (1 TBq.y^{-1}) for two finite times (40 years and 500 years). For each release of each radionuclide, the following quantities are calculated :

- the concentration of the radionuclide in the water column immediately above the dumpsite ;
- the concentration in the water column 500 m above the site ;
- the concentration at depths corresponding to the mid-points of the main compartments of the ocean model which are in the area of the ocean in which the dumpsite is located.

The UK/MAFF-NRPB model is described in full in the NEA report on the Site Suitability Review [4], and the US/TASC-SDP model in the Sandia Report "Preliminary NEA Dumpsite Safety Analysis" [5]. The three models have the same compartmental structure, but with different geometries, and then different flow patterns. For example, figure 3 enlights the different vertical geometries of the model near the site. One should note that only the F/CEA model includes coastal boxes. The three models have similar water/particles interaction submodels overlaid on them, except that the UK/MAFF-NRPB model takes into account the partial particle dissolution in bottom waters. Most of the parameters have the same value in the three models except for the following ones : the

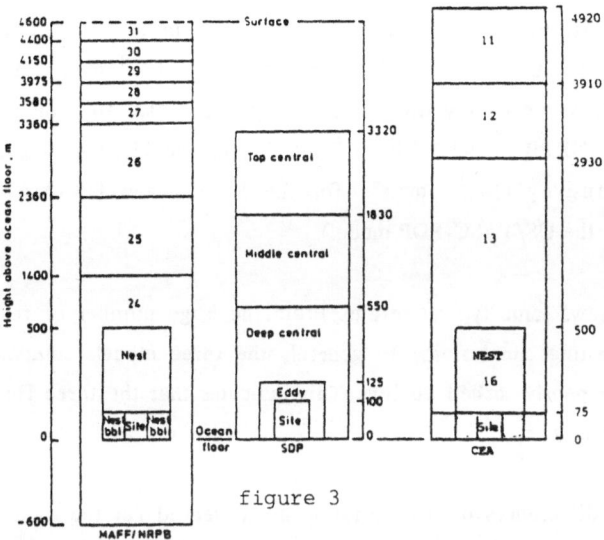

figure 3

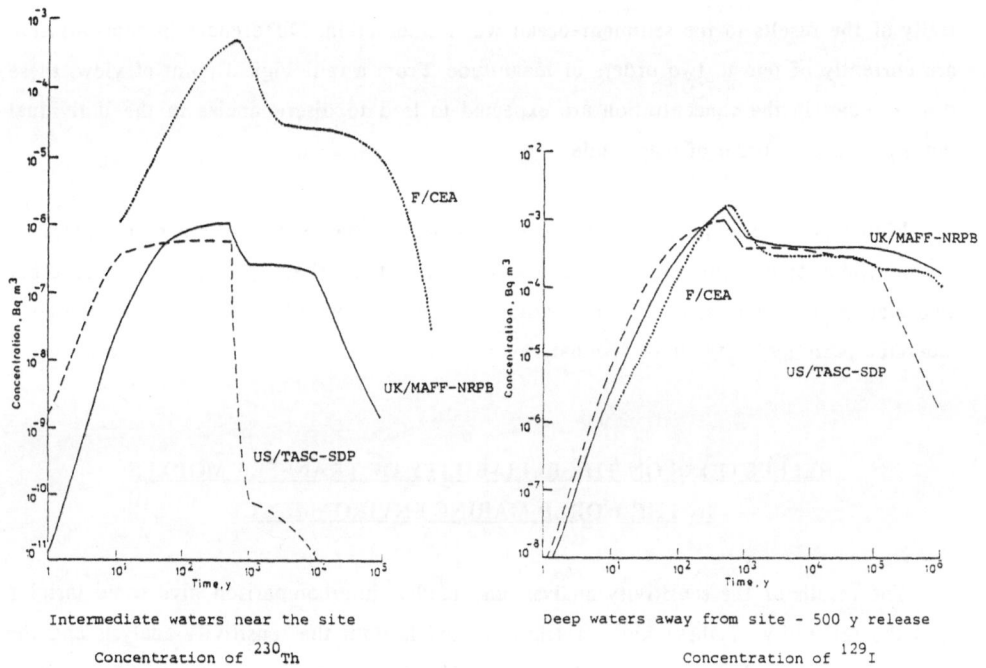

Intermediate waters near the site

Concentration of ^{230}Th

Deep waters away from site - 500 y release

Concentration of ^{129}I

figure 4

vertical eddy diffusivity in the BBL (2.5 10^{-3} $m^2.s^{-1}$ for the F/CEA model, 2.0 10^{-3} $m^2.s^{-1}$ for the UK/MAFF-NRPB model, 2.0 10^{-4} $m^2.s^{-1}$ for the US/TASC-SDP model), the vertical eddy diffusivity in the ocean (1.0 10^{-4} $m^2.s^{-1}$ for the F/CEA and UK/MAFF-NRPB models, 2.0 10^{-4} $m^2.s^{-1}$ for the US/TASC-SDP model) and the pore water diffusivity (10^{-10} $m^2.s^{-1}$ for the F/CEA and UK/MAFF-NRPB models, 10^{-9} $m^2.s^{-1}$ for the US/TASC-SDP model).

Figure 4 shows some typical results. From the large number of results obtained, we can draw interesting conclusions. In general, the three models compare rather well for ^{129}I, which is a poorly sorbed nuclide. This indicates that the three flow patterns do not exhibit large differences.

Significant differences of concentration are observed for the highly sorbed nuclides. They are mainly due to the differences in the parameterization of the sediment-ocean water interaction. This confirms the conclusion of the sensitivity analysis on the rather high sensitivity of the results to the sediment-ocean water interaction. Differences in concentration are currently of one to two orders of magnitude. From a radiological point of view, these discrepancies in the concentration are expected to lead to discrepancies in the individual doses of the same order of magnitude.

Moreover, it should be mentioned that none of the models produces estimates of radionuclides concentrations that are always higher than those predicted by the other two models. This implies that from a radiological angle, none of these models can be considered *a posteriori* as the most adequat.

REFLECTIONS ON THE RELIABILITY OF TRANSFER MODELS
IN THE WORLD MARINE ENVIRONMENT

The results of the sensitivity analysis and models intercomparison give some insights into the reliability of these kind of transfer model. Both the sensitivity analysis and the models intercomparison exhibit concentration differences of a few orders of magnitude which can be considered as a gross approximation of the uncertainty of the results.

Careful studies of the model accuracy should in fact take into account other modelling features : e.g. the scarcity of some key data, a certain subjectivity in the definition of the box geometry arrangement, or the influence of the numerics. One emphasizes in this section that the sensitivity analysis and the models intercomparison described above are both a very profitable but incomplete assessment of the reliability of these transfer models.

Nowaday, the assessment of the reliability of the transfer models remains the result of a synthesis of several techniques, which independently give insights to some aspects of the accuracy of the simulations : validation, sensitivity analysis, models intercomparison and uncertainty analysis. The validation of world ocean transfer models remains out of the science possibilities : no data set exists which is sufficiently extended in space and time to validate the world transport of nuclides in the ocean. One should note however that certain submodels of the composite model can be validated by field measurements. But no overall field-validation of the composite model is able at present to assess definitely the reliability of the results. The uncertainty analysis technique still faces the problem of the evaluation of uncertainty in parameter values. The specific problem of the composite world ocean transfer model is thus that uncertainty analysis and models intercomparison remain at the present time the only two reliable techniques available to assess the uncertainty of the dispersion simulations. The results described above are particularly profitable in this sense.

But one should also note that some inaccuracies of the models are not revealed by the sensitivity analysis or models intercomparison techniques. The following example enlights the inability of these techniques to identify the influence of the numerics in the results uncertainty. The three box models use an implicit or semi-implicit time scheme to allow a long enough time step. These schemes are known to create artificial computational diffusion. The results show for example that even after one time step (corresponding to one year in the F/CEA model) after the release starts, all the boxes are contaminated ! The time scheme is then expected to lead to inaccuracies in the time scales of the transfers, in the same way in the three models. Neither the sensitivity analysis nor the model intercomparison (limited here to the comparison of three models of compartmental type with similar space resolution) are able to identify this uncertainty of computational origin. This justifies the need of further research in that field and motivates the development of new strategies in the design of transfer model in the marine environment.

REFERENCES

1. Marietta,M.G., personal communication.

2. GESAMP, An Oceanographic model for the dispersion of wastes disposed of in the deep sea, Report N° 19 of the IMO/FAO/UNESCO/WMO/WHO/IAEA/UN/UNEP joint Group of Experts on the Scientific Aspects of Marine Pollution, International Atomic Energy Agency, Vienna, 1983.

3. Mobbs,S.F., Hill,M.D., Koplik,C.M. and Demuth,C., A preliminary comparison of models for the dispersion of radionuclides released into the deep ocean, NRPB-R194, National Radiological Protection Board, Chilton, Didcot, May 1986.

4. NEA, Review of the continued suitability of the dumping site for radioactive waste in the North East Atlantic, OECD/Nuclear Energy Agency, Paris, 1985, pp. 448.

5. Scott,J.I., Ensminger,D.A. and Koplik,C.M., Preliminary dumpsite analysis, SAND85-7157, Sandia National Laboratories, Albuquerque, September 1985.

MODELLING THE SEA-TO-LAND TRANSFER OF MARINE DISCHARGES FROM SELLAFIELD - VALIDATION AGAINST ENVIRONMENTAL MEASUREMENTS

J.M. Howorth and A.E.J. Eggleton
Environmental and Medical Sciences Division
Harwell Laboratory
Didcot, Oxfordshire OX11 ORA.
UK

ABSTRACT

Radionuclides discharged into the sea may return to land in the marine aerosol. The extent and significance of this transfer around Sellafield has been assessed using a series of computer models.

A marine model calculates the dispersion of radioactivity through the Eastern Irish Sea. Validation of the model is by comparison against the observed distributions of ^{241}Am. Transfer of radionuclides to the land is empirically modelled using measured deposition. This simple model reproduces both current and cumulative deposition well. Validation of airborne activity against time series data is less good.

The maximum dose to the average person in Seascale due to the transfer of actinides was calculated to be 24 μSv in 1973, which will reduce to 2 μSv in 2000. However, use of site-specific data may change these predictions.

INTRODUCTION

Measurements of radioactivity in Cumbria have shown the existence of a 'maritime effect' in which radionuclides, originally discharged into the sea from Sellafield, are resuspended from the sea surface and returned to the land in the marine aerosol [1,2]. The transfer causes enhanced deposition of some radionuclides along the coast and $^{239+240}$Pu deposits ten times higher than those expected from nuclear weapons fallout have been measured [1]. Approximately 0.07 TBq of marine-derived $^{239+240}$Pu has been deposited in a 5 km-wide coastal strip in W.Cumbria. Whilst this deposition is only 0.01% of cumulative marine discharges, it is of a similar size to routine atmospheric discharges from the site.

A series of computer models has been developed to estimate the radiological significance of the transfer [2,3]. These models calculate dispersion of radionuclides through the Eastern Irish Sea; the rates of deposition on land; airborne concentrations and the doses received by individuals in Cumbria. Validation of each model in this series is essential as the model output forms the input data for the next model. Validation by comparison against environmental data is described and the importance of time-series data for dynamic models is discussed.

STRUCTURE AND VALIDATION OF THE MARINE MODEL

The marine model is a 2-D, annually-averaged dispersion model. Radionuclides are assumed to be dissolved in seawater, adsorbed onto suspended particles or deposited on the seabed. The model uses a 6 km orthogonal grid which reduces to 1.5 km near the discharge point.

^{137}Cs is almost conservative in the water column and may be used as an effective tracer of water flows. The advection and diffusion rates were based, initially, on measured values and were adjusted until observed distributions of ^{137}Cs in the Irish Sea [4], calculated flushing times and ^{137}Cs budgets for the area [5] were reproduced. Measurements of ^{137}Cs reflect both the rapidly varying discharge rates [5] and time-varying flow rates. Comparisons between annually averaged calculated concentrations and spot measurements are, therefore, difficult. However, by repeating these comparisons with data collected over many years, some uncertainties may be reduced.

Unlike caesium, actinides are rapidly lost from the water column as they adsorb onto fine particles. Radionuclides dissolved in seawater and adsorbed on suspended particles are assumed to be in equilibrium and this assumption is supported by measurements of the partition of actinides between the two phases (the K_d) which are relatively constant across the Eastern Irish Sea [6].

A review of sedimentation in the Eastern Irish Sea concluded that evidence for erosion from or deposition onto the mud bank offshore from Sellafield is inconclusive and that the area is basically stable, but dominated by biological activity [7]. Thus, a second assumption is made that the fine mud particles are in a state of dynamic equilibrium with no net resuspension from or deposition onto the seabed.

Parameters controlling exchange rates between the three phases of the model were initially based on measured values and were adjusted until the model reproduced measurements of plutonium dissolved in seawater in

263

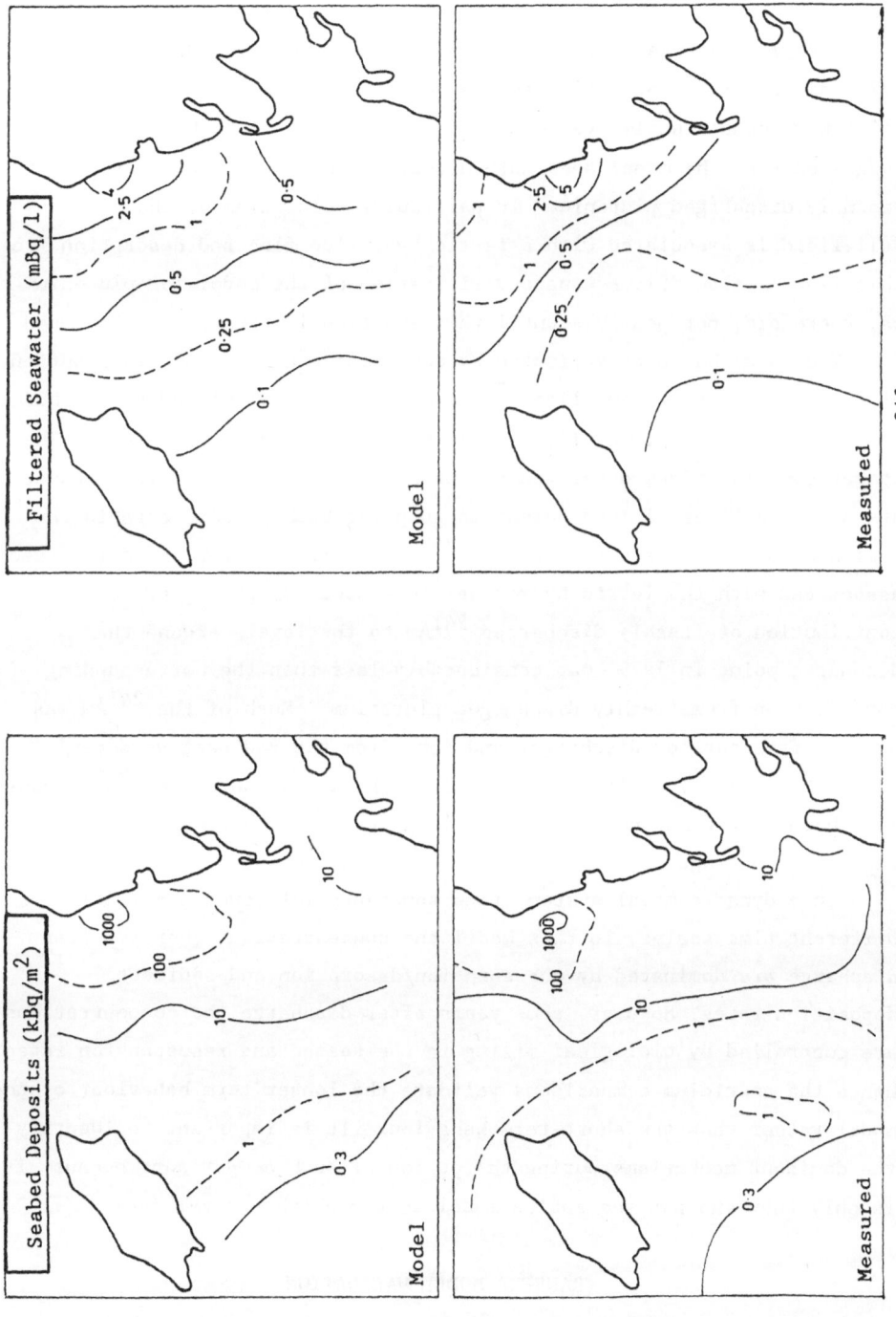

Figure 1. A comparison of calculated and measured [6] levels of ^{241}Am in the Eastern Irish Sea

1973, 1974 and 1979 and deposited on the seabed in 1978 [6].

Agreement between calculated and measured plutonium deposition in 1978 was good [1]. However, dissolved concentrations around the discharge point were consistently over-estimated by a factor of 5-10. It is suggested that the model does not adequately represent the behaviour of recently discharged plutonium. In particular, plutonium discharged from Sellafield is associated with a ferric hydroxide floc and desorption from this floc is slow [8]. Adequate verification of the model for plutonium is, therefore, not possible until this additional mechanism is included.

The model has been validated by comparing measured [6] distributions of an independent radionuclide, ^{241}Am, with those calculated by the model where only the experimentally determined K_d value has been changed. The comparisons for ^{241}Am deposited on the seabed and dissolved in seawater are shown in Figure 1. Agreement is good for both phases, even in the region around the discharge point. ^{241}Am is, like plutonium, discharged associated with the ferric hydroxide floc. However, the relative contribution of freshly discharged ^{241}Am to the levels around the discharge point in 1979, was considerably less than the corresponding contribution from freshly discharged plutonium. Much of the ^{241}Am was derived from earlier discharges and some from the radioactive decay of ^{241}Pu which was also discharged earlier. These two sources of ^{241}Am would have been in the marine environment long enough to have equilibrated with their surroundings.

In a dynamic model system, rate constants influence the model over different time scales. In this model the concentrations shortly after discharge are dominated by the sorption/desorption and sediment deposition rates. However, some years after discharge the concentrations are controlled by biological mixing in the seabed and resuspension rates. Hence the americium comparisons validate the longer term behaviour of the model rather than the short-term behaviour. It is important to identify the dominant mechanisms during the period of such comparisons because it is only they and not the entire model system which are validated.

TRANSFER MODEL VALIDATION

The rates of transfer of the marine aerosol have, to date, been reproduced by a simple empirical model. The calculated concentrations of radioactivity in the coastal grid cells of the marine model have been

related to measurements of cumulative deposition using deposition rates
which decrease exponentially with distance from the coast [2]. Although
simple, this model is able to reproduce 18 out of 21 measurements from an
independent survey to within a factor of two [2].

Continuous measurements of annual deposition have been made at
Eskmeals, a site 20 km south of Sellafield and 300m inland, since 1978
[9]. Figure 2 shows measured deposition rates and those calculated by the
transfer model. The agreement is generally good for an empirical model
and is within a factor of three for all years.

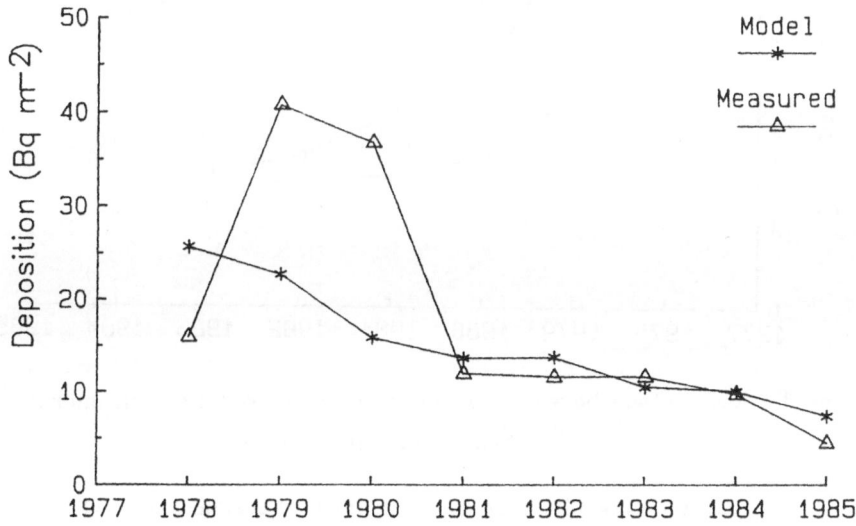

Figure 2. Comparison between modelled and measured [9] terrestrial
deposition of $^{239+240}$Pu at Eskmeals

The concentration of radionuclides in air has been derived from the
calculated deposition using effective deposition velocities to allow for
the changing aerosol size spectrum. These values were measured at varying
distances from the coast [2]. Calculated and measured concentrations of
plutonium at the Eskmeals site are within a factor of two between 1978
and 1985. However, examination of the data as a time series rather than
as individual points, as shown in Figure 3, shows different time-varying
behaviour. The calculated rate of decrease of the air concentrations is
less than that of the measured rate of decrease. Deposition at Eskmeals
is dominated by large spray droplets produced in the surf zone, but air
concentrations also contain a component from small spray droplets
produced offshore and capable of travelling long distances. The model

does not yet include these offshore sources and the relatively poor agreement between the time-varying behaviour of the measured and calculated data suggests either that this secondary source may be important, or that seasonal effects, for example meteorology, should be included in the model more explicitly.

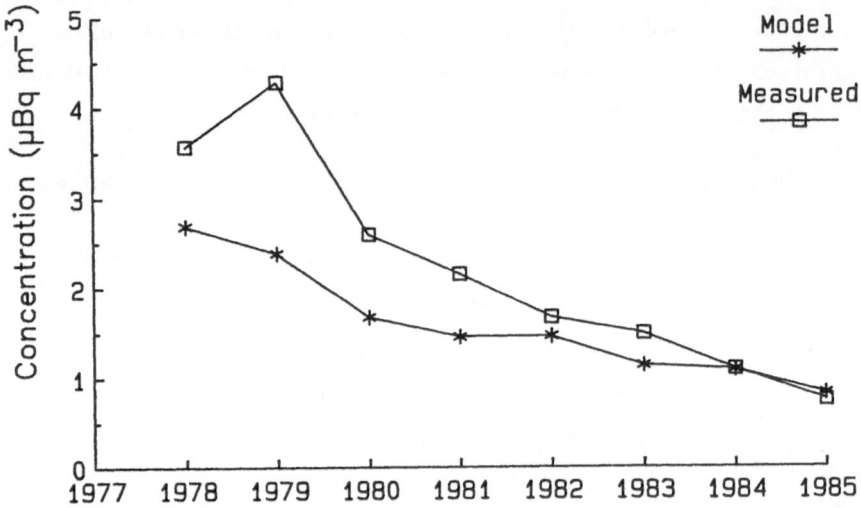

Figure 3.   Comparison between modelled and measured [9] concentrations of $^{239+240}$Pu in air at Eskmeals

USE OF SITE SPECIFIC DATA IN DOSE ASSESSMENT MODELS

The radiological significance of the transfer of actinides has been assessed for ingestion of terrestrial foodstuffs and inhalation [2]. The doses received by an average person living in Seascale resulting from the sea-to-land transfer of actinides are shown in Figure 4. The maximum dose was calculated to be 27μSv in 1973 and to reduce to 2μSv in 2000. The doses are dominated by the inhalation pathway.

However, root uptake of $^{239+240}$Pu and ^{241}Am in ryegrass grown in Cumbrian soil [10] may be significantly higher than standard literature values [2]. Preliminary calculations using higher root uptake values show the dose resulting from the ingestion of crops increasing to 10μSv in 2000 because of the effective reservoir of radioactivity in the soil. The use of site-specific uptake rates and validation of concentrations of radioactivity in foodstuffs by comparison with local measurements are both important in increasing the reliability of dose predictions.

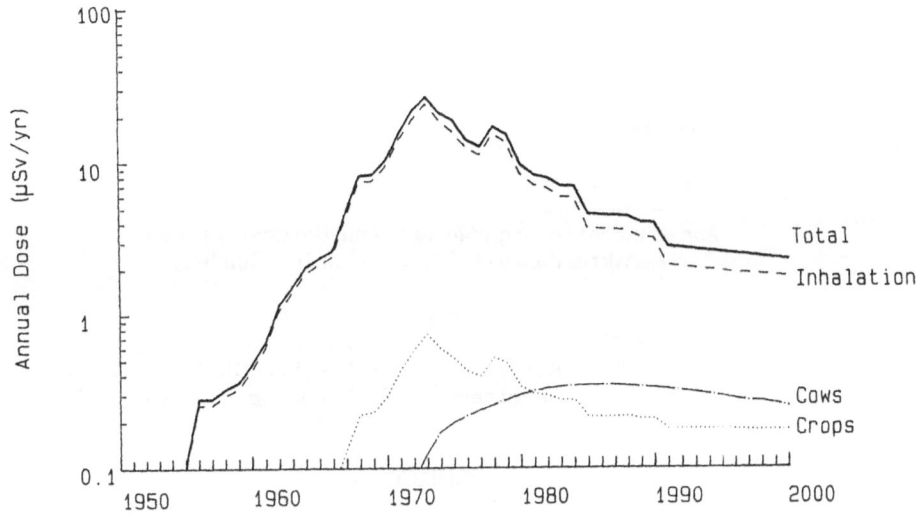

Figure 4. Model predictions of the variation, with time, of the dose received by an average person in Seascale arising from the sea-to-land transfer of actinides

DISCUSSION

A series of computer models has been used to assess significance of the 'sea-to-land transfer' in Cumbria. The models use annually averaged values, which reduces their complexity, but can make validation difficult. For example, measurements in the marine environment are frequently spot measurements and the deviation of these data from annually averaged values is not always clear.

Validation of dynamic models requires knowledge of the time-scale of application of the models. Methods must then be devised to validate each of the dominant mechanisms during that period. The use of time-series data is also essential when such models are to be used for predictive or retrospective analyses.

The maximum dose to the average person in Seascale due to the sea-to-land transfer was calculated to be less than 3% of the ICRP annual limit for members of the general public. This dose is small when compared to that received by the critical group of winkle eaters in Cumbria [4]. However, the potential for an increasing future dose when other pathways are decreasing indicates the long-term implications of this pathway and the need for site-specific validation of the terrestrial models.

COMPARISON OF TWO MODEL APPROACHES FOR THE GEOSPHERE/BIOSPHERE INTERFACE

Ann-Charlotte Argärde and Ann-Margret Ericsson
Kemakta Konsult AB, Stockholm, Sweden

and

Ulla Bergström and Björn Sundblad
Studsvik Energiteknik AB, Nyköping, Sweden

ABSTRACT

This paper presents a comparison between two approaches of modelling radionuclide transport in soil. A scenario within the BIOMOVS study is chosen for the comparison. The scenario treats soil as recipient for contaminated water from the geosphere. The groundwater, containing I-129 and Np-237, reaches the soil below the root zone. This paper compares and discusses the concentration of nuclides in the root zone.

One model is based on the compartment theory and is numerically solved by the code BIOPATH. Some input data to the model are obtained from a hydrological model (WATPATH) describing the dynamics of water in the unsaturated zone and saturated zone where important parameters are precipitation and mean air temperature. The BIOPATH approach considers the retention of nuclides in the different reservoirs. The uncertainty in the results due to the uncertainty in the parameter values are determined by the PRISM system.

The other model is based on the advection-dispersion equation which is numerically solved by the computer code TRUMP. The model takes into consideration the transport of nuclides with groundwater and percolating precipitation, and transport by dispersion and diffusion. It also considers the retention of radionuclides by sorption to the soil. The model gives the possibility of a detailed description of the flow and distribution of radionuclides (one or two dimensional) in a section of a soil profile as a function of time. Input data to the model is water flow, dispersivity, diffusivity and sorption characteristics in the different parts of the soil profile.

INTRODUCTION

In this paper a first comparison of modelling radionuclide transport with compartment theory and the advection-dispersion model is presented. The Scenario B6 within the BIOMOVS study [1] was chosen for the comparison. The scenario description is presented below.

The results will be treated within the BIOMOVS group.

In order to be able to compare the two modelling approaches a common data set was used when possible (see TABLE 1).

Presentation of the scenario taken from the BIOMOVS study

There is a flux of 1 Bq/a,hectare of Np-237 and I-129 in groundwater to soil below the rooting zone in a homogeneous farming area where there is no irrigation. Calculate concentrations in rooting-zone soil (Bq/kg dry wt.), root crops (Bq/kg fresh wt. prepared for human consumption) and in the atmosphere (Bq/m³ as inhaled) until a steady-state is reached. Uncertainties in the results should be estimated corresponding to 95% confidence interval.

In this paper the concentration of nuclides in the root zone is presented.

DESCRIPTION OF THE MODELS

Conceptual model and input data obtained for the comparison

Figure 1 presents the annual amount and schematic flow of water, the ground water level and the sizes of the different zones of the modelled area. For both models it was assumed that the nuclides were transported by the groundwater in a soluble form.

One of the models is based on compartment theory and the other model on the advection-dispersion equation. The hydrology is input data to the migration calculations for both models but the technical approach for obtaining the water flow is different.

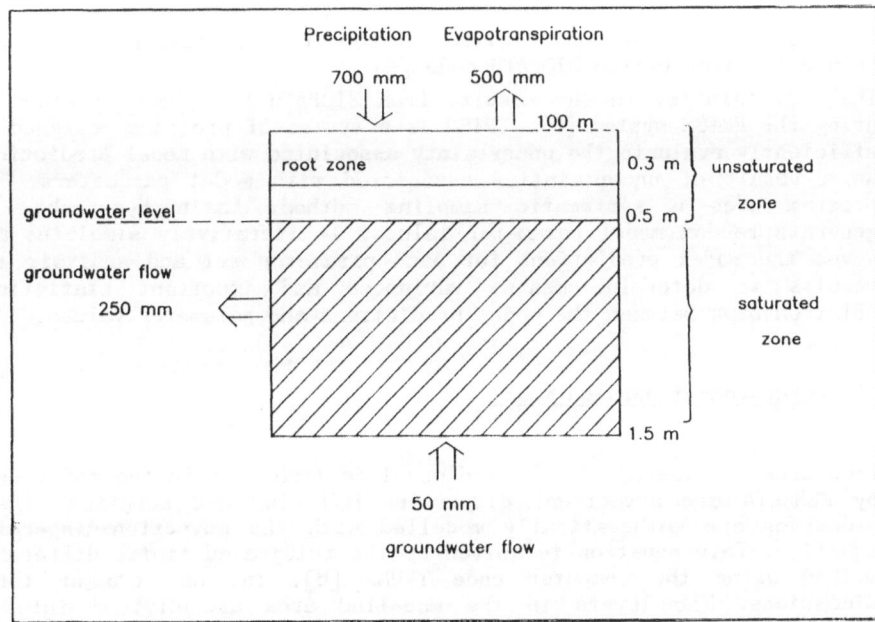

Figure 1. Schematic description of the modelled area

TABLE 1

Input parameter values

	Total porosity (%)	Effective porosity (%)	Density kg/m^3	Kd I* m^3/kg	Kd Np* m^3/kg
Unsaturated zone	50	30	2500	0.01	0.01
Saturated zone	40	30	2500	0.1	0.001

* [2]

The compartment model

The results from the STUDSVIK-group are based upon the use of three different codes. A hydrological model, WATPATH [3] has been used to obtain the water flow and the residence time in the different soil zones. A schematic picture of the model is shown in Figure 2. Processes which have been taken into account are precipitation, snow melt, interception, evapotranspiration, infiltration, ground water recharge/discharge and percolation of soil water. The model is driven by daily values of the air temperature and the precipitation, representative for the middle of Sweden. There are about twenty parameters included in the model, such as degree-day factor, interception capacity, altitude, latitude, dew-point depression, field capacity, permanent wilting point and recession constants.

The nuclide turnover in the different compartments (see Figure 3) was then calculated by the BIOPATH code [4].

The uncertainties in the results from BIOPATH have been obtained by using the PRISM system [5]. PRISM is a system of programs designed to efficiently evaluate the uncertainty associated with model predictions as a result of uncertainties associated with model parameters. The program uses a systematic sampling method, Latin Hypercube, to generate random model parameter values. It iteratively simulates and saves the model predictions for each parameter set and analysis the results to determine means, variances and important statistical relationships between the model predictions and parameter values.

The advection-dispersion model

Processes considered for the radionuclide transport in the model used by KEMAKTA are advection, dispersion/diffusion and sorption. These processes are mathematically modelled with the advection-dispersion equation. This equation is solved by the integrated finite difference method using the computer code TRUMP [6], in one, two or three dimensions. The layers in the modelled area are divided into 30 elements which are assigned different physical properties.

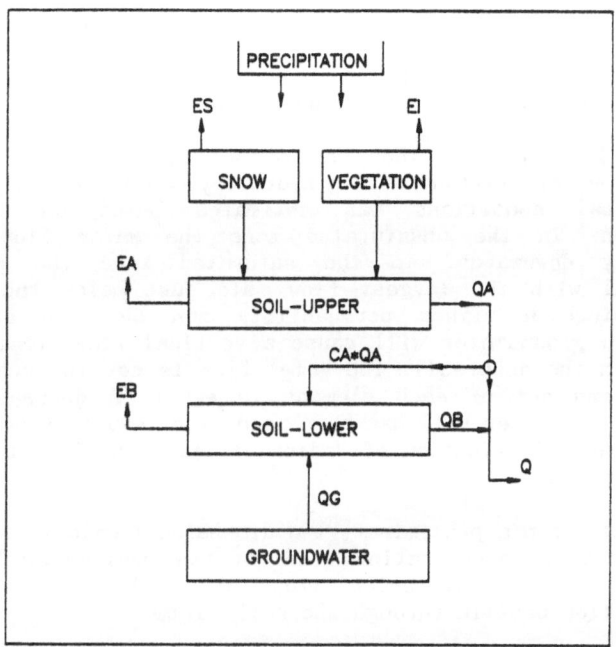

Figure 2. The hydrological model WATPATH used for this scenario

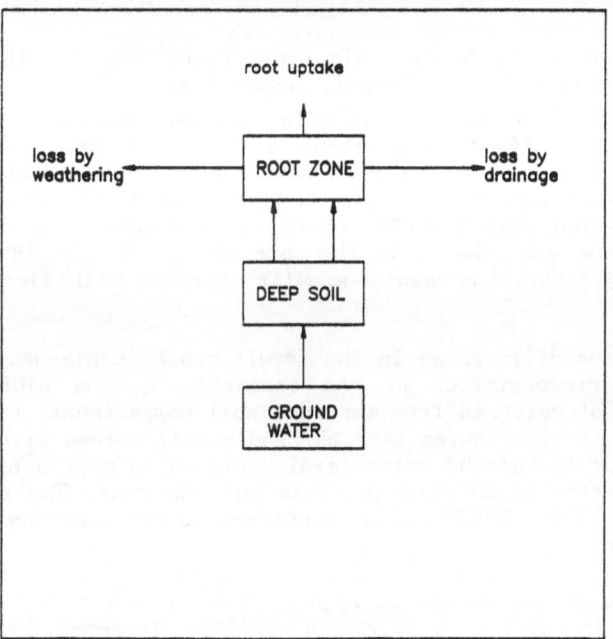

Figure 3. The compartment code BIOPATH used for this scenario

The mass transport is calculated from given initial conditions by keeping a record of the mass that enters, leaves, accumulates and reacts in each element. By the conservation of mass it is possible to obtain the concentration as a function of time.

The hydrological conditions can be set by the user of the code or obtained by calculations with a hydrology model. For this scenario the hydrological conditions was estimated based on the following assumption. In the unsaturated zone the water flow is directed vertically downwards. In the saturated zone the main flow is horizontal with the largest flow rate just below the ground water level, since a higher permeability can be expected there. The discharged groundwater will cause a vertical flow component directed upwards in the deep soil. The water flow is set in such way that the flow in and out of each element is egual. A decreased dispersion length in the upper soil has been used. The results from the model are the flow and distribution of radionuclides in different layers of the soil column.

The results from a preliminary two dimensional calculation showed that the horizontal concentration gradient was negligble. This indicated that one dimensional calculations could be made to study the concentration profile through the soil column.

RESULTS

In Figures 4 and 5 the calculated concentrations in the soil are presented for I-129 and Np-237, respectively.

The agreement between the results is good, though the time to reach steady-state differs with about a factor of two. Both models have about the same steady-state values, due to the same Kd values applied for the root zone for the two nuclides. For I-129, TRUMP gives higher concentration than BIOPATH for all time points, but the curves have roughly the same shape. In the case of Np-237, the TRUMP results are initially lower, but show a greater increase with time than those of BIOPATH.

Part of the differences in the result can be explained by differences in the interpretation of the scenario. In the BIOPATH model the drainage of water is from the root soil compartment, but in TRUMP the drainage is distributed over the whole soil column with the main part just below the ground water level. This would give a higher root soil concentration in BIOPATH which is not the case. The explanation for this is that TRUMP also considers dispersion as mechanism of transport.

The shape of the curves in Figures 4 and 5 shows that the two models react differently to a change in sorption. It seems that TRUMP is more sensible to a change in the Kd value of the deeper soil than BIOPATH. In BIOPATH the deep soil is modelled as one compartment and a average radioactive content is used, while in TRUMP the deep soil is divided into 20 elements. A high degree of discretization appears to be needed to describe the transient phase for strongly sorbed nuclides.

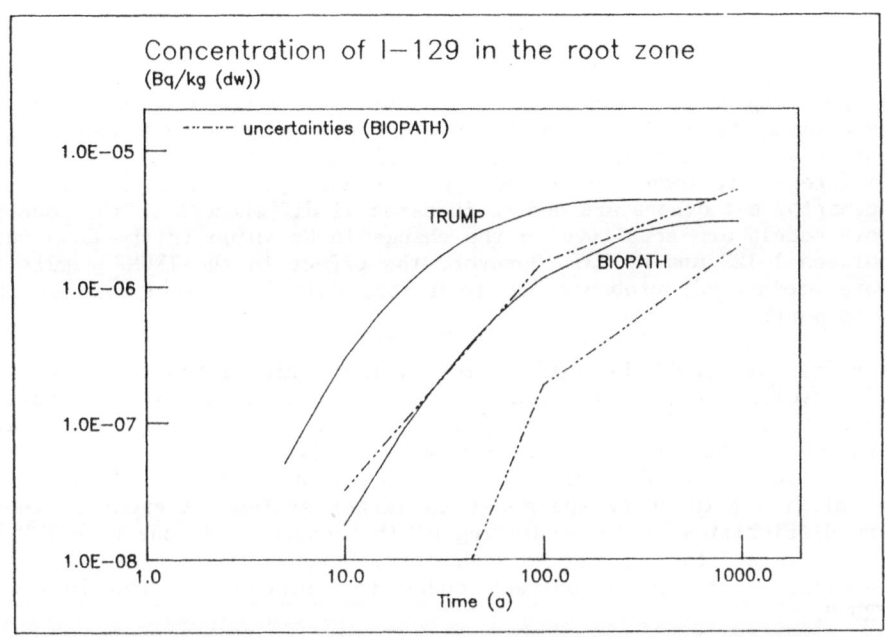

Figure 4. Concentration of I-129 in the root zone

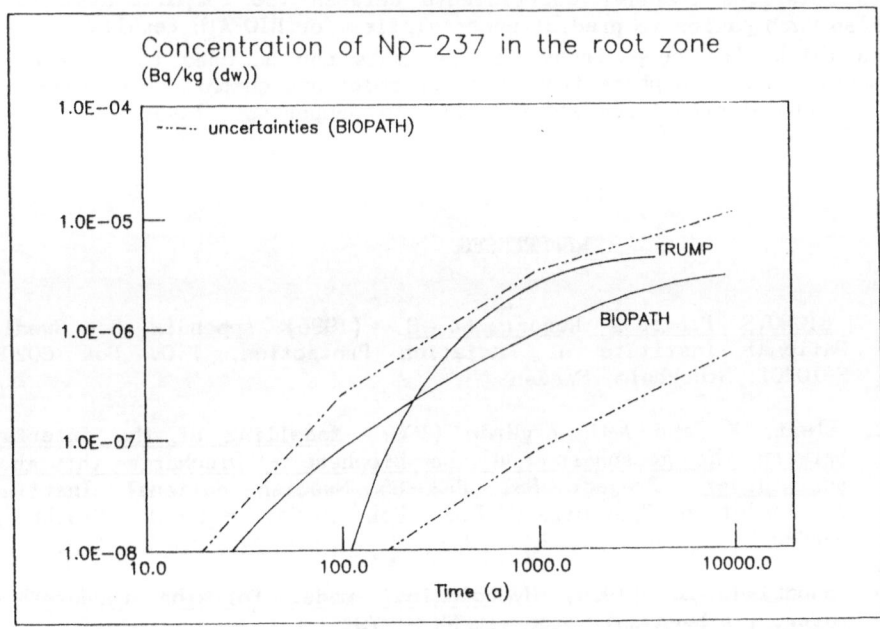

Figure 5. Concentration of Np-237 in the root zone

CONCLUSIONS

A good agreement was found in the steady-state concentrations calculated by BIOPATH and TRUMP. The differences that appear are mainly in the transient phase. The difference can be explained by a different treatment of hydrology and transport processes of the scenario, but others are due to fundamental differences in the models. Both models are sensitive to the change in Kd value in the deep soil between I-129 and Np-237. However, the effect in the TRUMP result is more pronounced, probably due to a more detailed description of the deep soil.

A model like TRUMP is capable to give more information on the effect of involved transport processes and might also give a better description of the transient phase, if the required input data are available. TRUMP is based on transport of solutes in water and it may be difficult to add other processes such as bioturbation. It may also be difficult to apply the model to larger systems. Presently, there are difficulties in the modelling of the unsaturated zone with TRUMP, mainly caused by the complex hydrology. However, work is going on to develope a connection between codes for unsaturated hydrology and TRUMP.

A compartment model like BIOPATH is good for assessment studies and can more easily be applied to larger systems. The main difficulty is to determine transfer coefficients between the compartments. It is also much easier to predict uncertainties for BIOPATH results.
We think that comparisons such as this can be used to improve the modelling of biosphere transport. Examples are comparison of different compartment structures and modelling of deeper soil layers.

REFERENCES

1. BIOMOVS Progress Report No 3, (1986) Appendix 3, Swedish National Institute of Radiation Protection, P.O. Box 60204, S-10401, Stockholm, Sweden

2. Elert, M. and A-C. Argärde (1987) Modelling of the interface between the geosphere and the biophere - Discharge through a soil layer, Project SSI P323-85 Swedish National Institute of Radiation Protection, P.O. Box 60204, S-10401, Stockholm, Sweden

3. Sundblad, B. (1984) Hydrological model for the turnover of water, (in Swedish), Studsvik/NW-84/794

4. Bergström, U., et al, (1982) BIOPATH - A computer code for calculation of the turnover of nuclides in the biosphere and the resulting doses to man. Basic description, Studsvik/NW-82/261

5. Gardner, R. H., B. Röjder, and U. Bergström (1983) PRISM - A systematic method for determining the effect of parameter uncertainties on model predictions, Studsvik/NW-83/555

6. Edwards, A.L. (1972) TRUMP: A computer program for transient and steady-state temperature distribution in multi dimensional systems, Nat. Tech. Inf. Sev., Nat. Bur. of Stand.; Springfield Va, USA

THE RELIABILITY OF ENVIRONMENTAL TRANSFER MODELS APPLIED TO WASTE DISPOSAL

H. Grogan
EIR, 5303 Würenlingen - SWITZERLAND

F. van Dorp
Nagra, Parkstrasse 23, 5401 Baden - SWITZERLAND

ABSTRACT

A large number of different radiological assessment models exist which vary in structure and degree of sophistication depending on the purpose for which they are designed. In this paper the various applications of environmental transfer models are outlined and the differences between models for present day releases and releases from radioactive waste disposal are highlighted. The second part of the paper addresses the uncertainties associated with both types of assessment. It is concluded that the long timescales involved are the major contribution to the overall uncertainty in the prediction of consequences of releases from a waste repository. Ideas are presented for tackling this problem.

INTRODUCTION

Performance assessment for nuclear waste disposal contains, in general, a part which describes the transfer of released radioactivity through the biosphere (Fig. 1). Many environmental transfer models exist which can be used for such performance assessments and their structure and degree of sophistication will vary depending on the purpose for which they were originally designed. Table 1 lists some characteristics which should be considered when comparing models and model results.

In this paper the interrelations of the following two topics are discussed:

- Geologic disposal of radioactive waste
- Modelling of radionuclide transfer through the biosphere.

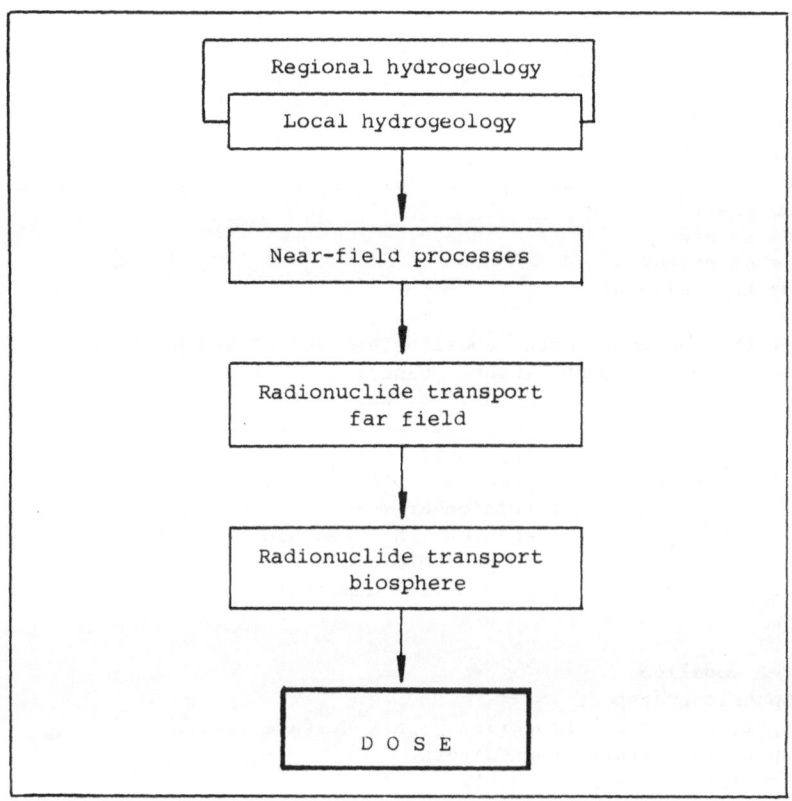

FIGURE 1: Model chain for safety analysis

Geologic Disposal of Radioactive Waste

A special feature of geologic disposal of nuclear waste is the long timescales involved: release of radioactivity from a repository is generally only expected to occur after a long time and to continue for long time periods.

In most cases the radioactivity is assumed to be released with (deep) groundwater. Release into the biosphere would then take place (primarily) in superficial groundwaters (treated as biosphere), rivers, lakes or seas. Secondary recipients might be soils or sediments.

Table 1:
Examples of Characteristics of Environmental Models
for the Transfer of Radioactivity

Field of modelling
Release from nuclear industry (including operational phase of waste repositories):
. Actual releases
. Potential releases
. Accidental releases

Releases from nuclear waste repositories (actual and potential releases during the post-closure phase):
- Repositories for   . high-level waste
                        . intermediate-level waste
                        . low-level waste
- Repositories in   . deep geologic formations
                        . rock caverns
                        . shallow land trenches
                        . seabed formations
- Repositories on   . seabed (sea dumping)

Processes modelled
. Atmospheric transport
. Transport in fresh / brackfish / salt surface waters
. Transport in surface groundwaters
. Transport in soils / sediments
. Biological transport

Aims of the model
. Definition of release limits
. Feasibility study
. Design, optimisation, cost/benefit estimate for a plant
. Licensing application

Results of the model
. Nuclide concentration in biosphere compartments or in foodstuffs
. Dose or risk to individuals or populations

Scale of the model
. Local
. Regional
. Global

Type of model
. Research / assessment
. Dynamic / steady state
. Compartmental / transport equation / finite difference / finite element
. Deterministic / stochastic
. Conservative / realistic (also applies to the input data)

Direct contamination of soil might result from erosion of the repository, or by human intrusion into the repository area and subsequent transfer of radioactivity to the surface. Whereas, atmospheric contamination could take place directly as a result of human intrusion processes which transfer dust into the air, or indirectly by the transfer of soils and sediments as dust into the air. It should be noted that atmospheric deposition of radioactivity generally plays a very minor rôle in biosphere modelling of waste repositories.

Environmental Modelling and Regulation

The radiological regulations for waste disposal are usually derived from general radioprotection criteria. Regulations may contain individual dose limits or individual risk limits. Risk includes consideration of the probability of a dose occurring. Collective doses are, in general, used for optimisation /1/, however, although radiological criteria must be used in this process they often have little impact on the final result. This is because the uncertainties in the radiological consequences of particular options (e.g. sites) might be larger than the differences between them.

Discussions are currently taking place on whether a dose or risk level can be defined, below which the effects can be considered of no concern /2/. This is especially important because "there is need to assign priorities so that limited resources are not wasted on trivial problems at the cost of neglecting major problems" /3/.

Several countries limit the time period over which detailed performance assessments and in particular biosphere modelling have to be carried out, since the development of the biosphere and human society is very uncertain over the expected lifespan of a repository /3/. In Sweden, for instance, only simplified performance assessment is required for periods beyond the next ice age (assumed to start in 10'000 years from now).

It is obvious that the type and details of the environmental models will depend on the regulation requirements. If, for example, it can be shown that releases into the biosphere will not occur before 10'000 years and the regulations only limit releases after 10'000 years, the modelling of biosphere transfer of radionuclides can be simple. If, however, regulations contain a dose limit for all times (as is the case in Switzerland) detailed biosphere calculations will have to be carried out for the total period of release.

Environmental Modelling

Instead of trying to predict the biosphere over the expected life-span of a repository another procedure can be used: the biosphere is modelled for the present-day situation assuming the predicted future releases occur today. It should be noted, however, that for the performance assessment of waste disposal, different aspects are of importance compared with the assessment of consequences of present-day releases. In contrast with the latter, water is thought to be the major release and transport agent for nuclear waste repositories.

In general, biosphere models can be sub-divided into description of the transfer of radionuclides through the environment or between compartments of the environment and into description of transfer through food-chains. Both parts can either be modelled as dynamic or steady-state processes.

For present-day releases of radionuclides, the following aspects are important as well as their **spatial** and **temporal** variability:
- Seasonal effects
- Agricultural (or fishing) practices
- Storage and transport time of food products
- Distribution pattern of food products
- Methods of food processing
- Consumption patterns.

For waste disposal the following have to be added to this list of important aspects:

- Climatic changes
- Changes in methods of agricultural food production
- Changes in consumption habits.

For waste disposal assessment the contribution of the last three aspects to the uncertainty in the predictions are more important than those from the first six aspects.

To evaluate changes in consumption habits, the model structure does not need to be changed. Total energy and nutrients could be balanced to avoid unrealistic or impossible diets (assuming no dramatic change in the size and physiology of man occurs). Changes in agricultural production methods could be evaluated by comparing different production methods practised at different locations in the world. So far only a few attempts have been made in this direction. Climatic changes have been taken into account by modelling e.g. tundra systems in arctic climates or hot climates /4/, /5/, /6/. The general data for these scenarios have been taken from literature describing these climates as they currently exist. The nuclide data for the tundra climates are based on real measurements of radionuclides in tundra areas, however, data for hot climates are more scant.

As indicated in the introduction, an important feature of waste disposal is the long timescale involved. This has particular consequences for modelling the transfer of radionuclides through the biosphere. Most environmental transfer models consider the biosphere compartments to be constant in time. A few models do consider processes which will influence the compartment sizes:

- Soil erosion/sedimentation processes - by wind and/or water action

- Sedimentation - which may change depth and size of lakes (BIOMOVS Scenario B6)

- Changes in the level of lakes and seas relative to the level of the land - which may cause land to become lake or sea and vice versa

The following topics require further consideration:

. Catastrophic erosion / sedimentation events like land and mountain slides (a question is whether modelling these events as continuous processes would produce different results to modelling them as single events)

. Glaciation, by transporting large amounts of material.

In addition, other soil-forming processes might also be important. For instance, transfer of materials (iron, clay, salt) within the soil profile, might have a considerable influence on the accumulation of nuclides in the soil. It is worth nothing that the period of release of radionuclides into the biosphere might be of the order of 10^4 to 10^6 years, whereas soil formation processes in moderate climates might have a time constant of the order of 10^3 years.

Uncertainties

Uncertainties can be divided into:

. Uncertainties concerning the models. For this the most important step is the derivation of the conceptual model(s), which is at the same time also the most difficult step

. Uncertainties concerning input data.

The uncertainties may result from:

. Inaccurate or insufficient data (these uncertainties could, in theory, be reduced (for instance: effects of season, agricultural practices, storage and transport of food materials, distribution pattern of foodstuffs, methods of food processing, consumption patterns on the transfer of nuclides in food chains).

. Natural variability (this uncertainty cannot be reduced, only properly accounted for, e.g. by probabilitstic modelling)

. Lack of knowledge and understanding of reality. This leads to uncertainties in the development of conceputal models and thus to uncertainties in the assessment results (it is virtually impossible to evaluate the consequences of such uncertainties)

. The development of environment and mankind in the (far) future (for instance: changes in climate, food production methods, and in consumption habits).

CONCLUSIONS

An important difference between the assessment of present-day releases of radioactivity and the assessment of releases from waste repositories into the biosphere is the long time periods involved in the latter. This causes the major part of the uncertainty in the predicted consequences of the release.

Whereas for the present-day case atmospheric deposition processes, seasonal effects, storage and transport time of food products cause uncertainties in prediction, uncertainties in the assessment of geologic waste disposal will mainly result from climatic changes, changes in agricultural practices and in consumption habits. Assessment modelling techniques are, in general, available, but data especially for warmer climates are scant. A possible solution would be to increase the exchange of information between moderate climate and more tropical countries.

Modelling of the effects of changes in environmental compartments, as caused by erosion and sedimentation and soil formation processes, on the behaviour of nuclides in the environment is still at a very early stage, and considerable effort is still required.

REFERENCES

/1/ Commission for Radiological Protection of the Federal Republic of Germany: Possibilities and limits for applying the concept of collective dose; A recommendation, Health Physics, Vol. 53, No. 1, pp. 9-10, 1987

/2/ van Dorp F., Grogan H., McCombie C.: Disposal of radioactive waste; Submitted for publication in Radiation Physics and Chemistry, Int. J. of Radiation Applications and Instrumentation, Part C, 198?

/3/ ICRP Publication 46: Radiation Protection principles for the disposal of solid radioactive waste; Annals of the ICRP, Vol. 15, No. 4, 1985

/4/ Nagra: "Biosphere modelling for a HLW repository - Scenario and parameter variations"; Nagra Technical Report Series, NTB 85-48, Nagra, Baden, Switzerland, 1985

/5/ Jones C.H.: "Basis of biosphere dose-conversion factors for a savanna environment"; ANS Report Series No. 736-1, ANS, Epsom, UK, 1986

/6/ Jones C.H.: "Basis of biosphere dose-conversion factors for a tundra environment"; ANS Report Series No. 736-2, ANS, Epsom, UK, 1986

SENSITIVITY ANALYSIS APPLIED TO THE DISPOSAL OF LOW AND MEDIUM-LEVEL WASTE

Ph. Guetat

Commissariat à l'Fnergie Atomique
Institut de Protection et de Sûreté Nucléaire
Département de Protection Technique
Fontenay-aux-Roses, France

and

Th. FOULT
Commissariat à l'Energie Atomique
Agence Nationale pour la gestion des Déchets RAdioactifs
Paris, France

RESUME

Au stade actuel des connaissances relatives au second site de stockage français, les calculs des conséquences radiologiques associés à une analyse de sensibilité mettent en évidence les principaux radionucléides à prendre en compte et font apparaître les paramètres sensibles de l'ensemble des barrières ainsi que les marges de sécurité résultant de la conception des ouvrages et de l'application des règles fondamentales de sûreté. Ceci permettra de mieux définir les capacités radiologiques et d'orienter les études expérimentales.*

SUMMARY

At the present development of knowledge of the second french disposal site, calculations of radiological consequences combined with a sensitivity analysis point out the main radionuclides to be taken into account, identify the sensitive parameters of the different barriers and evaluate the safety margins achieved through structural design and the application of general safety requirements. It will be possible to define the radiological capacity more accurately and derive guidance on the experimental studies.

* Version française : Rapport SEPD 88/02

INTRODUCTION

The second french center for disposal of low- and medium-level waste will become operational in the early 1990s. Setting up a disposal facility requires the radiological capacity of the site to be define in relation to the expected level of waste production and to the short- and long-term safety objectives. When the site characteristics are considered to be satisfactory and there is a firm commitment to set up a disposal facility, it is obviously desirable to determine how best to make use of the site. An attempt is therefore made, using an iterative approach between the study of the environment and the evaluation of the radiological consequences, to specify the site and engineering characteristics and the activity limits to be adopted for this disposal facility. The latter will ultimately be reflected in waste-package specifications for the producers. They are an important factor in the cost of disposal and of management of certain types of waste.

The present exercise should be considered as a generic study of the transfers via water pathway, which takes account of realistic geometry and flow characteristics.

METHOD

In order to approximate realistic conditions,the modelling of a site could be carried out in three dimensions,take account of the existence of an unsaturated zone, a zone of groundwater fluctuation and a saturated zone and, finally, integrate the physical, chemical and biological phenomena. Likewise, it would be necessary to know each type of waste: nature, activity distribution, chemical characteristics, etc.While increased knowledge in respect of each of these points is desirable,it is nonetheless necessary to estimate the importance of the mechanisms and to grade them in order of importance. Sensitivity analysis is an interesting means of dealing with that problem and can be carried out on basis far less complex.

A deterministic approach is used, and each result is associated to its particular set of data. This method does not provide a faithful image of realistic conditions, but it is easy to understand all the assumptions used and it does not include assumption concerning unknown data (law of distribution, point-by-point knowledge of the environment, etc.) Additionnally, it allows freedom of interpretation to the users of results, whether they are decision-makers or specific groups of the public.

This type of analysis is nowadays an integral part of safety reports.

We shall concentrate in the following text on the analytical solution of the mass transport equation in the saturated homogeneous environment in order to calculate radionuclide transport in the ground from a source emitting in short bursts and undergoing exponential decay. Subsequently, an analysis will be made of the influence of the various biosphere parameters and a low probability scenario.

RESULTS

The sensitivity analysis can be carried out on the basis of three periodes of time: the half-life of the radionuclide t_r, the travel time t_p between the point of emission and the point of escape and the leaching half-periode of the waste t_1 (deduced from the fraction of activity leached annually F.A.L. with $t_1 = 0.693/\text{F.A.L.}$).

The travel time takes account of all parameters relating to the site hydrology and the water-soil-radionuclide interactions. For a point-source emission in space and time, we have the relation : $t_p = L (\omega + \rho_s K_d)/\Phi_d$, where L is the distance travelled, Φ_d is Darcy's flux, ω is the kinematic porosity, ρ_s is the apparent dry density of the soil and Kd is the distribution coefficient. This expression can, in fact, be simplified when there is a certain retention of radionuclides in the solid phase, since ω then becomes negigeable. Thus we have $t_p = L.\rho_s.K_d/\Phi_d$ and the uncertainty relating to T_p is simply the sum of the uncertainties relating to each of the parameters. From experience, these parameters can be classified in decreasing order of accuracy as follows : Kd, Φ_d, ρ_s. The distance L can have a large influence in the calculation of the travel time. Thus,it is necessary to choose for a given site the longest possible distance between the disposal structures and the point of escape , and for the spatial representation of the site to be sufficiently accurate to reduce the associated uncertainty.

The leaching half-period takes account of all the parameters relating to seepage of water into the disposal facility and to the water-waste-radionuclide interactions.

For a point-source emission in space, the level of activity released at the point of escape lies between two extrem values :

- for leaching half-periods which are short in comparison with the travel time, the source can be considered as a point-source emission in time and the annual maximum activity released A, at the point a γ escape, can be deduced from the expression :

$$A \cong \frac{Ao}{t_p} \sqrt{\frac{L}{4 \pi \alpha_L}} \cdot 2^{-t_p/t_r} \tag{1}$$

Ao : activity disposed of

α_L : dispersivity

The associated uncertainty is : $\frac{\Delta A}{A} = (1 + 0.7 \frac{t_p}{t_r}) \frac{\Delta t_p}{t_p} + \frac{1}{2} \frac{\Delta \alpha_L}{\alpha_L}$ (2)

- for leaching half-periods which are long in comparison with the travel time, the annual activity emitted at the point of escape reaches a "steady" state, proportional to the activity emitted at the source :

$$A \cong \frac{Ln\ 2}{t_1} \cdot Ao \cdot 2^{-t_p/t_r} = FAL. \ Ao \cdot 2^{-t_p/t_r} \tag{3}$$

The associated uncertainty is : $\frac{\Delta A}{A} = \frac{\Delta\ t_1}{t_1} + (0.7\ \frac{t_p}{t_r}) \frac{\Delta\ t_p}{t_p}$ (4)

thus, according to equation (3), the activity released from the geosphere at the peak time is virtually proportional to the fraction of activity leached annually when this last is small. In the opposite case (1), the activity approaches a limit value depending principally on the radionuclide travel time. In equations (2) and (4), the ratio travel time/radioactive half-life is a factor multiplying the uncertainty associated to the travel time. In the first case (1), the activity released is always sensitive to the parameter travel time,while,in case (2), that sensitivity appears only when the travel time is equal to or greater than the radioactive half-life. In both cases, the sensitivity increases with increasing efficiency of the soil barrier.

This elementary mathematical development points out the determining factors in calculating the activities released without any need to use probabilistic methods which are generally costly both in time and money.

Table 1 summarized the various cases encountered in the generic study of the disposal site. The main parameters are sometimes the leaching half-periods and othertimes, the water-soil distribution coefficients and the distance from the disposal facility to the point of escape. It should be noticed that longitudinal dispersivity plays virtually no role in calculating the maxima. The influence of kinematic porosity could appear only in the case of very slight retention in the soil ($t_p = L/\Phi_d.w$). However, in that case, in which the travel time is short in comparison with the radioactive half-lives, the activities released do not depend on the travel time and consequently on the kinematic porosity.

Expressed in simple words, this means that, when the soil barrier is ineffective (H 3, C 14, U 238, Np 237, Ni 63, Nb 94), the sensitive parameters are those relating to the engineered barrier (package and structure), and, when the engineered barrier is ineffective, the sensitive parameters are those of the soil barrier (Co60, Pu239, Am241, Cs137, Sr90). This is the well-known law in ecology of the limiting factor.

In order to study the influence of these parameters in a less theorical way, use was made of the GEOLE code. In this code, it is taken account of the disposal facility geometry and of a more complex source term.This one results from the existence of a succession of periods (operational, survey, post-survey) and from a statistical law governing the deterioration of the engineered barriers and the subsequent leaching of wastes. The disposal is filled up gradually during the operational period. This corresponds only to an extension in space and time of the previous modelling and makes it possible to determine the influence of the engineered barriers which is now considered to be indispensable in most countries.

The characteristic parameters of the reference case are set out in tables 2 and 3, and the dose equivalent rates for the reference case and for the various sets of parameters are presented in figures 1 to 7.

The following comments can be done :

a) Reference case

Account is taken of the general safety requirements :

- The survey period cannot last more than 300 years.

- At the end of the survey period, the first and second barriers have lost their containment capability.

- The mean specific alpha activity in the disposal facility must not exceed 0.37 MBq.kg-1 at the end of the survey period.

Furthermore, no account is taken of the existence of an unsaturated zone between the base of the disposal and the watertable.

With these hypothesis, the dose equivalent calculated for a member of the critical group is always below 10^{-5} Sv.an^{-1}. This value corresponds to the exemption level proposed by ICRP and IAEA experts (2) (3) for the unrestricted use of solid waste.

The main radionuclides are H 3, C 14, Nb 94, Np 237 and Ni 63. For these radionuclides, the geosphere cannot be considered as a barrier. It should be pointed out that the best known elements Cs, Co, Sr and Pu do not contribute to the total dose, which is therefore insensitive to the variations of the associated parameters.

b) Sensitivity to the concrete ageing law

Deterioration of concrete is assumed to follow a normal law with a 500 years mean value and with a standard deviation of 170 years, such deterioration resulting in proportional leaching of the waste. In the reference case, as specified in the general safety requirements, the assumption of a total deterioration at the end of the survey period is considered. If this assumption is not made, results remain unchanged, since the first few hundreds of years have no influence on the long term dose caused by the long-lived nuclides C 14, Nb 94 and Np 237 and a very slight influence on Ni 63. Only Strontium gives rise to an appreciably smaller dose, but this has no effect on the total dose. Faster deterioration of the concrete (150 and 45 years) also results in a very minor change in the dose due to Ni 63. The absence of a concrete barrier results only in an increase in the dose rates due to Ni 63 and Sr 90 between 100 and 800 years without, however, exceeding the dose peaks of H 3, C 14, and Nb 94 + Np 237 obtained respectively at 40, 4000 and 20 000 years.

c) Sensitivity to the fraction of activity leached annually (FAL)

The FALs used are based on the existing technical specifications and on the safety criteria defined in the general safety requirements RFS III.2.e, which stipulate the maximum leaching rates from packages submerged in water, and on the assumption that the leaching rate is 100 times less during the operational and survey periods. The FALs during the post-survey period are defined on the basis of the leaching water – crushed concrete distribution coefficients.

The influence on the dose of the survey period FALs can be seen by comparison between figures 1 and 3. FALs higher by two orders of magnitude have been considered. These FALs play only a negligeable role owing to the intensity of the C 14, Nb 94, Np 237 and Sr 90 releases during the following period.We have moreover seen that the survey period FALs play no role at all in the case of Pu, Co, and Am. Only the dose peak of Ni 63 would be changed appreciably, without, however, exceeding the dose peaks of H 3 and C 14. The leaching rates required appear particularly severe. Indeed, values 100 times higher, corresponding for example, to submersion of the structures or to a lack of efficiency of the trench cap,would not change the dose calculated for the reference case.

The influence on the dose of the FALs of the post-survey period is shown in figure 4. The maxima are changed by the same factor as the FALs in the case of C 14, Np 237, and Nb 94.

Figure 5 presents the results obtained when engineered barriers are not considered.The FALs for the operational and survey periods are 100 times greater than those in the reference case,and a common value of 10^{-3} an^{-1} is adopted for the post survey period. This demonstrates the efficiency of the ground barrier itself.

Sensitivity to the soil parameters

It is shown in figure 6 that the results are sensitive to the water-soil distribution coefficient between 100 and 400 years (mainly Ni 63 and Sr 90 between 200 and 500 years and C 14 and Nb 94 subsequently). We here obtain the same results as in the previous analysis, and the use of a more complex model does not change anything.

The sensitivity to the disposal facility geometry is presented in the figure 7. There again, because of the dominance of radionuclides with low sorption, the influence is quite low.

Sensitivities to the other geosphere parameters have already been dealt with and are low (Darcy's velocity, porosity, dispersivity).

Sensitivity to the biosphere parameters

The calculations relating to the biosphere are,in most cases, products of transfer factors and sums of transfer pathways. Analysis of the results shows the clear predominance of the pathway "drinking water" except in the case of Nb 94 and C 14, for which the pathway "fish" predominates. For these radionuclides, it is highly probable that the transfer factors adopted have been considerably overestimated :

the Niobium water-fish transfer factor, previously estimated at 30 000 l.kg-1 (4), appears more likely to be of the order of 200 l.kg-1 (5). For this reason, transfer by water would also be dominant and the annual dose equivalent rate 40 times less. In the case of Carbon 14, the calculations are based on the isotopic ratio in the food chain and the dose factor for labelled organic compounds from ICRP (6). Most of the C 14 will be, in fact, in a mineral form and the isotopic ratio C14/C12 in the food intake of the fish will be much lower than that in the groundwater and in the river (water and sediment). Under these conditions, there is an over estimation of the transfers which may be very important. The dose equivalent rate for the drinking water pathway could also be overestimated by a factor of 100.

Tritium would then be the main nuclide and radiological consequences for Ni 63, C 14, Nb 94 and Np 237 would remain at the average level of 10^{-7} Sv.an^{-1} after the survey period.

The main biosphere parameter is the flow rate of the river at the water intake point for supplies of drinking water. The calculation was carried out for an intake point close to the site, whereas the first existing intake point is located further downstream where the average flow rate is 20 times greater.

For the present use of water, radiological consequences of the disposal would be at the level of 10^{-7} Sv.an^{-1} for Tritium.

Study of a scenario with low probability: use of water from a well located on the site after the survey period

Taking low probability scenarios into account may limit the radiological capacity of a disposal. In figures 8 and 9, results are given concerning the use of water from a well located at 10 m, 40 m and 120 m from the monoliths. For locations inside the disposal results are equivalent to those of the 10 m location. The dose equivalent rate is about 1 mSv.an-1. The presence of Sr 90 and Ni 63 can be explained by the assumption of total deterioration of the structures at the end of the survey period and the absence of an unsaturated zone ; results for C 14 depend on the use of the dose factor for the labelled organic compounds. The sensitivity study for Nb 94 and Np 237 shows that, in the case of pessimistic FAL values of the order of 10^{-3} an^{-1}, the dose equivalent rate would reach 10 to 20 mSv.an^{-1}. It hence seems that this scenario could be more restrictive than the basic scenario without, however, resulting in unacceptable exposures, even with very pessimistic assumptions.

CONCLUSIONS

This exercise shows the main features of the site at this first stage of investigation, for a disposal facility as conceived of in France at present.

It appears that, for a situation involving a realistic use of water immediately downstream of the site, the radiological consequences of a disposal of low- and medium-level waste corresponding to the waste production of 50 reactors for 30 years are below 10^{-5} Sv.an^{-1}.

The analysis of the reference scenario shows that the assumptions adopted lead to a result which is over-estimated. The safety margin appears large when studying the sensitivity of this global result to the different parameters as it is very difficult to obtain a dose exceeding the exemption level of 10^{-5} Sv.an^{-1} corresponding to a risk of 10^{-7} per year.

It also emerges that the level of confinment required seems, from the radio protection point of view, very high when compared to the real need.

In the case of Sr 90 and Ni 63, the total activity limits could depend on the well scenario during the post-survey period.

It seems, however, that, because of the average specific activities, the site volumetric capacity will lead to an activity inventory smaller than the radiological capacity.

Excepted for tritium, it appears that no limitation of the radioactive inventory should result from the study of the water pathway to man.

The studies concerning the migration of these radionuclides will probably confirm the results of this exercise. However, they will make it possible to demonstrate that the requirements relating to the F.A.Ls correspond to a safety redundancy which is not justified in the present case, owing to the retention quality of the ground barrier.

In order to define the radiological capacity for C 14, it is necessary to distinguish between organic labelled compounds (hospital waste) and graphite from the gaz-cooled reactors. Ingestion dose factors for carbonate and bicarbonate are needed in the latter case.

To demonstrate that using water from a well on the site is possible after the survey period, migration characteristics of Sr 90, Ni 63, C 14 and Nb 94 should be studied in both saturated and unsaturated zones.This point may be an important argument for public acceptance, but the interest is reduced from the point of view of safety.

	sensitive parameter	t_1	t_{pmin}	t_{pmax}	$\dfrac{t_{pmin}}{t_r}$	
$t_1 > t_p$ et $t_p \ll t_r$	t_1	1.4 E2	2.5 E0	9.5 E0	0.2	H 3
		6.9 E3	5.7 E2	2.2 E3	0.1	C 14
		3.2 E5	1.4 E3	5.4 E3	7 E-2	Nb 94
		1.4 E6	3.0 E3	1.1 E4	E-6	U 238
		3.2 E5	1.4 E3	5.4 E3	6 E-4	Np 237
$t_p \cong t_r$	t_1 et t_p	3.4 E4	2.2 E2	8.2 E2	2.3	Ni 63
$t_p \gg t_r$	t_p	3.4 E4	2.2 E2	8.2 E2	42.	Co 60
		5.8 E7	3.5 E5	1.3 E6	14.	Pu 239
		4.6 E5	7.1 E5	2.7 E6	2 E3	Am 241
$t_1 \ll t_p$ $t_p \gg t_r$	t_p	2.3 E2	5.7 E2	2.2 E3	20.	Sr 90
		2.3 E2	1.9 E5	7.2 E5	6 E3	Cs 137

TABLE 1

Value of the parameters t_1, t_p, t_r for various radionuclides and classification of those radionuclides according to the sensitive parameter
t_p min : travel time corresponding to the minimum distance
t_p max : travel time corresponding to the maximum distance

	FAL (an^{-1})		Half-life (an)	Activity (TBq)
	0 - 330 Yrs	after 330 yrs		0.00
H 3	5.00 E-03	1.00 E-01	1.23 E+01	4.00 E+03
C 14	1.00 E-04	1.00 E-04	5.70 E+03	2.00 E+02
Co 60	2.00 E-05	8.00 E-06	5.27 E+00	4.00 E+06
Ni 63	2.00 E-05	8.00 E-06	9.60 E+01	4.00 E+05
Sr 90	2.00 E-06	3.00 E-03	2.81 E+01	4.00 E+04
Nb 94	2.00 E-08	2.20 E-06	2.03 E+04	4.00 E+00
Cs 137	1.00 F-04	3.00 E-03	3.01 E+01	2.00 E+05
U 238	3.00 E-05	5.00 E-07	2.20 E+09	1.00 E+00
Pu 239	2.00 E-08	1.20 E-08	2.40 E+04	2.30 E+02
Pu 241	2.00 E-08	1.20 E-08	1.47 E+01	2.30 E+03
Am 241	2.00 E-08	1.50 E-06	4.32 E+02	3.50 E+02
Np 237	2.00 E-08	2.20 E-06	2.14 E+06	1.00 E+00

TABLE 2
Parameters of the source term

Longitudinal dispersion coefficient	(m)	5.00	E+01
Transversal dispersion coefficient	(m)	1.00	E+01
Apparent density (t/m^3)		1.70	E+00
Kinematic porosity of the soil		6.00	E-02
Darcy's velocity $(m.an^{-1})$		6.00	E+00

Distribution coefficients $(1.kg^{-1})$

H 3	0.00 E+00
C 14	8.00 E+00
Co 60	3.00 E+00
Ni 63	3.00 E+00
Sr 90	8.00 E+00
Nb 94	2.00 E+01
Cs 137	2.66 E+03
U 238	4.20 E+01
Pu 239	5.00 E+03
Pu 241	5.00 E+03
Am 241	1.00 E+04
Np 237	2.00 E+01

TABLE 3
Environmental parameters

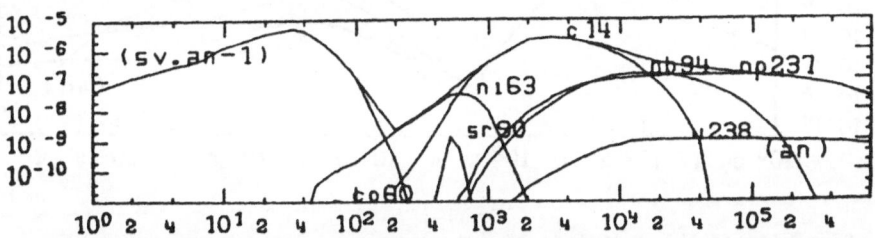

Graph n°1
Use of the river water - Reference case

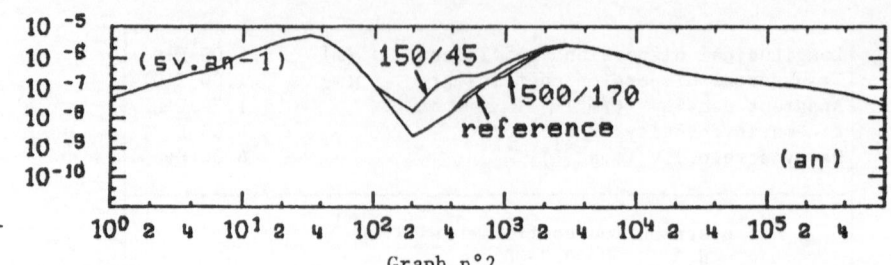

Graph n°2
Sensitivity of the radiological impact to the concrete ageing law

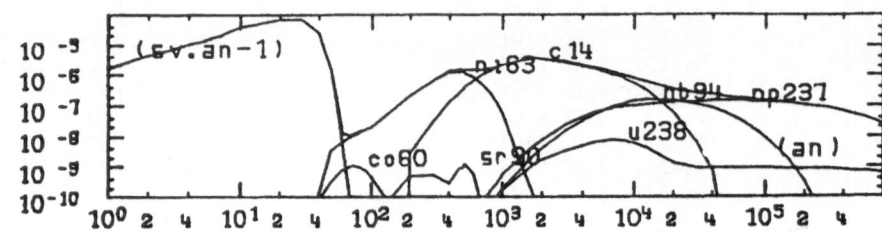

Graph n°3
Sensitivity of the radiological impact to the survey period FAL x 100

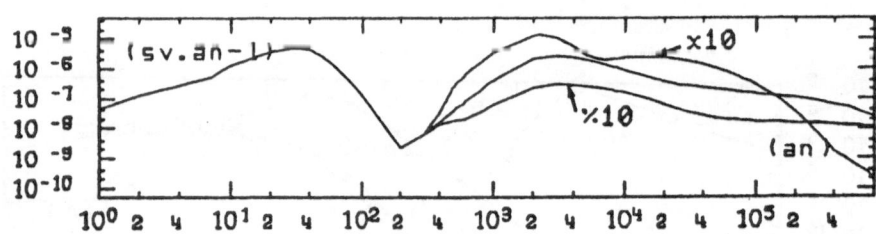

Graph n°4
Sensivity of the radiological impact to the post-survey period FAL

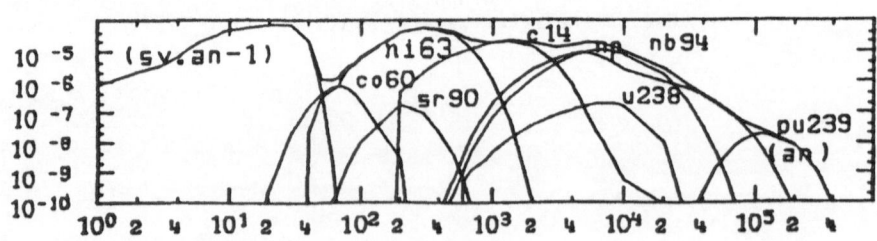

Graph n°5
Use of the river water – No concrete barrier

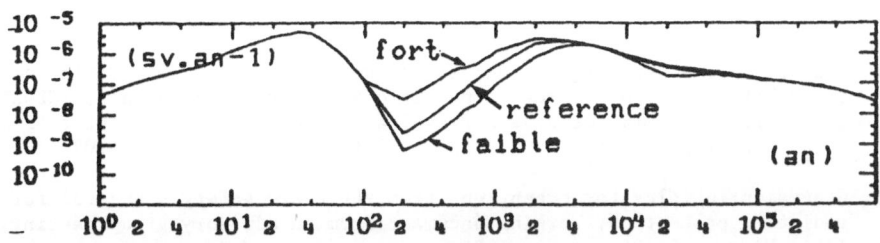

Graph n°6
Sensivity of the radiological impact to the water-soil
distribution coefficients

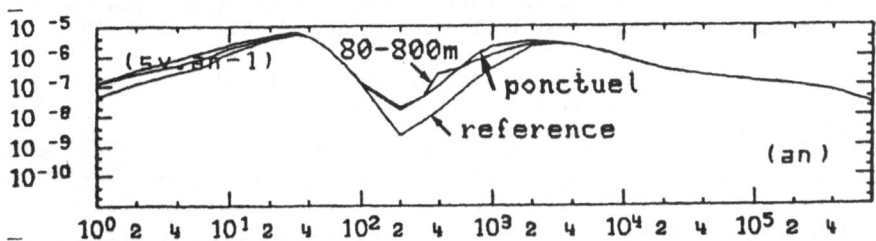

Graph n°7
Sensitivity of the radiological impact to the modelling of the
disposal facility geometry

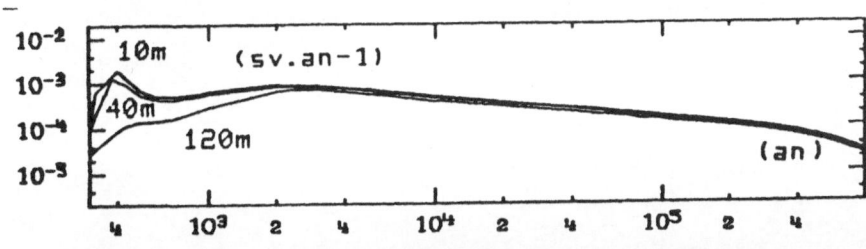

Graph n°8
Scenario with wells at 10 m, 40 m and 120 m - Influence of the
distance to the structures

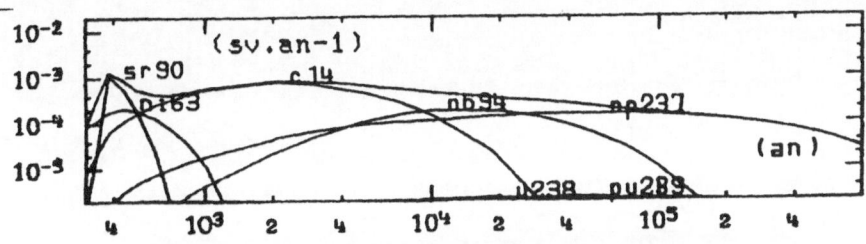

Graph n°9
Use of water from a well - Radionuclides contributions
Distance from monoliths to well 40 m

REFERENCES

1. Guetat, Ph., GEOLE version 1985 – Code de calcul pour l'évaluation des transferts de radionucléides à partir d'un stockage de surface. (GEOLE 1985 version – Computer code for evaluating radionuclide transfers from a surface disposal facility), Rapport interne ANDRA 754.NT.05.02.

2. General principles for exemptions from the basic safety standards for radiation protection. Working document from an advisory group meeting, IAEA, Vienna, 21-25 october 1985.

3. Radiation protection principles for the disposal of solid radioactive waste. ICRP Publication 46. Pergamon Press. Oxford. 1985.

4. Thompson, S., et al., Concentration factors of chemical elements in edible aquatics organisms. UCRL – 50564. Rev. 1, TID – 4500, 1972.

5. Coughtrey, P.J., Jackson, D., Jones, C.H., Thurne, M.C. Radionuclide distribution and transport in terrestrial and aquatic ecosystems. ANS Epson UK. Vol. 1, p.275.

6. International Commission on Radiological Protection. Limits for intakes of radionuclides by workers. ICRP Publication 30, Pergamon Press, Oxford UK, (1979 – 1982)

MODELLING THE RADIOLOGICAL IMPACT OF RELEASE OF RADIONUCLIDES INTO THE BIOSPHERE FROM SOLID WASTE DISPOSAL FACILITIES

G.M. Smith
National Radiological Protection Board
Chilton, Didcot, Oxon OX11 0RQ
UK

ABSTRACT

This paper describes the approach being adopted to improve confidence in biosphere modelling for the assessment of the post-disposal radiological impact of disposal of low and intermediate solid radioactive waste at alternative sites proposed by UK Nirex Ltd.

Release from a disposal facility may be precipitated by a number of mechanisms and ultimately result in release of activity into the surface environment, or biosphere. The range of potential receiving environments in the biosphere is discussed, and preliminary results for the consequences of unit release of relevant radionuclides are used to identify the most significant factors which lead to uncertainty or lack of confidence in results for biosphere modelling. The paper then goes on to consider the steps which might be taken to improve confidence in the predictions based on understanding of the relevant processes and mechanisms. These steps include the setting up of appropriate field and laboratory experiments and use of appropriate natural and other analogues to provide the best practicably obtainable database against which the biosphere model may be tested. The scope for such methods is discussed, noting the extended periods over which releases might occur, and the likely delays before they start.

This work is funded by UK Nirex Ltd.

INTRODUCTION

Movement of radionuclides into the surface environment, or biosphere, is only likely to occur some long time in the future, certainly on a long timescale compared with those usually considered in relation to the safety implications of other types of major project. In these circumstances predictive mathematical models are required to assess the radiological impact of such releases. It is also important that operators, regulators and the public should have the appropriate level of confidence in the predictions made with models. This paper discusses measures being adopted to improve confidence in models for the migration and accumulation of radionuclides released to the biosphere, and the calculation of associated doses to man, following the deep disposal of low and intermediate solid radioactive waste (LLW and ILW).

The process of assessment is essentially iterative; initial investigations are used to identify important transport processes and mechanisms, and these are then examined in more detail so as to feed back further information to the assessment. The experience gained to date and the implications for further investigations are discussed below, noting that the end-points of the biosphere modelling include predictions of maximum annual individual dose as a function of time, and of the temporal and spatial distribution of collective doses. It should be recognised that the different groups which have an interest in the assessment will have different attitudes to and interests in the results. Some will be convinced by more technical or analytical approaches, others by more demonstrative arguments.

EXPERIENCE FROM PRELIMINARY WORK

A considerable number of preliminary calculations have been made of the radiological impact of unit release in unit time of a wide range of radionuclides from groundwater into a number of potential biosphere receptors (see Table 1).

TABLE 1
Unit releases considered in preliminary calculations

Release of 1 Bq y^{-1} for 10^4 y to:-	Radionuclides
a, a small stream	H-3, C-14, Cl-36, Co-60, Ni-59, Ni-63 Se-79, Sr-90, Zr-93, Nb-93m, Nb-94,
b, adjacent sub-soil	Tc-99, Ru-106, Ag-108m, Sn-126, I-129 Cs-134, Cs-135, Cs-137, Ce-144, Sm-151
c, coastal marine waters	Eu-152, Eu-154 Pb-210, Po-210, Ra-226, Ra-228, Ac-227 Th-228, Th-229, Th-230, Th-232, Pa-231 U-233, U-234, U-235, U-236, U-238 Np-237, Pu-238, Pu-239, Pu-240, Pu-241 Pu-242, Am-241, Am-243, Cm-244

The calculations were made using the model BIOS [1] which is
fairly comprehensive in its consideration of all parts of the
environment and all the potential pathways for the exposure of
man. Three different types of biosphere receptor have been
considered for activity released from the geosphere, noting that
groundwater flows near the surface at the time of release may be
quite different from those obtaining currently. These receptors
are freshwater streams, soils, and the marine environment. The
long list of radionuclides examined arises from the consideration
of radionuclide content of LLW and ILW streams to be disposed of.
The radionuclides include fuel and structural activation
products, and fission products, as well as some others arising
outside the nuclear power industry. In addition, since a
disposal facility will be in operation for some considerable time
it is appropriate to try to anticipate what radionuclides might
arise in future wastes. However, for the purpose of prioritising
future effort it should be noted that long-lived radionuclides
which are not significantly sorbed by geological media are more
likely to reach the biosphere via migration through the geosphere
in groundwater.

The numerical results of these calculations show that
according to the different biosphere receptor, the period of
release, and the individual radionuclide the range of potentially
important transport processes and mechanisms is very large. Just
what is critical in respect of a particular site and disposal
will depend on what is predicted for release from the geosphere.
Significant factors which lead to uncertainty or lack of
confidence in the results for biosphere modelling fall into
several broad categories. These are:

- long term evolution of the surface geological features at
  and around the site,

- long term behaviour of the relevant radioelements in soil
  sediments and aerosols, and their uptake into foodstuffs,

- long term assumptions for human behaviour,

- lack of understanding of the interface between the
  geosphere and biosphere transport models.

The important geological features include changes in the
surface hydrogeology (eg through climatic changes, erosion of
surface soils by wind, water, and in the longer term, by
glaciation) and, particularly at coastal sites, the potential for
inundation by the sea.

The processes relating to individual radionuclides which are
both important and poorly understood include the long term
accumulation and migration (vertical and horizontal) in soil and
sub-soil, their soil chemistry which affects root uptake into
crops as well as migration, their accumulation and subsequent
transport in freshwater sediments, and their potential for
becoming attached to suspendable particulates thus giving rise to

an inhalation hazard. The chemistry of radionuclides at the time of intake is also important for dosimetry. In addition, in order to investigate the consequences of contaminated waste or geological media being deposited on the surface following intrusion, the release of activity from such material should be examined.

The long term assumptions for human behaviour cannot be evaluated in this way. Control of the environment may be maintained as at present, for example through dredging of water courses, and maintenance of sea walls. Use of land for farming is another important example, but farming practices can change considerably. These controls could be relaxed, but this would probably be associated with a lower population density.

Alternatively control over the environment could be increased. Notwithstanding all these possibilities for change, the periods of major release from the geosphere for many radionuclides are very long compared with the timescales over which effects of man tend to operate. Thus all the possible scenarios for human behaviour will probably obtain during the period of significant release. The same could be said in respect of many of the naturally occurring changes.

The problem of interfacing the geosphere and biosphere models concerns the use of appropriate boundary conditions. The relevant questions are:

- what is the area of sub-soil receiving discharge from below?

- if the release is controlled by diffusion, what does the geosphere model need to know about concentrations in the biosphere?

- how do radionuclides passing into the aqueous environment from the geosphere interact with freshwater or marine bed sediments?

IMPLICATIONS FOR IMPROVING MODEL RELIABILITY

Many of the problems noted above are associated with long term considerations, and it is for this reason that models for the transport of radionuclides in the biosphere following release from waste repositories are indicated [2] at the qualitative rather than the quantitative end of the ranking for potential for validation. The issues fall into three broad categories:

a) natural short-term effects,
b) natural long-term effects,
c) human influences.

Processes in category a, will generally be considered in sub-model within the BIOS (or equivalent model) framework, and there is considerable scope for validation of such sub-models. An example might be investigation of the distribution of radionuclides among surface soil particulates and the potential for suspension of these particulates to form a respirable aerosol. Further examples may be found in the bibliography of the Nirex safety assessment research programme [3]. Clearly a wide range of field and laboratory experiments can provide relevant information in this area.

For category b, the difficulty is that the relevant experiments or field measurements require impracticably long timescales. In this case natural analogues can provide useful information for assessments. Even if an analogue is insufficiently precise to provide detailed model data, nevertheless it may support background suppositions and assumptions. In addition, examination of the biosphere transport of appropriate analogues can result in identification of relevant processes which might otherwise be omitted.

A further approach to examining natural long term effects is through model comparison exercises, such as BIOMOVS [4]. In the context of long term biosphere releases the participants are asked to predict concentrations of radionuclides in relevant environmental media on the basis of only a loosely described biosphere. Comparisons of the results and discussion of the differences in a workshop type environment provide valuable information exchanges and offer the prospect of an international consensus on important issues.

Human behaviour, category c, is inherently unpredictable, so in order to build a high degree of confidence in an assessment it is important to demonstrate that a full range of potential exposure pathways has been examined. This might involve consideration of some possibilities which, from a technical standpoint, do not appear to be warranted.

Given the particular difficulties of validation in all these areas a further method of improving the degree of confidence is to allow for peer review by all interested parties at each iteration of the assessment process (see INTRODUCTION).

CONCLUSIONS

A number of issues relevant to confidence in results of long term biosphere modelling have been presented. Overall it seems that the extent to which the models can be validated is limited to qualitative considerations, although some important sub-models may be validated quantitatively. If a safety case is to be convincing to a wide audience then it must be robust. In essence this means that it must be demonstrated that safety targets can be met for a wide range of possible scenarios, processes and parameter assumptions. It is a corollary that a robust safety case will rely on quite general considerations and not an excessive esoteric detail.

This work is funded by UK Nirex Ltd.

REFERENCES

1. Lawson, G and Smith, G M. BIOS: A model to predict radionuclide transfer and doses to man following releases from geological repositories for radioactive wastes. NRPB-R169, London, HMSO, 1985.

2. Hill, M D. Verification and validation of NRPB models for radionuclide transfer through the environment. These proceedings.

3. Cooper, M J and Hodgkinson, D P. Nirex safety assessment research programme bibliography 1987. AERE-Bib206 Harwell Laboratory 1987.

4. BIOMOVS. Progress Report No. 4. Swedish National Institute for Radiation Protection. Stockholm 1987.

DIVIS

A PROGRAMM PACKAGE TO SUPPORT

THE PROBABILISTIC MODELLING

OF

PARAMETER UNCERTAINTIES

by

E. NOWAK and E. HOFER

Gesellschaft für Reaktorsicherheit

D-8046 Garching, F.R.G.

Abstract

This paper presents details of the program package DIVIS which is to support the construction of subjective probability distributions and the specification of correlation coefficients for uncertain parameters of computational models. DIVIS provides graphical representations of selected distributions and of specified measures of correlation interactively so that the model expert may quickly judge how well they represent his knowledge base on the uncertain parameter.

1. Introduction

Environmental transfer model predictions are subject to parameter uncertainties. Quantitative information on the influence of parameter uncertainties on model predictions is obtained through a parameter uncertainty
analysis which is generally probabilistic. In the first step of such an
analysis subjective probability distributions are constructed for each
potentially important uncertain parameter and dependences between parameters are specified either deterministically (for instance, via value
restrictions) or probabilistically via correlation coefficients. The subjective probability distribution is to quantitatively represent all relevant
knowledge about the value of the parameter in the context of the question
asked of the model.

DIVIS supports the construction of the distributions and the specification
of correlation coefficients. For each parameter it is assumed that an
expert has specified the maximum conceivable range of possibly applicable
alternative values together with his degrees of belief that the appropriate
parameter value is not larger than specific values selected from this
range. These values are known as fractiles, quantiles or percentiles of
the subjective probability distribution with the percentage given by the
quoted degree of belief. Often few (two or so) quantiles are provided.
Additionally, the expert may have qualitative information concerning the
shape of the distribution that best represents his knowledge base. The
problem is to select a probability distribution that fits the quoted degrees
of belief and accounts for the qualitative information on shape.

2. Summary of DIVIS

The program package DIVIS is an interactive computer code. It offers 18
different types of probability distributions that may be interactively fitted
to the information provided by the expert. Some of these distributions
can be used not only in their standard version but also in truncated
form.

Table 2.1

Probability distributions
Nonparametric Distributions
- Histogram
- Histogram with Triangular Endings

- Logarithmic Histogram
- Discrete Distribution

Parametric Distributions with Compact Support
- Uniform Distribution
- Logarithmic Uniform Distribution
- Triangular Distribution
- Logarithmic Triangular Distribution
- Beta Distribution

Parametric Distributions with Infinite Support
(Truncation possible)
- Logarithmic Normal Distribution
- Gamma Distribution
- Exponential Distribution
- χ^2 Distribution
- F Distribution
- Normal Distribution
- Extreme Value Distribution Type I
- Extreme Value Distribution Type II
- Weibull Distribution

Different measures of correlation may be specified to account for dependences between uncertain parameters.

Table 2.2

Correlations
- Ordinary Population Correlation
- Population Quadrant Measure
- Population Kendall´s τ
- Population Spearman´s ρ
- "Complete Dependence"
- Sample Rank Correlation

On input DIVIS accepts various kinds of data (see Table 2.3) whichever the expert may find convenient to quantitatively express his knowledge about the uncertain parameter. DIVIS prompts step by step for the necessary distribution data to construct a probability distribution which can be fitted to the expert's input. Where necessary it suggests a list of possible distributions or it determines the parameter values for the selected distribution type such that the distribution satisfies the input data sufficiently well.

On output probability density, distribution function and complementary distribution function may be viewed on the monitor or may be obtained as hardcopy or printout. Scatter plots are offered to study the effect of the specified correlations.

To make different plots comparable DIVIS supports several plot specifications such as plot intervals, type of axis (linear/logarithmic), etc..

Table 2.3

Input/Output Data
Distribution Input Data
- support and/or parameters
- support and/or quantiles
- support, mean value and standard deviation
- quantiles and modal value
- median and factor $q_{0.95}/q_{0.50}$
- correlation measure

Plot Input Data
- plot interval
- scaling (linear, logarithmic)
- plot device

Output Data
- parameters and/or support
- list of possible distributions
- distribution plot and output protocol
- scatter plot of a random sample from two random variables with given correlation

All the distributions generated by DIVIS may be stored in a distribution table to use them, for instance, in the subsequent construction of an experimental design. A DIVIS session may be interrupted and continued lateron without loss of the previously inserted distribution data.

3. Methods
The two most important parts of the numerical calculations done by DIVIS are parameter search and correlation calculation.

3.1 Parmeter search

The problem of parameter search arises whenever the distribution parameters of any selected distribution type cannot be calculated by an analytical formula. In this case DIVIS starts an iteration process to obtain a
set of parameter values such that the selected type of distribution complies with the given input data. This iteration procedure is formulated in
terms of a nonlinear constraint optimization problem which is solved
through a random search method. The used random search method is a
modification of a procedure described by Tarasenko (1980) which does not
require any evaluation of the derivatives of the objective function.

Table 3.1

Parameter Search

Optimization Problem

$$\cdot \quad |\ F_{\alpha,\beta}(a) - p\ | + |\ F_{\alpha,\beta}(b) - q\ | = \min_{\alpha,\beta \varepsilon A} !$$

Starting Values
- parameters of related distributions
- parameters of limit distributions

Optimization Procedure
- random search method

In table 3.1 $F_{\alpha,\beta}$ denotes a distribution function with parameters
$(\alpha,\beta)\ \varepsilon\ A\ c\ R^2$ and p and q denote the given cumulative probabilities to
the quantiles a and b of the unknown distribution.

3.2 Correlation Calculations

DIVIS uses the computer program MEDUSA (Krzykacz 1987) to generate
random samples from subjective probability distributions of dependent
uncertain parameters. This is achieved by using MEDUSA subsequently to
the interactive generation of a MEDUSA input file under the control of
DIVIS.

4. DIVIS Environment

The numerical parts of DIVIS are written in FORTRAN 77 using some
standard routines from the IMSL-Library. The graphics procedures are
supported by the SAS/GRAPH software package. Different parts of the

DIVIS code are connected by a command procedure, so called CLIST, running on an IBM-compatible mainframe computer (e.g. AMDAHL 5870) under the operating system MVS/XA using ISPF and ISPF/PDF.

ISPF is a dialog manager for interactive applications. It supports the PDF facility which aids in the development of dialogs and other types of applications. The CLIST invokes ISPF dialog services to display panels and messages, build and maintain tables, generate output files and control operational modes (IBM 1984).

Table 4.1

Environment

Hardware

- IBM-compatible mainframe computer (e.g. Amdahl 5870)
- graphics terminal, plotter, printer

Software

- operating system MVS/XA
- interactive dialog manger ISPF/PDF
- SAS/Graph
- IMSL-Library
- Fortran 77

5. Example 1

In this section we illustrate the use of DIVIS with the aid of a sample session. Let q_α denote the α-quantile of a probability distribution and let X and Y denote two random variables. Table 5.1 gives a summary of the initial data (distributions of X and Y) and some illustrations generated by DIVIS.

Table 5.1

Example 1

Initial Data

- X is logarithmic normal distributed with quantiles
  $q_{0.10} = 5.64E-04$
  $q_{0.90} = 1.52E-03$
- Y is logarithmic uniform distributed with support
  minimum = 1.00E-03

maximum = 2.00E-03

Illustration of Different Distributions

- distribution of X
- distribution of X truncated at quantiles $q_{0.05}$ and $q_{0.95}$
- logarithmic normal distribution truncated at quantiles $q_{0.05}$, $q_{0.95}$ of X with $q_{0.10}$, $q_{0.90}$ as given above

Illustration of Different Ordinary Population Correlation Coefficients between X and Y

- 0.00
- 0.50
- 0.90

Figure 5.2 and figure 5.3 show the distribution functions and probability density functions of the three logarithmic normal distributions specified in table 5.1:

Fig. 5.2

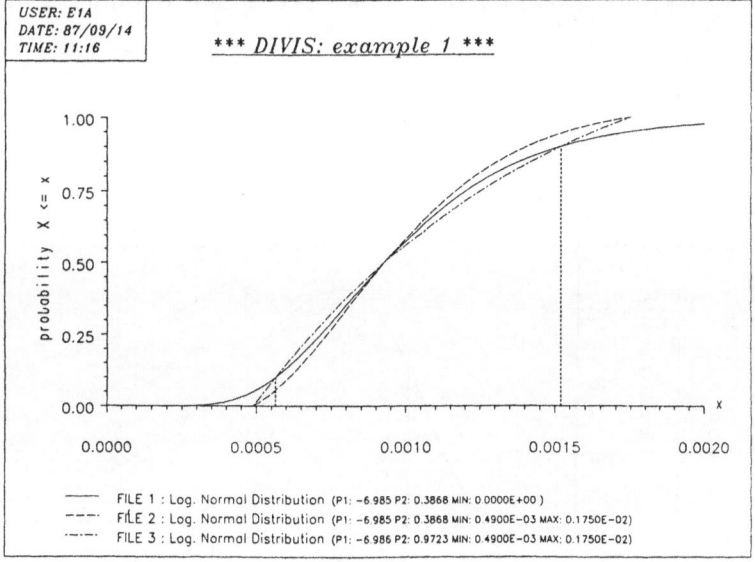

Fig. 5.3

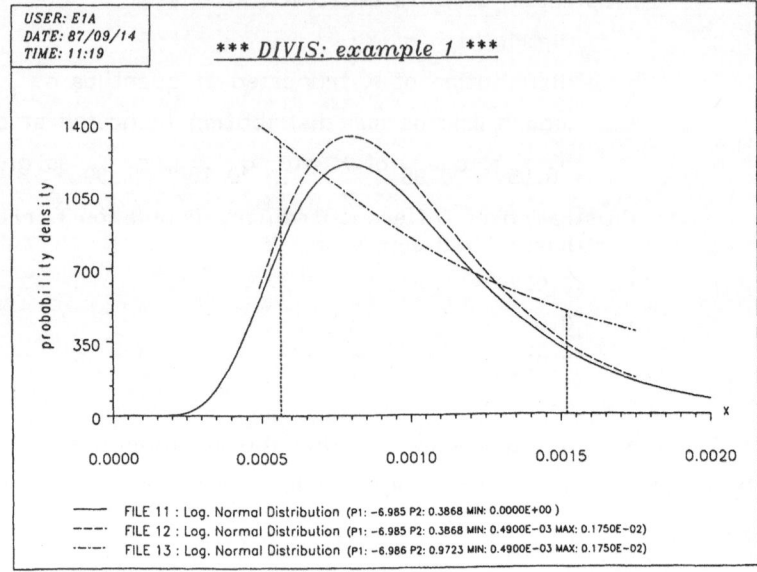

The density functions of X and Y are presented in figure 5.4:

Fig. 5.4

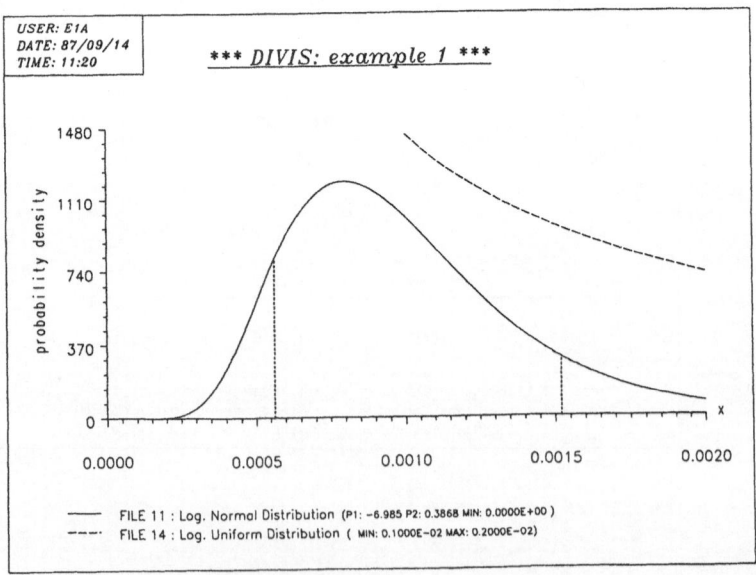

Illustration of the three different correlation coefficients of table 5.1 are given in the figures 5.5, 5.6, 5.7:

Fig. 5.5

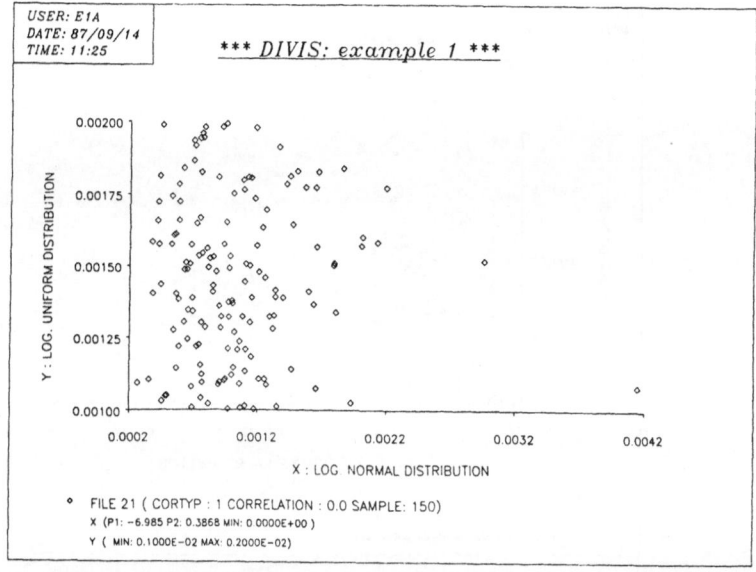

Fig. 5.6

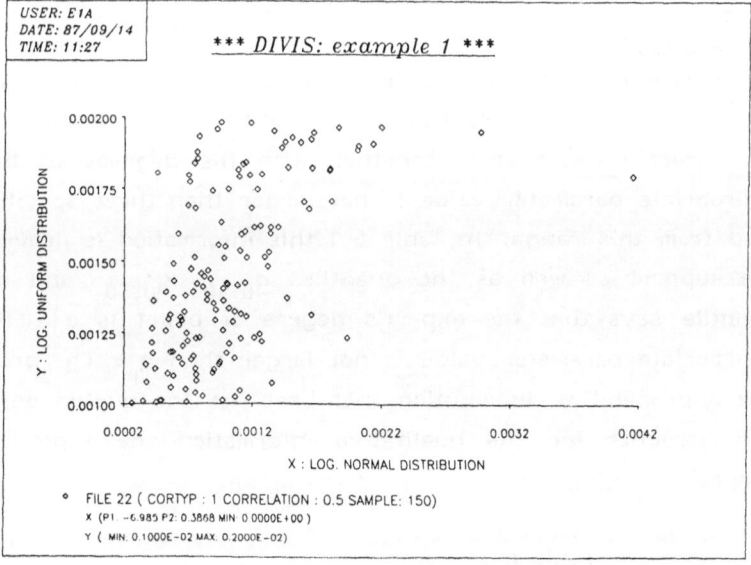

314

Fig. 5.7

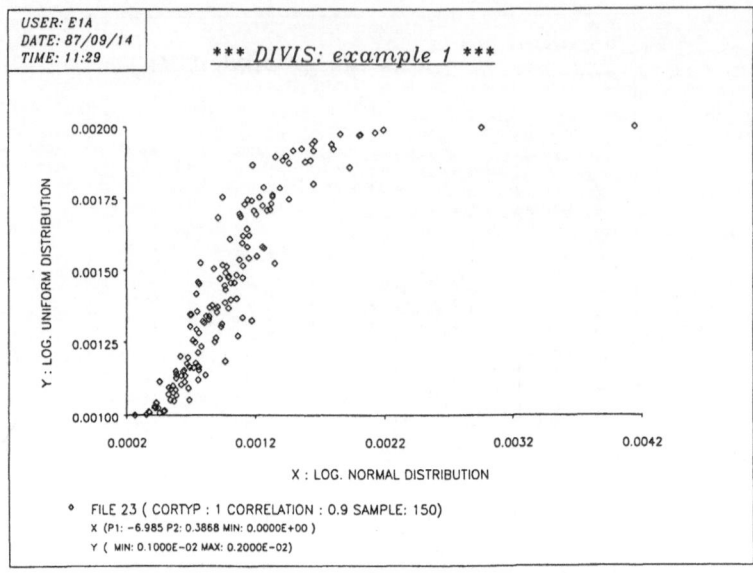

The main panels of an example of a DIVIS session were part of the presentation and are contained in the version of this paper distributed at the workshop.

6. Example 2

As a second example we consider the case where an expert has specified the maximum conceivable range of possibly applicable alternative values of an uncertain parameter, together with his degrees of belief that the appropriate parameter value is not larger than three specific values selected from this range. In table 6.1 this information is provided in form of the support as well as the quantiles $q_{0.10}$, $q_{0.50}$ and $q_{0.90}$. The q_α quantile says that the expert's degree of belief is $\alpha \cdot 100$ % that the appropriate parameter value is not larger than q_α. The problem is to select a probability distribution that best fits the quoted degrees of belief and accounts for the qualitative information the expert may have on shape.

Table 6.1

Initial Data

support of unknown distribution
minimum = 1.0E-05, maximum = 2.0E-03

quantiles of unknown distribution

$$q_{0.10} = 6.0E-04, \quad q_{0.50} = 8.0E-04, \quad q_{0.90} = 1.0E-03$$

Illustration of Different Distributions

Histogram

interval	probability
[1.0E-05, 6.0E-04]	0.1
[6.0E-04, 8.0E-04]	0.4
[8.0E-04, 1.0E-03]	0.4
[1.0E-03, 2.0E-03]	0.1

Distributions Fitted to Quantiles $q_{0.10}$ and $q_{0.90}$

- beta distribution
- logarithmic normal distribution
- normal distribution
- extreme value distribution (type I)
- extreme value distribution (type II)
- Weibull distribution

Logarithmic Normal Distribution Fitted to Quantiles

- $q_{0.10}, q_{0.50}$
- $q_{0.50}, q_{0.90}$
- $q_{0.10}, q_{0.90}$

Beta Distribution Fitted to Quantiles

- $q_{0.10}, q_{0.50}$
- $q_{0.50}, q_{0.90}$
- $q_{0.10}, q_{0.90}$

For inspection by the expert the alternative distributions listed in table 6.1 are generated. Additionally a histogram is plotted that completely agrees with the expert's quantitative information.

Figure 6.2 can be used to compare quantiles of the different distributions. For instance, the quantiles $q_{0.50}$ can be compared with the given 0.50-quantile. Figure 6.3 permits comparisons of the shapes, particularly the distribution behaviour at large or small values of the uncertain parameter may be of interest.

Limitation to the beta distribution and the logarithmic normal distribution and generation of all beta distributions and logarithmic normal distributions that may be fitted to two of the three given quantiles (cf. table

6.1) provides the probability densities shown in figures 6.4, 6.5.

A comparison of the densities shows that the differences between the distributions fitted to any two of the three given quantiles are smaller in the
case of the beta distribution. Thus it seems reasonable to choose one of
the beta distributions provided it complies with the expert's idea of what
the relevant tail end of the distribution should look like.

Fig. 6.2

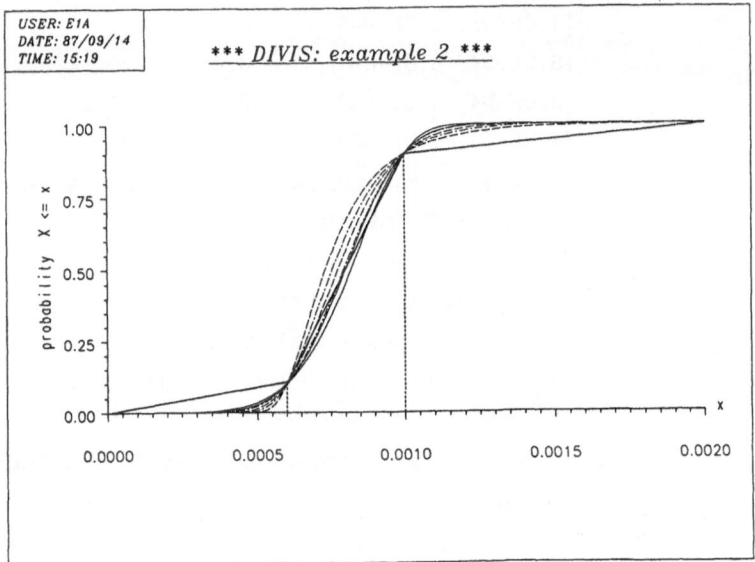

Fig. 6.3

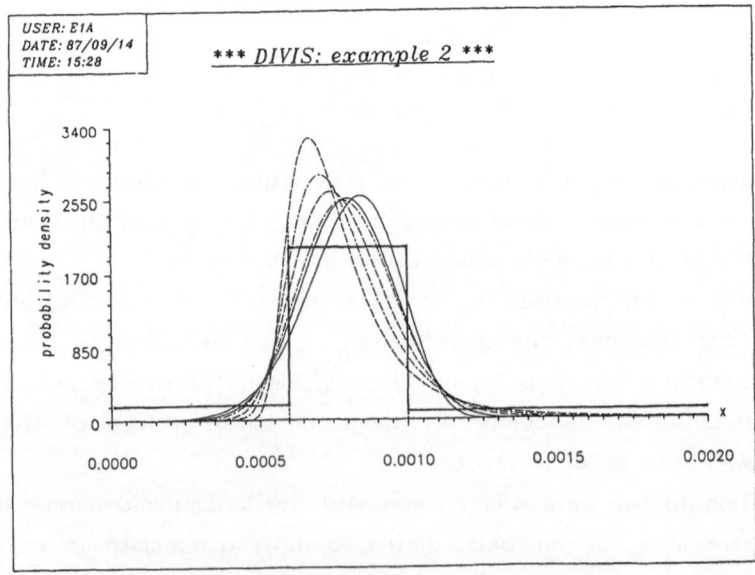

Fig. 6.4

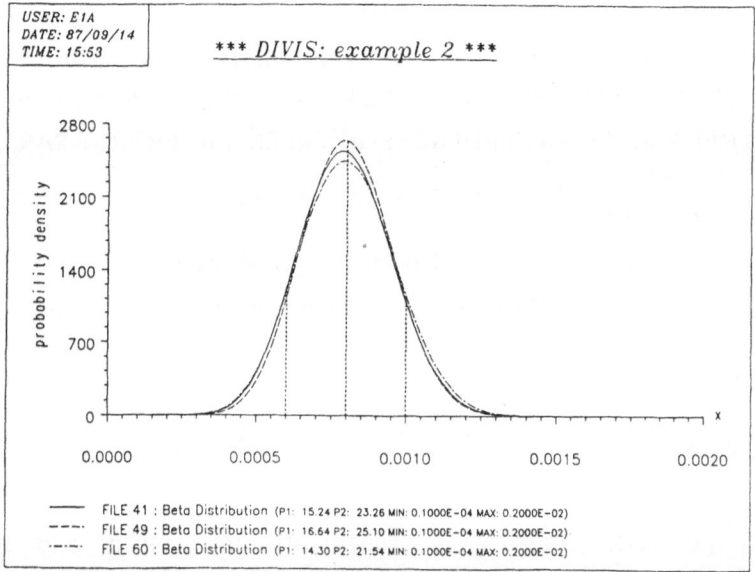

Fig. 6.5

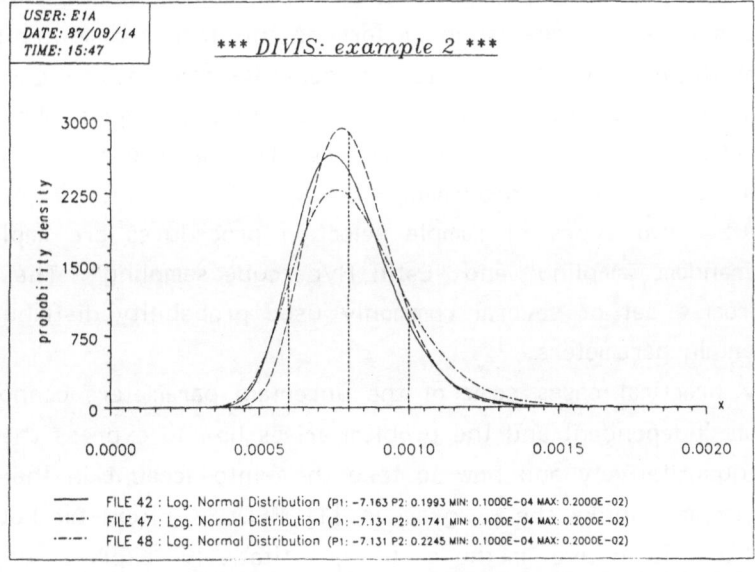

References

IBM: Interactive System Productivity Facility, Version 2, Dialog Management Services, MVS, 1984.

Krzykacz, B.: "MEDUSA: Ein Programm zur Generierung von Simple-Random- und Latin-Hypercube-Stichproben, GRS-A Report, 1987.

Tarasenko, G.S.: "Über die Konvergenzgeschwindigkeit der adaptiven zufälligen Suche" (in Russisch), Problemy slucajnogo poiska, 1980.

THE GENERATION OF EXPERIMENTAL DESIGNS
FOR UNCERTAINTY AND SENSITIVITY ANALYSIS OF MODEL PREDICTIONS
WITH EMPHASIS ON DEPENDENCES BETWEEN UNCERTAIN PARAMETERS

B. KRZYKACZ and E. HOFER

Gesellschaft für Reaktorsicherheit

(GRS) mbH

D-8046 Garching, F.R.G.

Abstract

One of the major steps of a probabilistic uncertainty and sensitivity analysis of model predictions is the generation of an experimental design, i.e. the selection of a multivariate sample of parameter values suitable to study the influence of parameter uncertainties on model predictions. In order to support the analyst in performing this task a computer program, named MEDUSA, has been written to generate the desired design after having received the necessary input data from the experts who are familiar with the various uncertain parameters. The paper presented describes the main features of this program.

In MEDUSA two types of sample selection procedures are implemented: "simple random sampling" and "Latin Hypercube sampling". The user can select from a set of several commonly used probability distributions for his uncertain parameters.

In many practical cases some of the uncertain parameters cannot be regarded as independent and the problem arises how to express these dependences quantitatively and how to take them into account in the selection of the sample. Unlike the program in [1] MEDUSA offers for both design types two alternative methods to quantitatively express dependence between uncertain parameters. One of these methods, first introduced in [2], employs empirical Spearman's correlation coefficients to account for parameter dependence within the design sample. The other employs the following quantities to measure the degree of association between random variables:

ρ: ordinary correlation coefficient

q: quadrant measure

τ_K: Kendall's τ

ρ_S: Spearman's ρ

These quantities are defined and interpreted in terms of population properties of bivariate distributions which seems to be more natural than a definition and interpretation in terms of empirical properties of bivariate samples.

Additionally, MEDUSA contains a method to represent a kind of total dependence where there is some deterministic functional relationship between two parameters which complies with their marginal distributions.

Introduction

The main purposes of a parameter uncertainty and sensitivity analysis of model predictions are

a) to quantitatively express the uncertainty of model predictions in terms of suitable probability statements

b) to identify and to rank the most important contributors to that uncertainty.

To perform this task several steps have to be taken (cf.[5]) that in case of complex computer models can be summarized as follows:

- identify the relevant uncertain parameters of the model and specify probability distributions to express their uncertainty by the "degree of belief" probability interpretation

- generate an experimental design based on the parameter distributions i.e. select a suitable sample of parameter vectors from the corresponding distributions as input to the model

- perform the model predictions with this design as input data and analyse the results of these computations in order to

    a) quantify the uncertainty in the model prediction that is due to the uncertainty of the parameters

    b) rank the parameters according to their contribution to the uncertainty in the model prediction.

This paper is concerned with the generation of experimental designs. Emphasis is on the treatment of dependences among uncertain parameters.

The program MEDUSA

Since the computer codes to be analysed are usually complex and very costly to run, only a small and carefully selected number of runs can be executed for the purpose of uncertainty and sensitivity analysis.

To support the analyst in the selection of runs a computer program named MEDUSA (Multivariate Experimental Designs for Uncertainty and Sensitivity Analysis) has been written. It generates an experimental design, i.e. a multivariate sample

$$
\begin{pmatrix}
x_1^{(1)}, \ldots, x_K^{(1)} \\
\cdot \\
\cdot \\
\cdot \\
x_1^{(N)}, \ldots, x_K^{(N)}
\end{pmatrix}
$$

of N K-dimensional parameter vectors necessary for uncertainty and sensitivity calculations.

The main features of MEDUSA are:

- two alternative types of sample selection procedures
   (1)   simple random sampling (SRS)
   (2)   latin hypercube sampling (LHS)

- two alternative methods to quantify dependence between uncertain parameters by specifying for all pairs of dependent parameters either
   (a)   one of the four <u>populational</u> measures of
          association ρ, q, τ_K, ρ_S
   or
   (b)   the <u>empirical</u> rank correlation $\hat{\rho}_S$ for the design sample.

Additionally a method is offered to take into account a kind of complete positive or negative dependence between uncertain parameters.

In the following the main features of MEDUSA are described more precisely (for more details cf. [3]).

Sample selection procedures in MEDUSA

Two types of sample selection procedures are implemented in MEDUSA. Both have the common property that the parameter values are varied randomly and simultaneously according to the corresponding method. This is in contrast to some old fashioned selection types like e.g. one-at-a-time design where one parameter is varied while the others are held fixed. This method cannot account for dependence among parameters, cannot reveal combined effects of parameter uncertainties and provides information only around the selected nominal point in the parameter space.

(1)   SRS (simple random sampling)
A simple random sample of size N of K uncertain parameters is obtained by sampling the K components of the parameter vector independently N times according to the distributions of the parameters specified previously. The sampling is performed in the usual way using a uniform number generator and the inverse distribution method. If dependences are to be taken into account the SRS-method will be modified in a certain way (cf. models of dependence).
Although relatively simple this selection procedure has many advantages and useful properties (cf. final table).

(2)   LHS (latin hypercube sampling)
This selection procedure is designed to ensure that the sample values evenly (with respect to probability) cover the range of each parameter. It consists of the following steps:
- the range of each parameter is divided into N equiprobable intervals (N = desired sample size)
- from each interval of each parameter one value is selected
- the N values for each of the K parameters are permuted randomly and finally they are combined to N K-tuples, the resulting design-matrix.

So, the value from each interval of each parameter is represented in the selected sample once and only once.
If dependences are to be taken into account the random combination of the N values of the K parameters will be modified in a certain way (cf. models for dependence and final table for further properties of LHS).

Models of dependence in MEDUSA

Each of the two types of sample selection procedures can be combined with one of the following two methods to model dependence between uncertain parameters. Both methods have the common feature that for each pair of dependent parameters with given marginal distributions some coefficient of association, i.e. a number between -1 and +1, must be specified in order to quantify the degree of dependence between them. The first method employs some <u>population</u> measures the second an <u>empirical</u> measure of association. In the following these methods are explained in more detail.

(a) <u>Population measures of association</u>

For each pair of dependent parameters one of the following four coefficients of association

ρ = correlation coefficient

q = quadrant measure

τ_K = Kedall's tau

ρ_S = Spearman's rho

can be specified. The first dependence model implemented in MEDUSA defines a multivariate distribution of the parameter vector $(X_1,...,X_K)$ which is completely characterized by the marginal distributions of $X_1,...,X_K$ and the corresponding coefficients of association. When selecting this method the user specifies the properties of the multivariate population where the desired design sample should come from. To the authors' opinion this way to quantify parameter dependence has a more direct and more natural interpretation than the second method where some empirical properties of the future design-sample must be specified.

Definitions and interpretations of the four coefficients of association (cf [4]).

1. <u>Ordinary population correlation coefficient ρ</u>

$\rho(X,Y):=$cov $(X,Y)/$SQRT(var $X \cdot$ var Y)

$\quad\quad = E\ ((X-EX)(Y-EY))/$SQRT(var $X \cdot$ var Y)

ρ is a very popular measure of association since it completely charac-
terizes the bivariate normal distribution which is the most frequently used
bivariate distribution.

Some well known properties of ρ:
- $-1 \leq \rho (X,Y) \leq +1$
- X and Y stoch. independent => $\rho (X, Y) = 0$
  (the reverse does not hold generally)
- $\rho(X,Y) = 1 (-1) <=>$ there is a linear increasing (decreasing) relation
  between X and Y.

A practical interpretation of a specific value of ρ, other than -1, 0, 1, is
not easy. A special case of an overlapping additive structure of X and Y
may be helpful:
- if $X = Z + X'$
        $Y = Z + Y'$
  where the components Z, X', Y' are independent and have equal variances, then
        $$\rho(X,Y) = 1/2$$

Due to this property the value $\rho = 1/2$ is frequently used in practical
situations. Except for the normal case, $\rho = 1/2$ does, however, not always
imply the above overlapping additive structure of X and Y.
Another unfavourable property of ρ which is also inconsistent with the
intuitive concept of association is the fact that ρ is not ordinally
invariant, i.e. not invariant under monotone transformations of X and Y.
Hence, if $X' = g(X)$ and $Y' = h(Y)$ with both increasing or both decreasing functions g and h, then generally $\rho(X',Y') \neq \rho(X,Y)$.

The remaining three coefficients of association q, τ_K, ρ_S are all ordinally
invariant. They also have more natural and direct interpretations in terms
of certain simple probabilities that are more basic than the moments
involved in the definition of ρ.

2. Population quadrant measure q

Let m_X, m_Y be the medians of the distributions of the parameters X and
Y (assumed to be continuous).

$q(X,Y) := $ prob((X,Y) will fall into quadrant I or III around (m_X, m_Y))

$\quad\quad\quad\quad$ - prob((X,Y) will fall into quadrant II or IV around (m_X, m_Y))

Some properties of q

$\quad -1 \leqq q(X,Y) \leqq +1$

$\quad X,Y$ independent $\Rightarrow q(X,Y) = 0$

$\quad Y$ is increasing (decreasing) function of $X \Rightarrow q(X,Y) = 1$ (-1)

$\quad q$ is ordinally invariant

Elementary manipulations give some useful alternative expressions for q:

$\quad q(X,Y) \quad = \quad$ 4 prob ((X,Y) will fall into quadrant I around $(m_X,$

$\quad\quad\quad\quad\quad\quad\quad m_Y$)) -1

$\quad\quad\quad\quad = \quad$ 2 prob ($(X - m_X)(Y - m_Y) > 0$)) -1

$\quad\quad\quad\quad = \quad$ 2 prob ($Y < m_Y | X < m_X$) -1

Thus according to the last relation the quadrant measure q can be expressed by a simple probability statement on the parameters X, Y and hence it can be interpreted and specified by the model expert more easily than the ordinary correlation coefficient ρ. (cf. [4] for a thorough discussion of q and the other measures of association.)

3. Kendall's tau τ_K

The definition of τ_K is very close to that of q but instead of the medians (m_x, m_y) another random point is used, i.e. τ_K involves two hypothetical independent "observations" (X_1, Y_1) and (X_2, Y_2) of the bivariate random variable (X, Y).

$\tau_K (X, Y) := $ prob $((X_2, Y_2)$ will fall into quandrant I or III around

$\quad\quad\quad\quad (X_1, Y_1))$

$\quad\quad\quad\quad$ -prob $((X_2, Y_2)$ will fall into quadrant II or IV around

$\quad\quad\quad\quad (X_1, Y_1))$

$\quad\quad\quad = \quad$ 2 prob $((X_1 - X_2)(Y_1 - Y_2) > 0) - 1$

$\quad\quad\quad = \quad$ 2 prob $(Y_2 < Y_1 | X_2 < X_1) - 1$

τ_K, too, has the properties that are listed for q but its interpretation via the above probability statements is not as direct as that of q. It can be considered as the population analogue of the well known Kendall's sample rank correlation coefficient.

4. Spearman's rho ρ_S

Unlike q or τ_K the definition of ρ_S involves three hypothetical independent "observations" (X_1, Y_1), (X_2, Y_2) and (X_3, Y_3) of the bivariate random variable (X, Y) and hence its interpretation is not as direct as that of q or τ_K.

$$\rho_S (X, Y):= \; 3 \; \text{prob} \; ((X_2, Y_3) \; \text{will fall into quadrant I or III around}$$
$$(X_1, Y_1))$$
$$-3 \; \text{prob} \; ((X_2, Y_3) \; \text{will fall into quadrant II or IV around}$$
$$(X_1, Y_1))$$
$$= \; 6 \; \text{prob} \; ((X_2 - X_1) (Y_3 - Y_1) > 0) - 3$$
$$= \; 6 \; \text{prob} \; (Y_3 < Y_1 \mid X_2 < X_1) - 3$$

ρ_S, too, has the properties that are listed above for q. It is sometimes called the grade-correlation-coefficient between X and Y and can be considered as the population analogue of the well known Spearman's sample rank correlation coefficient.

Internal realisation of the dependence model (a) in MEDUSA:

After having received the values of the corresponding coefficients of association for all dependent parameter pairs, MEDUSA internally establishes a joint multivariate distribution of the entire parameter vector with the desired properties, i.e. with the given marginal distributions and the required measures of association. This multivariate distribution of the parameter vector $(X_1, ..., X_K)$ is obtained via the transformation

$$X_i = F_i^{-1} (\phi(Z_i)), \; i = 1, ..., K$$

of a certain multivariate normal vector $(Z_1, ..., Z_K)$ using the following relations between ρ, q, τ_K and ρ_S in the bivariate normal case:

$$q = \tau_K = (2/\pi) \arcsin \rho$$
$$\rho_S = (6/\pi) \arcsin (\rho/2)$$

(F_i is the marginal distribution function of X_i, ϕ is the distribution function of a standard normal variate).

Such a multivariate distribution of the entire parameter vector may be useful in some practical situations, e.g. when tolerance limits for the model prediction should be determined.

(b) Empirical measure of association

If this type of dependence quantification is selected the empirical rank correlation coefficients $\hat{\rho}_S$ must be specified for all pairs of dependent parameters. That means that for each dependent pair (X, Y) the user has to specify a (correlation) value $\hat{\rho}_S \ \varepsilon \ (-1, +1)$ that should be attained by the correlation coefficient of the rank transforms of the corresponding parameter values in the design sample ($\hat{\rho}_S$ is ordinally invariant too). Unlike dependence model (a), model (b) asks the user to require some empirical properties of the future design sample. In model (a) he requires properties of the population the design sample should come from. Hence, when selecting this method, the user wishes that the design sample to be generated should strictly reproduce the degree of population association he has in mind.

Internal realisation of dependence model (b) in MEDUSA:
First an independent design matrix is generated from the marginal distributions of the parameters (regardless of the desired dependences). Then the values of each column of this matrix are permuted (applying some rank transformations and matrix operations, cf [2]) so that the rank correlation coefficients between the columns of the new design matrix are very close to those specified by the user, i.e.

$$1 - 6 \sum_{i=1}^{N} (r(x) - r(y))^2 / N(N^2 - 1) \approx \hat{\rho}_S (X,Y)$$

where $r(x_i)$ = rank of the i-th value in the sample of X and $\hat{\rho}_S(X,Y)$ = specified value of the empirical rank correlation between parameters X and Y. Hence, with this dependence model no multivariate structure of the parameter vector is obtained.
Other properties of this model are summarized in the final table.

An additional method to model complete dependence

A certain type of total positive (negative) dependence is incorporated in MEDUSA that allows to take into account an often occuring situation: Two parameters X_1 and X_2 are highly associated in the sense that the values of one parameter are completely determined by the values of the other, i.e. the uncertainty in X_1 may completely be reduced to the uncertainty in X_2 but the functional relationship between X_1 and X_2 is not given explicitely. For such cases where this ralationship is strictly montone MEDUSA offers a method that can be used for both sample selection procedures and both dependence models.

Intuitively motivated we call a pair of parameters X_1, X_2 with given marginal distributions F_1, F_2 "completely positively (negatively) dependent" if there is a strict monotone increasing (decreasing) functional relationship between X_1 and X_2, e.g. X_2 = strictly monotone increasing (decreasing) function of X_1.

Since the marginal distributions F_1, F_2 of X_1 and X_2 are already given (and assumed to be continuous) , the corresponding monotone increasing (decreasing) relationship between X_1 and X_2 can be chosen as

$$X_2 = F_2^{-1} (F_1(X_1)) \quad (X_2 = F_2^{-1}(1 - F_1(X_1))$$

The following properties can be shown to hold for "completely dependent" parameters X_1, X_2:

1. X_1, X_2 "completely positively (negatively) dependent"
   $\rightarrow q(X_1, X_2) = \tau_K(X_1, X_2) = \rho_S(X_1, X_2) = 1 \ (-1)$
   (the reverse doesn't hold generally)

2. $\rho(X_1, X_2) = 1 \ (-1) \rightarrow X_1$, X_2 are "completely positively (negatively) dependent".
   (The reverse doesn't hold generally)

3. X_1, X_2 "completely dependent"
   $\rightarrow$ the sample rank correlation coefficient of any bivariate sample of (X_1, X_2) equals 1 (-1).

Hence, "complete dependence" implies that all ordinally invariant measures of association equal ± 1.

"Complete dependence" between two parameters X_1, X_2 is realized in MEDUSA by simply taking the same uniform random number R (resp. 1 - R for negative dependence) from the random number generator for both distributions.

Comparison of the designs in MEDUSA

The combination of the two sample selection procedures with the two models of dependence described before results in four design options. Which one should the user select in a specific application? Usually, the selection of a design is determined primarily by the type of statements on model prediction uncertainty to be made and by the sample size that can be afforded.

In order to support the user to select the suitable design the following table shows the principal properties of the four design options available in MEDUSA.

Principal properties of the design options in MEDUSA:

design option	(1) SRS (a) popul. meas.	(1) SRS (b) empir. meas.	(2) LHS (a) popul. meas.	(2) LHS (b) empir. meas.
direct and natural interpretation of dependence	●		●	
confidence statements can be made (tolerance conf. limits etc.)	●			
sample size can be increased by the addition of new runs	●			
number of parameters can exceed the sample size	●		●	
empirical correlations are guaranteed to be close to the required correlations		●		●
estimates of mean value and cumul. probabilities have smaller variance [1]			●	●

[1] has been shown in [6] for independent parameters and monotone relationship between model prediction and each of the uncertain parameters.

Example of input data for MEDUSA:

```
**        INPUT-DATA FOR MEDUSA  ,SEPTEMBER 1987
**
**================= 1. GENERAL INPUT INFORMATIONS =======================
**
**                          TITLE:
**
**        SRS - EXPERIMENTAL - DESIGN  ( EXAMPLE )
**
**        TYPE OF DESIGN = 10 :  SRS, POPUL. MEASURE OF ASSOC. GIVEN
**                        11 :  SRS, SAMPLE GRADE CORRELATION GIVEN
**                        20 :  LHS, POPUL. MEASURE OF ASSOC. GIVEN
**                        21 :  LHS, SAMPLE GRADE CORRELATION GIVEN
**
**
** TYPE OF | NUMBER OF  | NUMBER OF  | SAMPLE | NUMBER OF   | INITIAL
** DESIGN  | PARAMETERS | FULLY DEP. |  SIZE  | DESIGN-     | DSEED
**         |            | PARAMETERS |        | REPETITIONS |
**
**    10        9            2         100         1         123457.D0
**
**..............NEXT INPUT FOR DESIGNS 20,21 ONLY....................
**
**       TYPE OF POINT  = 1 :  CONDITONAL MEDIANS OF EACH SUBINTERVAL
**       -SELECTION       2 :  RANDOM SAMPLING FROM EACH SUBINTERVAL
**
**              TYPE OF POINT SELECTION
**
**                        1
**
**
**                   OUTPUT-CONTROL
**
**          PRINT    PRINT    PRINT    WRITE    PRINT
**          INPUT    CORR.    TRIANG.  DESIGN   DESIGN
**          DATA     MATRIX   DECOMP.  ON OUTP.
**                            MATRIX   FILE
**
**           +1       +1       -1       +1       +1
**
**
**        PRINT    PRINT    PRINT    PRINT    PRINT   PRINT    PRINT
**        SAMPLE   SAMPLE   SAMPLE   SAMPLE   RANKS   ORDERED  CORR.
**        CORR.    RANK     RANK     QUADR.           SAMPLES  DIFF.
**                 CORR     CORR.    MEAS.
**                 (SPEAR.)(KENDALL)
**
**         +1       +1       +1       +1       -1       -1       +1
**
**================= 2. DISTRIBUTIONAL INFORMATIONS ====================
**
**                 AVAILABLE DISTRIBUTIONS:
**
** NO. | DISTR. TYPE | NO. OF PAR.|     PARAMETERS
**
**  1 =  DISCRETE       2*N + 1     N,N POINTS,N PROBABILITIES
**  2 =  HISTOGRAM      2*N + 2     N,(N+1) INTERVAL BOUNDS,N PROBA-
**                     ( N <= 20 )                        BILITIES
**  3 =  NORMAL (TRUNC.)   4        MY,SIG,ALFA1,ALFA2 (SIG>0,0<=ALFA1
**  4 =  LOGNORMAL(TRUNC.) 4        MY,SIG,ALFA1,ALFA2  <ALFA2<=1)
**  5 =  UNIFORM           2        A,B              (A<B)
**  6 =  LOG-UNIFORM       2        A,B              (0<A<B)
**  7 =  TRIANGULAR        3        A,B,C            (A<B<=C)
**  8 =  LOG-TRIANGULAR    3        A,B,C            (0<A<=B<=C)
**  9 =  WEIBULL (TRUNC.)  5        XI ,ALFA, C, ALFA1   ,ALFA2
**                                 (A(FA>0, C>0, 0<=ALFA1<ALFA2<=1)
** 10 =  BETA              4        A,B,P,Q          (A<B, P,Q>0)
** 11 =  GAMMA (TRUNC.)    4        B,P,ALFA1,ALFA2 (B>0,P>0.5, 0<=
**                                                   ALFA1<ALFA2<=1)
** 12 =  EXTR.I  (TRUNC.)  4        XI,THETA ,ALFA1,ALFA2 (THETA>0,
**                                             0<=ALFA1<ALFA2<=1)
** 13 =  EXTR.II (TRUNC.)  5        XI,THETA,K,ALFA1,ALFA2 (THETA>0,
**                                         K>0,0<=ALFA1<ALFA2<=1)
**
** PARAM. | NO. OF | LIST OF
** NO.    | DISTR. | DISTRIBUTIONAL
**        | TYPE   | PARAMETERS
**
**     1      3      0. 2.  0.0  1.0
**     2      5      0. 10.
**     3      4      0. 0.5 0.0  1.0
**     4      8      1. 2. 5.
**     5     10     -2. +4. 3.  4.
**     6      2      3.  0. 1. 2. 3.    0.2 0.5 0.3
**     7      6      1. 5.
**     8     13      0. 3. 2. 0.1 0.9
**     9      7      0. 3. 5.
**
**================= 3. DEPENDENCE INFORMATIONS =======================
**
**                   FULL DEPENDENCE
**
**          # | FULLY DEP. | +/- CORRESPONDING
**            | PARAMETER  |     FREE PARAMETER
**
**          1      4              7
**          2      8             -2
**
**.............................................................
**
**        MEASURES OF ASSOCIATION BETWEEN FREE PARAMETERS
**
**  FOR DESIGN-TYPES 10,20 FOUR TYPES OF MEASURES ARE AVAILABLE:
**
**             1 = ORDINARY POPULATION CORRELATION
**             2 = POPULATION QUADRANT MEASURE
**             3 = POPULATION KENDALL'S TAU
**             4 = POPULATION SPEARMAN'S RHO
**
**  FOR DESIGN-TYPES 11,21 SAMPLE RANK CORRELATIONS (SPEARMAN) MUST
**                                                    BE GIVEN
**
**      FREE PAR.1 | FREE PAR.2 | CORRESPONDING | TYPE OF
**                 |            | MEASURE       | MEASURE
**                 |            | OF ASSOCIATION | (FOR DESIGNS
**                 |            |                | 10,20 ONLY)
**
**          1           2           0.5            1
**          2           3           0.6            4
**          1           3           0.4            2
```

Programming notes

MEDUSA is written in FORTRAN 77 and uses some standard routines from the IMSL-library [7]. To run MEDUSA the user has to create an input data file where he must provide all necessary information in a prescribed format (cf. input example before). The output of MEDUSA consists of the generated design matrix which can be written on a particular output file. On request the ordered parameter values, the sample rank-correlation coefficients and other sample statistics are printed.

References

[1]   Iman, R.L., Shortencarier, M.J., "A Fortran 77 Program and User's Guide for the Generation of Latin Hypercube and Random Samples for Use with Computer Models", NUREG/CR-3624, Sandia National Laboratories, Albuquerque (1984).

[2]   Iman, R.L., Conover, W.J.,
"A Distribution-Free Approach to Inducing Rank Correlations Among Input Variables"
Commun. Statist.-Simula. Computa., 11 (3), 311-334 (1982)

[3]   Krzykacz, B.,
"MEDUSA - Ein Programm zur Generierung von Simple Random- und Latin Hypercube-Stichproben für Sensitivitäts- und Unsicherheits- analysen von Ergebnissen umfangreicher Rechenmodelle"
GRS-Report, to be published

[4]   Kruskal, W.H.,
"Ordinal Measures of Association"
J. Amer. Statist. Assoc. 53, Dec. 1958, 814-861

[5]   International Atomic Energy Agency, Division of Nuclear Fuel Cycle
"A working document on Procedures for Assessing the Reliability of Radionuclide Environmental Transfer Model Predicitions"
To be published

[6]   McKay, M.D., Beckman, R.J., Conover, W.J., "A Comparison of Three Methods for Selecting Values of Input Variables in the Analysis of Output from a Computer Code", Technometrics, Vol. 21, No. 2, 239-245 (1979).

[7]   IMSL, International Mathematical and Statistical Libraries, Inc., Houston (1984).

UNCERTAINTY ANALYSIS FOR RANKING PARAMETERS
IN ENVIRONMENTAL MATHEMATICAL MODELLING

A. S. Paschoa
Pontifícia Universidade Católica do Rio de Janeiro
Departamento de Física, C.P. 38071
Rio de Janeiro, RJ 22453
BRASIL

and

M. E. Wrenn
University of Utah, Radiobiology Laboratory
Salt Lake City, Utah 84112
USA

ABSTRACT

Each parameter entering a model may have a different impact on the model outcome, depending on its intrinsic uncertainty. Therefore, ranking parameters according with their relevancy to the model outcome is an useful exercise for environmental modellers and model users as well. A sensitivity index is used to rank all parameters entering a model designed specifically to interpret the atom ratios I-129/I-127 in terms of the retrospective dose commitment due to I-131 absorbed in human thyroids as a consequence of nuclear tests in the atmosphere.

INTRODUCTION

Models used in radiological assessment can be of several types. Thus, for example: models can involve several radionuclides and be applicable to a number of distinct locations [1-3]; be designed for site specific cases dealing with only one particular radionuclide and a straightforward pathway [4]; or concerned with the global cycling of radionuclides [5-7]. However, the uncertainties associated with each particular parameter entering a model will have an outcome as reliable or unreliable as the quantification of the uncertainties associated with the parameters. In addition, the amount of effort, time and money to be spent in improving the quality of any parameter value should be balanced by the overall effect that such parameter would have in the model outcome.

This work will focus on the discussion and application of a simple criterion to rank parameters in environmental models with the objective to help optimizing the use of human and material resources in the attempt to improve the quality of values intended to be used in radionuclide environmental transfer model predictions.

RANKING PARAMETERS

A sensitivity analysis convenient to rank parameters has been suggested, based on the Tomovic sensitivity [8,9]. This analysis introduces a sensitivity index that can vary between zero and the unity [8].

The sensitivity index is used here defined as follows:

$$S = 1 - \frac{\chi_{min}(p)}{\chi_{max}(p)} \tag{1}$$

where: $\chi_{min}(p)$ and $\chi_{max}(p)$ are the minimum and maximum values, respectively, of the output of an equation (or a model) χ, as a function of a single parameter p entering the equation or (model).

The sensitivity index defined by equation (1) is particularly easy to interpret when χ increases or decreases monotonically with p, because in this case χ_{min} or χ_{max} will correspond to either the lower or upper acceptable (or realistic) value of p. There are, however, many cases in which the extreme values of p do not correspond to neither χ_{min} nor χ_{max}, as for example, when χ oscillates as p increases or decreases. In such cases the sensitivity index S can still be used for ranking parameters, but care should be exercized to avoid misinterpretation of the values obtained with the extreme values of p.

The maximum theoretical sensitivity corresponds to S = 1 (i.e., $\chi_{min} \ll \chi_{max}$), while the minimum theoretical sensitivity would be attained at S = 0 (i.e., $\chi_{min} = \chi_{max}$). Thus, the most sensitive (or relevant) parameter in a model should be the one that makes S the closest to the unity, while the one making S nearest to zero should be the less sensitive (or relevant).

The final output of a model, however, is usually the important quantity to be considered as far as the sensitivity to any particular parameter in the model is concerned. Accordingly, the parameters entering a model should be ranked by their effect in the final model output. Thus, considering for example the model developed to interpret the atom ratios $^{129}I/^{127}I$ in terms of the retrospective ^{131}I dose commitment to deceased individuals whose death occurred at a known time after a particular nuclear weapons test, one should analise the outcome of the time dependent function 1/Q for a particular time t, which is the outcome χ in this case.

The function Q can be expressed as [10]:

$$Q = \frac{(A_T/M)}{D_c} \tag{2}$$

where: A_T/M is the ^{129}I concentration measured in the thyroid, given in $1/(^{129}$I atoms$/^{127}$I atoms$)$; D_c is the estimated dose commitment, in Gy.

The expression for Q can be written more explicitly as [10]:

$$Q = \frac{C(^{129}I,0)}{FC(^{131}I,0)} \cdot \frac{(\lambda'_m - \lambda'_v)(\lambda'_T - \lambda'_m)(\lambda'_v - \lambda'_T)}{(\lambda_m - \lambda_v)(\lambda_T - \lambda_m)(\lambda_v - \lambda_T)}$$

$$\times \frac{\left[(\lambda_T - \lambda_m)e^{-\lambda_v t} + (\lambda_m - \lambda_v)e^{-\lambda_T t} - (\lambda_T - \lambda_v)e^{-\lambda_m t}\right]}{\left[\frac{\lambda'_m - \lambda'_T}{\lambda'_m}(1 - e^{-\lambda'_m t}) + \frac{\lambda'_m - \lambda'_v}{\lambda'_T}(1 - e^{-\lambda'_T t}) - \frac{\lambda'_m - \lambda'_T}{\lambda'_v}(1 - e^{\lambda'_v t})\right]} \tag{3}$$

where the nominal values for the decay constants λ'_i and λ_i, i = i, m, t are listed in Table 1; $C(^{129}I,0)/C(^{131}I,0) = 2.0 \times 10^{-10}$; F is a dose factor expressed numerically as:

$$F = \frac{2.6 \times 10^{-9}(Gy/d)/(Bq/kg) \cdot 7.2 \times 10^{14} \ (atoms \ ^{129}I)/(Bq \ ^{129}I)}{10^3(g/kg) \cdot 0.6(mg \ ^{127}I/g) \cdot 4.78 \times 10^{19} \ (^{127}I \ atoms)/(mg \ ^{127}I)}$$

$$= 0.104 \ (Gy/d)/(^{129}I \ atoms)/(^{127}I \ atoms) \tag{4}$$

The measured ^{129}I concentrations in the thyroids are reported in ^{129}I atoms$/^{127}$I atoms, so is more convenient to express F in (Gy/day)$/^{129}$I atoms $/^{127}$I atoms).

Figure 1 shows the graphical representations of the time varying function 1/Q, for a time interval from zero to 100 days. The value for the function 1/Q at 100 days is 2.85×10^9 cGy/(^{129}I$/^{127}$I) in atom helio. The function 1/Q can also be used for times considerably higher than 100 days, however, this particular time interval is adequate to represent the function graphically in a linear scale.

TABLE 1
Decay constants for ^{131}I and ^{129}I

Meaning of the subscript i	Subscript i	^{131}I $\lambda_i (d^{-1})$	^{129}I $\lambda_i (d^{-1})$
Physical	-	8.62×10^{-2}	1.20×10^{-10}
Vegetation *	v	1.39×10^{-1}	5.33×10^{-2}
Milk	m	3.65×10^{-1}	2.77×10^{-1}
Human thyroid	T	9.12×10^{-2}	6.93×10^{-3}

*This decay constant includes radioactive decay, and losses due to rain wind, etc.

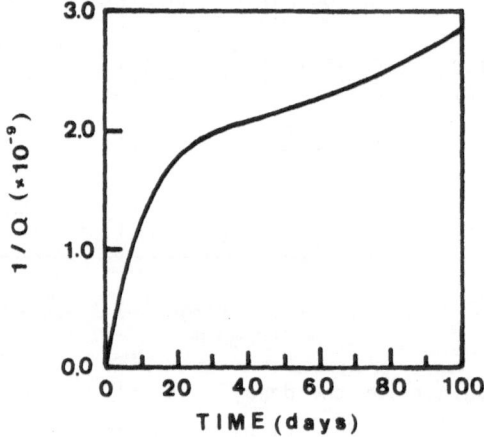

Figure 1 - Function 1/Q from zero to 100 days.

Figure 2 represents the same function 1/Q for a time interval extended from zero up to 500 days. One can see from Figures 1 and 2 that the doubling time for the function 1/Q after about 30 days remains constant and equal to 100 days.

Considering, for example, an individual that had died at a date corresponding to 100 days after a nuclear test, and that the thyroid of this individual was recently measured for ^{129}I concentration, with a result of $(3.16 \pm 0.39) \times 10^{-9}$ ^{129}I atoms/^{127}I atoms, the function 1/Q would allow one to estimate that the committed dose to the thyroid from absorbed ^{131}I associated with that particular test was 9.0 ± 1.0 cGy. One needs to be aware, when making this kind of estimate, that the uncertainty reported here is associated only with the experimental procedures involved in the measurement of the atom ratio $^{129}I/^{127}I$ in thyroids. However the actual uncertainties associated with the model will be mostly due to the uncertainties associated with those parameters entering the model calculations.

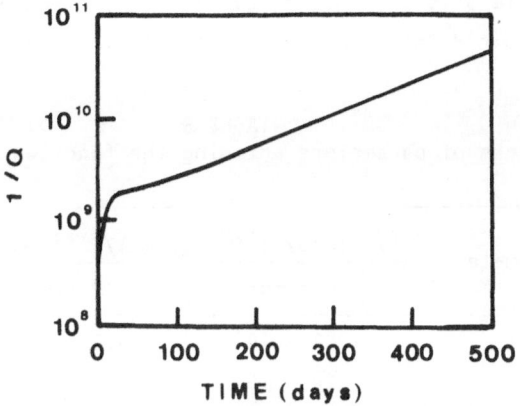

Figure 2 - Function 1/Q from zero to 500 days.

Table 2 lists the estimated thyroid dose ranges based on a plus or minus variation of 20% for each parameter entering the function 1/Q, for a measured concentration of 3.16×10^{-9} atoms $^{129}I/^{127}I$ in the thyroid of an individual whose death occurred at 100 days after a particular nuclear test. The estimated doses listed in Table 2 range from 4.2 to 26 cGy. This range corresponds to the variation of a single parameter λ_v', (i.e., the decay constant of ^{131}I in the vegetation). Accordingly, the uncertainties associated with parameters entering the function 1/Q are reflected in the fact that the earlier estimated dose of 9.0 cGy may be affected by factors that range from about 0.5 to 3.0, as the parameters vary for only 20%.

TABLE 2
Range of estimated retrospective ^{131}I thyroid dose commitments.

Parameter	Estimated dose range (cGy)	
	0.8 p	1.2 p
λ_v'	4.2	26
λ_v	7.1	11
λ_m'	7.5	11
λ_m	7.1	11
λ_T'	7.5	11
λ_T	8.1	10
$\dfrac{C(^{129}I,0)}{C(^{131}I,0)}$	7.5	11

Table 3 shows the ranks of the parameters entering the function 1/Q, based on the sensitive index defined by equation (1). The highest ranking parameter in Table 3 is λ_v'. This result could be expected from examining the data on estimated dose ranges presented in Table 2.

TABLE 3
Ranks of parameters entering the function 1/Q.

Rank	Parameter	1/Q (cGy/ (^{129}I atoms)/(^{127}I atoms))		S
		Minimum	Maximum	
1	λ_v'	1.32×10^9	8.23×10^9	0.846
2	λ_v	2.24×10^9	3.47×10^9	0.354
3	λ_m	2.26×10^9	3.43×10^9	0.341
4	$\dfrac{C(^{129}I,0)}{C(^{127}I,0}$	2.37×10^9	3.56×10^9	0.334
5	λ_m'	2.37×10^9	3.56×10^9	0.334
6	λ_T'	2.37×10^9	3.55×10^9	0.332
7	λ_T	2.56×10^9	3.16×10^9	0.190

CONCLUSIONS

1. The most important uncertainties associated with a model outcome, even when the model is designed to interpret measured results, are likely to be associated with uncertainties of the parameters entering the model.

2. Each particular parameter entering a model will have a corresponding effect in the model outcome, irrespective one deals with specific or generic models, and both modellers and model users should always be aware of this fact.

3. Ranking parameters is an exercise particularly useful when one needs to know the most relevant parameters entering complex mathematical models.

4. Human and material resources should not be spent in improving the values of low ranking parameters (like the ^{129}I decay constant for the thyroid λ_T that appears in the 1/Q function) at the expense of more relevant parameters (like the decay constants λ_v' and λ_v of ^{131}I and ^{129}I, respectively) when the objective is to improve the quality of the overall outcome of a model.

5. Model users and developers, particularly from developing countries, may find useful to give priority in the allocation of meager resources to improve parameter values entering a model, by ranking the parameters in accordance with their relevancy to the model outcome.

ACKNOWLEDGEMENT

This work has been developed with support received under contract DNA 00818C0138. One of us (ASP) acknowledges the partial support of FINEP, CNPq and CNEN.

REFERENCES

1. U.S. Nuclear Regulatory Commission (USNRC), Liquid Pathway Generic Study, NUREG-0440, 1978.

2. U.S. Nuclear Regulatory Commission (USNRC), Draft Generic Environmental Impact Statement on Uranium Milling, NUREG-0511, Vols. 1 and 2, 1979.

3. International Atomic Energy Agency (IAEA), Generic models and parameters for assessing the environmental transfer of radionuclides from routine releases, IAEA Safety Series 57, 1982.

4. Evans, A. G., New dose estimates from chronic tritium exposures. Health Phys., 1969, 16, 57-63

5. Koheler, D. C., A dynamic model of the global iodine cycle and estimation of dose to the world population from releases of I-129 to the environment. Environ.Int., 1981, 5, 15-31.

6. Soldat, J. K. and Baker, D. A., Worldwide population doses from tritium released from nuclear facilities. In The Behaviour of Tritium in the Environment, ed. S. Freeman, International Atomic Energy Agency, Vienna, 1979, 575-581.

7. Killough, G. G. and Till, J. E., Scenarios of C-14 releases from the world nuclear power industry from 1975 to 2020 and estimated radiological implications. Nuclear Safety, 19, 1978, 602-617.

8. Hoffman, F. O. and Gardner, R. H., Evaluation of uncertainties in radiological assessment models. In Radiological Assessment: A Textbook on Environmental Dose Analysis, NUREG/CR-3332, 1983, 11-1 - 11-55.

9. Tomovic, R., Sensitivity Analysis of Dynamic Systems, McGraw-Hill, New York, 1963.

10. Paschoa, A. S., Singh, N. P., Torrey, J. A., and M. E. Wrenn, Retrospective dosimetric modeling for I-129 and I-131. To appear in Health Phys.

Session 1: Overview of Model Reliability and Testing Studies
(Chairman: C. MYTTENAERE)

We have been asked by the organizers to give our personnal point of view on the papers presented and the following discussion.

The main question which was raised was : How to increase the reliability of the models and consequently how to decrease the uncertainty level of calculated doses ?

Most of the models which have been developed must be considered as real deskworks. So the best method to increase their reliability is to know better the "environment"! Part of the information may be found in the non nuclear current litterature and works developed recently in different fields (conventionnal pollution) may be considered as very useful.
As soon as the model fits the environmental conditions more knowledge has then to be introduced in the scenario in order to reduce its uncertainty. Two different ways are then offered to the "team" in charge of the calculation : to use more brain and to dispose of more data regarding the behaviour of the radionuclides in the environment. Initiatives from the "International Union of Radioecologists" as well as efforts produced by Universities will surely increase the intellectual potential of future radioecologists. Exploration of the litterature and a better experimentation thanks to coordinated programmes will help us in getting the necessary information.

Nothwithstanding these efforts part of the uncertainty will be "irreducible" and will have to be quantified.
Intercomparison of models shall thus help the radioecologists to select the best scenario. Criteria of selection will have to be defined.
Shall we keep the model which gives the lowest uncertainty or the more conservative model ! These questions are now asked to the "Biomovs" gestion Committee and will have to be solve very soon.

Finally let us give to the radioecologists the necessary time to analyse their data. Such a transition period would allow to avoid duplications and the "ageing" of the papers presented during scientific manifestations.

Session 2: Transfer Air—Land
(Chairman: F. LUYKX)

This was a very short session and the papers presented, while each of excellent scientific value, did not taken together cover all aspects of the problems of reliability of models describing the transfer of radionuclides from air to soil.

One paper dealt with carbon-14 in the environment, two with tritium, another covered iodine from the Nevada bomb tests, and finally one discussed atmospheric dispersion. While there were no presentations on experience gained after the Chernobyl accident, several papers in Session 3 'Transfer in the terrestrial environment' did include air soil processes and gave results, derived from post-Chernobyl measurements, on dry and wet deposition velocities, crop interception factors, losses by weathering etc.. Data have been shown comparing predicted to observed results obtained with the different models in use in Europe and elsewhere.

It should be recognized, however, that most laboratories are still analyzing environmental data collected after the accident and have not yet finalized their conclusions. Moreover, contamination is still present in several environmental vectors and measurements continue.

It is probably too early, therefore, to draw final conclusions from the Chernobyl "experiment". In about one year, possibly longer, a more complete picture of the air-soil transfer parameter values for the different radionuclides detected after the accident will be available. At that time another seminar certainly would be justified to examine how well the environmental models perform.

Session 4a: Transfer in the Aquatic Environment Fresh Water
(Chairman: D. ROBEAU)

Three kinds of papers were presented and discussed in the session 4a devoted to the transfer of radioactivity in fresh water :

- firstly, a paper on intercomparisons of models on dynamic food chain and geosphere transfers (U. Bergström : intercomparisons of models calculations of the turnover of Ra 226 within an aquatic ecosystem)

- secondly, two papers on uncertainties and sensitivity analysis (A.L. Brenkert : incertainties associated with estimates of radium accumulation in lake sediments and biota ; Th. Zeevaert : dose assessment and uncertainty with respect to liquid effluent discharges).

- thirdly, an comparison between results of modelisation and measurements at the oportunity of the Chernobyl accident (R. Korhonen: Biosphere model validation by intercomparison to observed behaviour of fall-out radionuclides in the environment).

The first paper presented by Dct. Bergström is a very interesting intercomparison of seven models on dynamic food chain and geosphere transfers : AMURAD, BIOPATH, BIOS, DETRA, ECOS,LASER,NCRP.

This presentation explains main differences between these seven codes and recommands priorities, but also the difficulties to compare models : each model having special particularities preventing a very rigorous comparison.

The second and third papers of this session are based on use of uncertainties and sensitivity analysis methods. The methods employed are, in the first case, a ranking method and, in the second case, a classical statistical method of confidence interval following a Latine Hypercule Sampling procedure.

These two presentations have indicated that uncertainties and sensitivity analysis are very important points, because the study of environment implies a lack of knowledge or bad knowledge of fundamental parameters, and that the use of these methods are absolutely necessary.

Mathematical and statistical methods, as Latin Hypercule Sampling, Monte Carlo or Rank correlation are good methods and must not be discussed even if these methods must be applied to fast run computer programs. The most important point is the determination of available probabilitic distributions of values of parameters, this determination having a great influence on results.

The last paper presented was a very realistic comparison between results of measurements of Cs 137 and Sr 90 in the environment at the oportunity of the Chernobyl accident, and results of modelisations using the DETRA Code. Whole deposition, deposition on soil and deposition in water and sediments are studied around three Finish lakes. The main interest of this work is the large surveyed area, and the long period of survey. The first results of this work shows that the agreement between measurements and theoritical results are good.

Session 4b: Transfer in the Aquatic Environment Marine Ecosystem
(Chairman: A. AARKROG)

May I start by making a general remark:

My friend Elis Holm once told me that when he was a soldier
in the Swedish Army his officer taught him that when a map
didn't agree with the terrain, it was the terrain that was
right.

The situation is not that simple for models and observations.
Observations are based on measurements of samples, and this
inevitably entails both sampling and analytical errors.

In the session on modelling in the marine ecosystem this
problem was evident in several of the papers. Coming from the
data side rather than the data model side I would like to
elaborate a little on the problem of acquiring proper data for
the model validation.

In the first three papers by Nielsen, Robeau and Johnson, bio-
indicators such as seaweed were applied. As mentioned by René
Kirchmann in the discussion, this procedure introduces certain
difficulties. For example, you have to be sure that the concentration factors you are using for the calculation of water
concentration are applicable to the biota you have collected;
in other words that the collection was made properly. However,
I also think it is worthwhile to mention that the bioindi-
cator collected at a given location has integrated the water
concentration over a prolonged period; a water sample, on the
other hand, is transient.

The paper of Chartier, where three independent box models
from the CRESP ocean dumping programme were compared, may have

shocked some of the participants. The difference found between
the models amounted to as much as three orders of magnitude.
In this case it is nearly impossible to validate the models.
If we should use fallout plutonium, for instance, for a partial
evaluation we would find that the data were very inadequate.
As far as I can recall, less than 100 vertical transects of
plutonium the world oceans have been performed until now. The
reason is that both the sampling and making of measurements
are extremely costly.

I would also like to comment upon Dr. Howorth's paper. Here
we saw an excellent demonstration of what Dr. Paretzke has
mentioned earlier at this meeting, namely the importance of
not forgetting those major pathways that are not immediately
obvious to a researcher. In particular we should be aware of
pathways between the terrestrial and aquatic environments and
vice versa. Dr. Pattenden stressed that such pathways may
not be taken note of in that neither terrestrial nor marine
radioecologists may feel any obligation to care about them.

I would also like to remind you of what we can learn from Peter
Kershaw's work showing the importance of burrowing animals
in accounting for the distribution of radionuclides in sediments. This is clearly not a simple diffusion process.

In conclusion I think it is fair to say that marine modelling
work for dose estimates has been based upon box models to a
great extent. Furthermore, it is a very difficult to validate
those models that include the deep ocean, whereas we are in a
far better position to represent coastal (shallow) waters by
models. To accomplish this work it is of the utmost importance
to have good measurements, but equally important to gain
access to reliable source term estimates.

Session 5: Transfer in the Biosphere from Waste Repositories
(Chairman: M. HILL)

Many people are highly sceptical about this area of environmental
transfer modelling because they doubt the meaning of predictions of doses
arising thousands and tens and hundreds of thousands of years in the
future. Such scepticism is based on a misunderstanding of the object of
the exercise. Radiological protection standards for the disposal of solid
radioactive waste are set in terms of the dose or risk to people who have
the same habits and metabolic characteristics as we do today. This is
done because it is impossible to predict how human beings will develop in
the far future, so the best that can be achieved is to ensure that the
risks of waste disposal do not exceed those that would be considered
acceptable today. To estimate these risks there is no alternative but to
use mathematical models.

The papers in this session fell into two groups: those dealing with
the results of specific modelling exercises, and those of a more general
nature which identified some of the outstanding issues in biosphere
modelling for waste disposal assessments. From the specific papers, the
main conclusion I drew was that one should use the simplest model which is
adequate for the purpose. It is interesting to develop more complex
models, but this should not be undertaken if nothing is gained in terms of
the accuracy of predicting doses and risks, and estimating the
uncertainties associated with these predictions.

From the more general papers, I drew two conclusions. The first is
that much more effort needs to be devoted to formulating conceptual models
for biosphere transfer, and to clearly explaining the meaning of the
parameters used in such models. At present, most attention seems to be
directed to calculations and model-model comparisons, with the results
that not all potentially important exposure scenarios are being addressed
and the description of parameters is not sufficiently detailed to allow
experts outside the modelling community to make judgements about
probability distributions of values.

The second conclusion concerns the extent to which uncertainties
about human behaviour should be included in analyses of the uncertainties
in waste disposal assessments. To attempt to include all these
uncertainties would, in my view, be fruitless and inconsistent with the

philosophy underlying radiological protection standards. It would also lead to endless speculation and argument about the completeness of the assessments and the uncertainty analyses. Those carrying out such studies should make it clear at the outset that they are trying to predict the doses and risks to people who live as we do now. There is no need to consider possible changes in agricultural practice and food consumption in the future per se; such changes should only be considered as part of broader scenarios representing possible changes in biosphere conditions (eg glaciation, general rises in temperature). Similarly, uncertainties in the parameters related to agricultural practice and food consumption should be derived from knowledge of present day activities.

Conclusions

Summary of a panel discussion.
Members : M. Hill, A. Hofer, F.O. Hoffman, J. Kollas, H. Paretzke.

This workshop was unusual for two reasons. Firstly, it provided a rare opportunity for people involved in modelling radionuclide transfer through the terrestrial, freshwater and marine environments to meet each other and discuss problems of common interest. Secondly, it was the first European meeting entirely devoted to methods for assessing the reliability of radionuclide transfer model predictions. Although most of the time was taken up by relatively formal presentations rather than the discussion typical of workshops, this was appropriate in view of the differing interests of the participants and the unfamiliarity of some of them with the principles and methods of uncertainty analysis and the latest thinking on model validation.

The five technical sessions of the workshop covered the following areas of radionuclide transfer modelling : air to land, terrestrial environment, aquatic environment (freshwater and marine), biosphere modelling for assessments of the radiological impact of disposal of solid radioactive wastes and uncertainty analysis. A majority of the participants were strongly in favour of carrying out uncertainty analyses, but there were a few people who clearly felt that quantitative statements of the uncertainty in model predictions would confuse rather than clarify issues. There was also general agreement that for quantitative model validation it was preferable to compare probability distributions of model predictions with corresponding distributions of measured data, rather than to compare one model prediction with a range of measurements. However, there was disagreement about the role of uncertainty analysis in model development. Those who were used to dealing with relatively simple models which can be fairly directly validated by comparing predictions with independent sets of measured data (e.g. terrestrial foodchain models) were of the opinion that uncertainty analysis should be an integral part of model development.

Those whose expertise lay primarily with more complex models which cannot be broken down into smaller parts, or which are not amenable to very direct quantitative validation (e.g. models of radionuclide dispersion in the deep ocean, models for biosphere transport of radionuclides following release from deep geological repositories) felt that uncertainty analysis was something to be carried out after a model has been developed, because it may entail simplification of the original model.

One of the points of criticism which clearly emerges from this workshop, but also from previous ones, is about the adequacy and value of monitoring data in model validation exercises. Many of the results obtained by monitoring suffer very often from the fact that data collected in one particular area or time are compared with other data collected in a distant area or time to calculate e.g. transfer or concentration factors. Such manipulations of "data collection" do certainly not improve the quality of the output ; they do not very much contribute to the comprehension of the phenomena, they do increase instead the uncertainties of the models.

A kindred discussion arose concerning the difference in model predictions due to so called stochastic variability ("Type 1 uncertainty") and uncertainties due to lack of knowledge ("Type 2 uncertainty"). Some participants were quite clear that the two types of uncertainties should be treated separately in assessing the radiological impact of nuclear facilities, the first being included in "best estimate" assessments of radiological risks and the second in analyses of the uncertainties associated with the results of these assessments. Such distinctions seem very artificial. A lot of these so-called stochastic variability in the environment are in many cases artificially created, being the result of the mismatching of data which are simply not related at all. A better understanding of the facts would certainly, in a number cases, have lead to a description of a mathematical functional relationship with still its own statistical variation instead of having brought about a set of figures with a very large uncertainty just because they are basically only very loosely connected. More intensive consultations between scientists working in the field and modellists have to be stimulated.

In general, modelling is an activity which is an intrinsic part of the scientific method and not something which can be considered as a separate operation. Modelling perhaps is a relatively new word, but still it has be to remembered that the scientific understanding and description of natural processes frequently come about by first the collection of data from experimental measurements, followed by their analysis in terms of mechanisms, ending up in a synthesis. It owes moreover nothing to the recent focussing of interest on environmental science and radioactivity, nor is it dependent on the advent of computers nor to elaborate or simple programs. Neither computers nor the elaborate or simple programs which they may use are essential features of modelling or synthesis. What is essential is scientific understanding and imagination.

In conclusion, it may be said that this workshop brought together both modellists, being involved with the "cleaning up and fashioning" of data, and those scientists who are dealing with "bringing data". The gap was very obvious and should be redeemed as quickly as possible, if models predictions are hoped to remain reliable.

Dr. A. AARKROG
Health Physics Department
Risø National Laboratory
Postbox 49
DK-4000 ROSKILDE

Mr. A. ALBERGEL
DER/MAPA
E.D.F.
Quai Watier 8
F-78400 CHATOU

Dr. E.C.S. AMARAL
Inst. de Radioprotecao e Dosimetria
Avenida das Américas km 10.5
Barra da Tijuca
22700 RIO DE JANEIRO, Brasil

Mrs. A.C. ARGARDE
Kem Akta Konsult AB
Pipersgatan 27
S-112 28 STOCKHOLM

Dr. P.A. ASSIMAKOPOULOS
Nuclear Physics Laboratory
The University of Ioannina
GR-45332 IOANNINA

Dr. R.S. ATHERTON
British Nuclear Fuels
Risley
GB- WARRINGTON, Cheshire WA3 6AS

Mr. P.J. BARRY
Chalk River Nuclear Laboratory
CRNL/AECL
CHALK RIVER, Ontario K0J IJO, Canada

Dr. J.G. BARTZIS
Nuclear Technology Department
N.C.R.P.S. Demokritos
Aghia Paraskevi
GR-15310 ATTICA

Dr. U. BERGSTRÖM
Studsvik Energiteknik AB
S-611 82 NYKOPING

Mrs. F. BOURDEAU
E.D.F.
Avenue Wagram 20-30
F-75008 PARIS

Mrs. A.L. BRENKERT
Environmental Sciences Division
Oak Ridge National Laboratory
P.O. Box X
OAK RIDGE, TN 37831, U.S.A.

Mr. A. BROGUEIRA
DPSR
LNETI
Estrada Nacional 10
P-2685 SACAVEM

Mrs. J. BROWN
N.R.P.B.
Chilton
Didcot
GB-OXON OX11 ORQ

Mrs. G. CAMPOS VENUTI
Instituto Superiore Di Sanita
Viale Regina Elena 299
I-00161 ROMA

Mr. D. CANCIO
Area Protec. Radio. y Medio Ambiente
C.I.E.M.A.T.
Avenida Complutense 22
E-28040 MADRID

Mrs. A.M. ERICSSON
Kem Akta Konsult AB
Pipersgatan 27
S-112 28 STOCKHOLM

Mr. P. CARBONERAS
ENRESA
Paseo de la Castellana 135
E-28020 MADRID

Dr. A. ERIKSSON
Radioecology Department
University of Agricultural Sciences
P.O. Box 7031
S-75007 UPPSALA

Dr. N. CATSAROS
N.C.R.P.S. Demokritos
Aghia Paraskevi
GR-15310 ATTICA

Dr. S. FINZI
DG XII
C.E.C.
Rue de la Loi 200
B-1049 BRUXELLES

Mr. A. CERNES
ISPN/CEA
B.P. 6
F-92265 FONTENAY-AUX-ROSES

Mr. J.L. FONT
Area Protec. Radio. y Medio Ambient
C.I.E.M.A.T.
Avenida Complutense 22
E-28040 MADRID

Mr. M. CHARTIER
IPSN/CEA
B.P. 6
F-92265 FONTENAY-AUX-ROSES

Dr. H. FORSTEL
Institut für Radioagronomie
KFA Jülich GmbH
Postfach 1913
D-5170 JÜLICH

Dr. S. DANALI-COTSAKI
N.C.R.P.S. Demokritos
Aghia Paraskevi
GR-15310 ATTICA

Mr. T. FOULT
CEA ANDRA
Rue de la Fédération 31-33
F-75015 PARIS

Dr. G. DESMET
DG XII
C.E.C.
Rue de la Loi 200
B-1049 BRUXELLES

Dr. M.J. FRISSEL
Laboratory for Radiation Research
R.I.V.M.
P.O. Box 1
NL-3720 BA BILTHOVEN

Dr. M.J. FULKER
Environmental Protection Group
B433 British Nuclear Fuels plc
Sellafield, Seascale
GB-CUMBRIA CA20 1PG

Dr. E. HOFER
Gesellschaft für Reaktorsicherheit
Forschungsgelände
D-8046 GARCHING

Mrs. E. GARCIA MONTANO
Dpto. Fisica Atomica y Nuclear
Facultad de Fisica - Universidad de
Sevilla - Apdo 1065
E-41080 SEVILLA

Dr. F.O. HOFFMAN
Environmental Sciences Division
Oak Ridge National Laboratory
P.O. Box X
OAK RIDGE, TN 37831, U.S.A.

Mr. G. GOUVRAS
C.E.C., DG V/E/1
Bât. Jean Monnet C4/46
Rue Alcide de Gasperi
L-2920 LUXEMBOURG

Dr. J.M. HOWORTH
Environmental & Medical Sciences Div
B551 Harwell Laboratory
U.K. Atomic Energy Authority
GB- OXFORDSHIRE OX11 ORA

Dr. H. GROGAN
Swiss Federal Institute
for Reactor Research
CH-5303 WÜRENLINGEN

Dr. K.G. IOANNIDES
Nuclear Physics Laboratory
The University of Ioannina
GR-45332 IOANNINA

Mr. P. GUETAT
IPSN/CEA/DPT/SEPD
B.P. 6
F-92265 FONTENAY-AUX-ROSES

Mr. G. JOHANSSON
National Institute of
Radiation Protection
Box 60204
S-10401 STOCKHOLM

Dr. C. HAEGG
National Institute of
Radiation Protection
Box 60204
S-10401 STOCKHOLM

Dr. C.E. JOHNSON
Environmental & Medical Sciences Div
B551 Harwell Laboratory
U.K. Atomic Energy Authority
GB- OXFORDSHIRE OX11 ORA

Mrs. M.D. HILL
Head of Assessments Department
N.R.P.B.
Chilton, Didcot
GB-OXON OX11 ORQ

Dr. C.H. JONES
Associated Nuclear Services
Eastleigh House
60 East Street
GB- EPSOM, Surrey KT17 1HA

Dr. S.R. JOSHI
Canada Centre for Inland Waters
National Water Research Institute
P.O. Box 5050
BURLINGTON Ontario LTR 4A6, Canada

Dr. J. KOLLAS
Nuclear Technology Department
N.C.R.P.S. Demokritos
Aghia Paraskevi
GR-15310 ATTICA

Dr. I. KALEF-EZRA
Medical Physics Laboratory
The University of Ioannina
GR-45332 IOANNINA

Dr. L.A. KONIG
Schüttelkopf S. Diabatè
K.F.K.
Postfach 3640
D-7500 KARLSRUHE 1

Dr. B. KANYAR
National Research Institute for
Radiobiology & Radiohygiene
Pentz Karoly u. 5.
H-1221 BUDAPEST, XXII.

Dr. P.M. KOPP
Swiss Federal Institute
for Reactor Research
CH-5303 WÜRENLINGEN

Mr. P.J. KERSHAW
Fisheries Laboratory
M.A.F.F.
Pakefield Road
GB- LOWESTOFT, Suffolk NR33 0TH

Mrs. R.I. KORHONEN
Nuclear Engineering Laboratory
Technical Research Centre of Finland
P.O. Box 169
SF-00181 HELSINKI

Dr. R. KIRCHMANN
Département de Radiobiologie
SCK/CEN
Boeretang 200
B-2400 MOL

Mr. B. KRZYLACZ
Gesellschaft für Reaktorsicherheit
Forschungsgelände
D-8046 GARCHING

Dr. J. KOCH
Soreq Nuclear Research Centre
Israël Atomic Energy Commission
70600 YAVNE, Israël

Mr. G.S. LINSLEY
Waste Manag.Sect.-Div.Nuc.Fuel Cycle
I.A.E.A.
Wagramerstrasse 5, P.O. Box 100
A-1400 VIENNA

Mr. H. KOHLER
Institut für Strahlenhygiene
GSF
Ingolstädter Landstrasse 1
D-8042 NEUHERBERG

Dr. F. LUYKX
C.E.C., DG V/E/1
Bât. Jean Monnet C4/48
Rue Alcide de Gasperi
L-2920 LUXEMBOURG

Dr. M.R. MALISAN
Servizio di Fisica Sanitaria
Ospedale Civile
Via G. Pieri
I-33100 UDINE

Dr. Y.C. NG
Lawrence Livermore National Lab.
University of California
P.O. Box 808
LIVERMORE, CA 94550, U.S.A.

Mr. A. MANOUKAS
Biology Department
N.C.R.P.S. Demokritos
Aghia Paraskevi
GR-15310 ATTICA

Dr. S.P. NIELSEN
Health Physics Department
Risø National Laboratory
Postbox 49
DK-4000 ROSKILDE

Dr. H. MAUBERT
IPSN/DERS/SERE
CEN de Cadarache
B.P. 1
F-13108 ST PAUL-LEZ-DURANCE

Dr. E. NOWAK
Gesellschaft für Reaktorsicherheit
Forschungsgelände
D-8046 GARCHING

Mr. N. MOUSSIOPOULOS
Inst. für Technische Thermodynamik
Universität Karlsruhe
Postfach 6980
D-7500 KARLSRUHE 1

Dr. E. OLYMPIOS
N.C.R.P.S. Demokritos
Aghia Paraskevi
GR-15310 ATTICA

Dr. H. MÜLLER
Institut für Strahlenschutz
GSF
Ingolstädter Landstrasse 1
D-8042 NEUHERBERG

Dr. A.A. PAKOU
Nuclear Physics Laboratory
The University of Ioannina
GR-45332 IOANNINA

Dr. C. MYTTENAERE
Unité de Physiologie Végétale
U.C.L.
Place Croix du Sud 4
B-1348 LOUVAIN-LA-NEUVE

Dr. G. PANTELIAS
N.C.R.P.S. Demokritos
Aghia Paraskevi
GR-15310 ATTICA

Dr. S. NAIR
Berkeley Nuclear Laboratories
CEGB
Dursley
GB- BERKELEY, Gloucestershire GL13 9PB

Dr. D. PAPADOPOULOS
Hauptabteilung Sicherheit
K.F.K.
Postfach 3640
D-7500 KARLSRUHE 1

Dr. N. PAPADOPOULOS
N.C.R.P.S. Demokritos
Aghia Paraskevi
GR-15310 ATTICA

Dr. J.E. PINDER III
Savannah River Laboratory
E.I. du Pont de Nemours & Co.
AIKEN, SC 29808, U.S.A.

Dr. I. PAPAZOGLOU
N.R.C.P.S. Demokritos
Aghia Paraskevi
GR-15310 ATTICA

Dr. K. PSARRAKOS
Medical School
University of Thessaloniki
GR-54006 THESSALONIKI

Dr. H. PARETZKE
Institut für Strahlenschutz
GSF
Ingolstädter Landstrasse 1
D-8042 NEUHERBERG

Dr. Y.Z. QI
Department of Biology
I.U.R.
Jinan University
GUANGZHOU China

Dr. A.S. PASCHOA
Departamento de Fisica
Pontificia Universidade Catolica
Cx. Postal 38071
22453 RIO DE JANEIRO, Brasil

Dr. J.M. QUINAULT
DERS/SERE
CEN de Cadarache
B.P. 1
F-13108 ST PAUL-LEZ-DURANCE

Dr. N.J. PATTENDEN
International Union Radioecologists
Essex Street 73 B
GB- NEWBURY, Berkshire RG14 GRA

Mr. F. RICCI
Physics Department
University of Rome
Pzale Aldo Moro
I-00161 ROMA

Dr. PERSSON
Radiation Physics Department
University of Lund
Lasarettet
S-221 85 LUND

Mr. D. ROBEAU
Département de Protection Sanitaire
IPSN/CEA
B.P. 6
F-92265 FONTENAY-AUX-ROSES

Dr. S. PIERMATTEI
ENEA-DISP, Casaccia
Via Vitaliano Brancati 48
I-00144 ROMA

Dr. U. SANSONE
ENEA-DISP, Casaccia
Via Vitaliano Brancati 48
I-00144 ROMA

Dr. E.G. SIDERIS
N.C.R.S. Demokritos
Aghia Paraskevi
GR-15310 ATTICA

Dr. J. van den HOEK
Vakgroep Dierfysiologie
Landbouwuniversiteit
Haarweg 10
NL-6709 PJ WAGENINGEN

Dr. S.L. SIMON
Dept.Environmental Sciences & Engin.
The University of North Carolina
Rosenau Hall 201 H
CHAPEL HILL, NC 27514, U.S.A.

Dr. F. VAN DORP
Nation.Genossenschaft für die Lager.
Radioaktiver Abfälle NAGRA
Parkstrasse 23
CH-5401 BADEN

Dr. J. SINNAEVE
DG XII
C.E.C.
Rue de la Loi 200
B-1049 BRUXELLES

Mrs. M. VARVAYANNI
N.C.R.P.S. Demokritos
Aghia Paraskevi
GR-15310 ATTICA

Dr. H. SMITH
Biology Department
N.R.P.B.
Chilton, Didcot
GB-OXON OX11 ORQ

Dr. C.R. WILLIAMS
HM Inspectorate of Pollution
Mitre House
Church Street
UK- Lancaster LAI IBG

Dr. P. STEGNAR
Jozef Stefan Institute
LJUBLJANA JAMOVA 39, Yugoslavia

Dr. L.G. WILSON
Radiation Protection Supervisor
Food R.A.
Randalls Road
GB- LEATHERHEAD, Surrey KT22 7RY

Mrs. M. TOMBROU
Nuclear Technical Section
DEME/DEH
Navarinou Street 10
GR-10680 ATHENS

Dr. E. WIRTH
GSF
Ingolstädter Landstrasse 1
D-8042 NEUHERBERG

Dr. J. URBANCIC
Joseph Stefan Institute
LJUBLJANA JAMOVA 39, Yugoslavia

Mr. T. ZEEVAERT
Health Physics Department
SCK/CEN
Boeretang 200
B-2400 MOL